W9-ADM-677

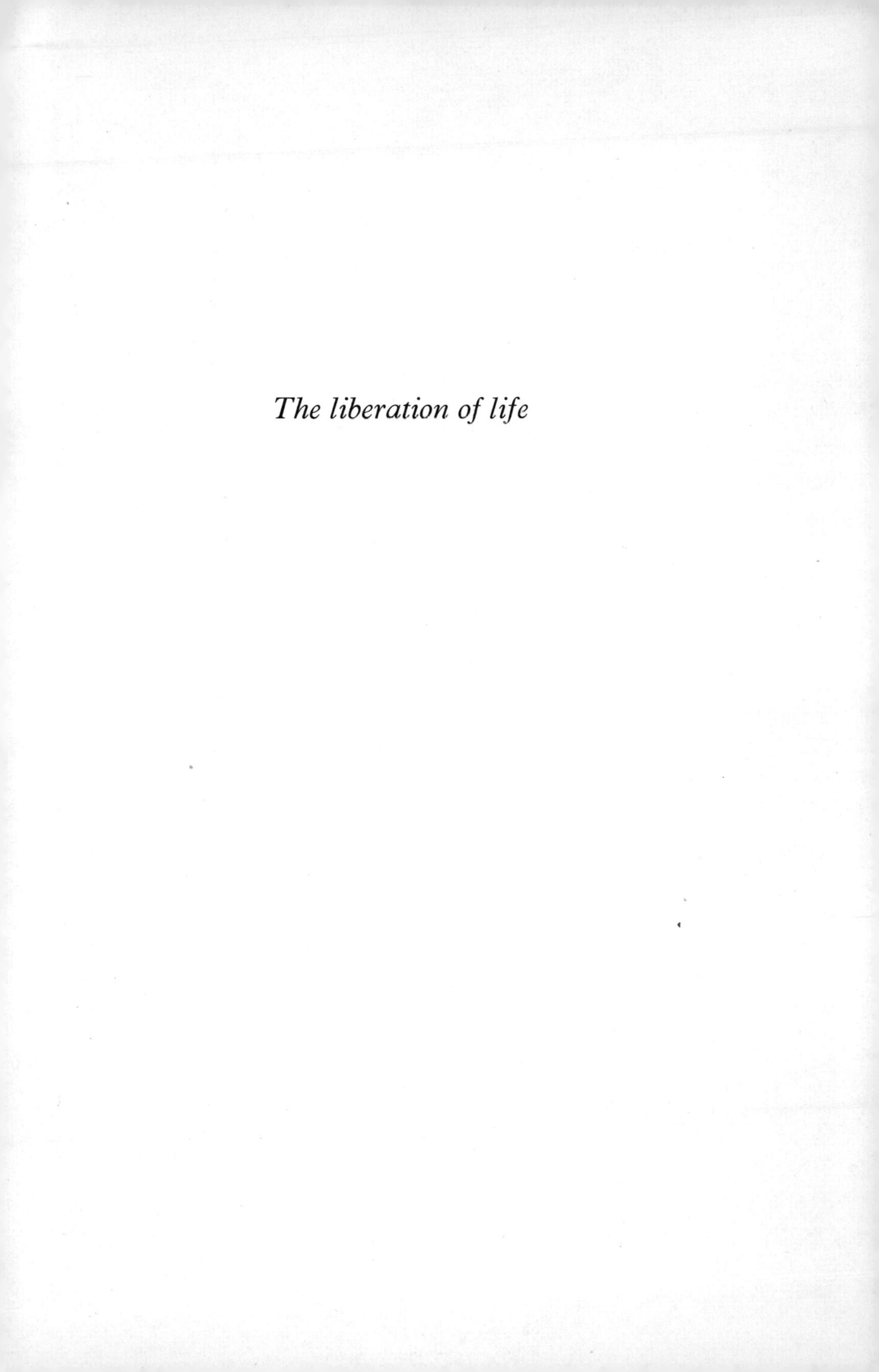

The liberation of life

The liberation of life

FROM THE CELL TO THE COMMUNITY

CHARLES BIRCH
Challis Professor of Biology, University of Sydney
JOHN B. COBB, JR
Ingraham Professor of Theology and Director of the Center for Process Studies
School of Theology, Clarement, California

CAMBRIDGE UNIVERSITY PRESS
Cambridge
London New York New Rochelle
Melbourne Sydney

Published by the Press Syndicate of the University of Cambridge
The Pitt Building, Trumpington Street, Cambridge CB2 1RP
32 East 57th Street, New York, NY 10022, USA
296 Beaconsfield Parade, Middle Park, Melbourne 3206, Australia

First published 1981

Printed in Great Britain at the
University Press, Cambridge

British Library Cataloguing in Publication Data

Birch, Charles
The liberation of life.

1. Human ecology
I. Title II. Cobb, John B.
304.2 GF31 80-42156

ISBN 0 521 23787 4

TO CHARLES HARTSHORNE

CONTENTS

ACKNOWLEDGEMENTS

We gratefully acknowledge the advice and criticism of a number of our colleagues who read the manuscript in earlier drafts. Clifford Cobb, David Griffin and David Vergin read the entire manuscript and gave much helpful advice. The following helped greatly with comments and criticism of particular chapters: Chapters 1 and 2 – Clive Crossley, Darcy Gilmour and Geoffrey Grigg; Chapter 4 – Vaughan Hinton and Graham Pyke; Chapter 5 – William Godfrey-Smith; Chapter 6 – Gary Watson; Chapter 8 – Paul Ehrlich and Anne Ehrlich; Chapters 9 and 10 – Gordon Douglass, Jane Douglass, Charlotte Ellen and Dean Freudenberger. Sung Pak checked most of the chapters for errors and helped greatly with checking references. The whole manuscript in a number of drafts was typed by Sylvia Warren whose skill and patience have been indispensable. We are grateful for the helpful advice of Cambridge University Press which enabled us to make a number of critical concepts clearer than they were.

INTRODUCTION

Liberation is a powerful word. It has caught the imagination of many. It is the slogan of oppressed human groups of all sorts. It has been extended to animals by their human defenders.

As applied to human beings liberation has both an outer and an inner meaning. Outwardly it is the call to throw off the economic, political and ideological yoke of the oppressor. Inwardly it is the call to be freed from the internalisation of the oppressive ideology.

We affirm all these movements for liberation. Each has its necessary specificity reflecting the particular experience of oppression of particular groups such as women, children, the aged, colonised people, Jews, peasants, homosexuals, labourers, Blacks and Indians. Yet there are common themes in all liberation movements, and as they go to the deepest roots of their oppression they find there images and paradigms so basic to our society that they tend to go unrecognised. There is a danger that oppressed groups may share these paradigms which, if not exorcised, will distort their own liberation if they find it. These paradigms function also in the human treatment of animals and there too they need to be exorcised if there is to be true liberation.

The paradigms with which this book is chiefly concerned are the exemplars of life. What people think of life affects the way they deal with their fellow humans and other living organisms. Following the usual paradigm of life most people treat living organisms as objects to be manipulated rather than subjects that experience. We agree with Passmore that the understanding of the living being as primarily a means rather than an end found expression 'in a metaphysic, for which man is the sole finite agent and nature a vast system of machines for man to use and modify as he pleases. That is the metaphysic that ecologists are particularly, and rightly, rejecting' (Passmore, 1974, p. 27).

Although the dominant paradigm of life theoretically treats all human beings as subjects, and thus as standing outside of this objectified nature, in fact those who have accepted this paradigm tend to treat other human groups as part of the world of objects. Those without power, be they women or Blacks or other oppressed groups are too easily denied agency as subjects and are treated as mere means to the ends of those who have power. It is from the results of just such objectification, of regarding people and other living organisms as objects only, that many are now seeking liberation.

The objectification of the natural world, including human groups, has been all too dominant in the modern Western tradition. It has been closely related to scientific models and scientific ways of thinking. But we agree with Passmore that 'this metaphysics by no means constitutes the entire Western tradition. Nor ... does its rejection entail the rejection of the science with which it has so often been associated' (Passmore, 1974, p. 27). We *can* free ourselves from the tyranny of this metaphysics and from the personal and institutional habits it has encouraged.

We call for the liberation of life in this twofold sense. There is the liberation of the conception of life from its objectifying character right through from cell to human community, for the concept of life itself is in a bondage fashioned by interpreters of life ever since biology and allied sciences began. Secondly, there is the liberation of social structures and human behaviour such as will involve a shift from manipulation and management of living creatures, human and non-human alike, to respect for life in its fullness.

Like many powerful words 'liberation' by its extensive use in the past decade is in danger of wearing itself out. That would be unfortunate, for it would be difficult for any other word to capture its full connotations. The important and needed movements for which it has functioned as a slogan and guide would be hampered, their unity obscured in the midst of their mutual controversies. It is our hope that the use we make of this word is not part of that overuse which deadens our language but a serious contribution to its enrichment and strengthening beyond the level of slogans.

This is particularly important in one respect. The rhetoric of liberation is largely negative. That is, it is oriented to the removal of oppression. There is an underlying but inadequately developed assumption that when people are freed from oppressive constraints something very positive happens. But this positive result of liberation is not often considered critically. Because it is lacking, liberation can seem to be the call for sheer individualism, licence or the competition of interest groups for power. Protest movements sometimes destroy without providing any guiding images for reconstruction, apart from which the destruction of oppressive structures can result in even greater evils filling the vacuum.

The reason that advocates of liberation are confident that the removal of oppression will lead to good is that they have faith in life itself. When oppressive forces are removed, both outwardly and inwardly, people become more alive. We share that faith. But when the meaning of the life in which we trust is not clarified, the faith can be romantic in the sense of failing to recognise our capacities for evil. It can be a faith that turns away from what is central in human experience, the individual being confronted by the complexity and ambivalence of desires and being driven by impulses which are neither controlled nor understood. Some of these impulses are nasty, brutish and destructive. Such a faith turns a blind eye to the negative emotions of rancour, envy and vengeance, not to mention pride and vanity that are all a part of human life. It refuses to recognise that civilisation is a precious balance between liberty and authority.

When life is rightly understood, to trust life is to be responsibly open and sensitive to its call, not simply to let things happen without vision and planning. Clarification of what life is and what it means to trust life will support the confidence that liberators long for. It will also show their critics the deeper justification for that confidence and it will provide guidance to the free action which liberation from oppression makes possible.

This book falls in a tradition of writing about human problems from a perspective informed by biology. This is a respectable tradition, which since Darwin's discovery of the basic principles of evolution has dealt with problems from which many people cannot escape. If human beings are a part of the natural, evolutionary process, this cannot but affect the way we understand ourselves. One response to the implications of evolutionary theory is a bifurcation of reality that tries to create a protected sphere for the human wholly separate from the rest of nature. It has profound effects on human self-understanding which we find incredibly damaging. We take evolutionary history seriously in the effort to understand humanity, its future, its relations to other living things and to the life that is inwardly experienced.

This way of thinking is related to that of the great French visionary scientist Teilhard de Chardin. In his numerous books he too took evolutionary history as a basis for understanding humanity, its destiny and its relations to all things. He deserves gratitude for reopening in a commanding way a discussion which advocates of bifurcation had tried to shunt aside. There is much in his vision that we share and we have no desire to engage in a detailed critique. Yet his account of evolution is one-sided. It is insufficiently sensitive to the close connection between ecology and evolution, whereas evolution should be seen ecologically through and through. This one-sidedness expresses itself in a lack of appreciation for the positive value of other forms of life and even for other cultures and religions. His extrapolation from the evolutionary process to a single goal and inevitable destiny for the whole is unconvincing. The future is truly open in the sense that the human species may extinguish itself and destroy much of the rest of life on the planet; yet such an outcome is not inevitable. Hence there is great urgency about establishing the right direction through the current maze of problems and dangers.

While we do not follow Teilhard de Chardin at some fundamental points, we believe that Jacques Monod has gone too far in his attack. He pictures Teilhard's position as an 'animist ideology' which considers evolution to be 'the majestic unfolding of a programme woven into the very fabric of the universe'. Against this he insists:

> Pure chance, absolutely free but blind, at the very root of the stupendous edifice of evolution; this central concept of modern biology is no longer among other possible or even conceivable hypotheses. It is today the *sole* conceivable hypothesis, the only one compatible with discovered and tested fact. And nothing warrants the supposition (or the hope) that conceptions about this should, or ever could, be revised (Monod, 1974, p. 11).

Monod adds to this principle of chance that of necessity and entitles his most famous book accordingly, *Chance and Necessity*. The necessity of which he speaks is that of selection of those genetic characteristics which best adapt creatures to reproduce. To these two principles he seems to attribute the entire evolutionary process. Since the chance mutations which are at the root of the 'edifice of evolution' are themselves caused by natural processes, in the end he appears to support a totally deterministic and mechanistic view of life.

Although Monod is right as to the importance of chance and necessity, he exaggerates and, indeed, provides evidence against the exclusivity which he apparently claims for them. He is not wrong in rejecting a single cosmic purpose which works its inexorable way out in the evolutionary process, but he does not clarify the role of intelligent problem solving and purposeful behaviour in the evolutionary process. The future of the planet is not in the hands of chance and necessity alone. Human decisions are extremely important, and these can be influenced by thought, a process poorly illumined by deterministic and mechanistic models. Further, Monod himself, and others, provide evidence that purposive actions did not originate with human beings and that they played a role in evolutionary selection long before *Homo sapiens* arrived on the scene. We are not sure that in any of this we disagree with Monod's actual views, but Monod's comments about purposive action are not integrated into his dominant models and stand in tension with much of his rhetoric.

Sir Peter Medawar may be correct when he writes: 'The only such overall purpose, Monod believes, is the amplification or expansion of DNA, a conception summed up by the schoolboy aphorism "a chicken is merely the egg's way of making another egg!"' (Medawar & Medawar, 1977, p. 169). In any case, Monod does not sufficiently protect himself from this interpretation. If Monod had entitled his book 'Chance, Necessity and Purpose', and if he had elaborated the interweaving of purpose with chance and necessity more thoughtfully, we would be more satisfied with his account. We believe that he could have then moved on to his suggestions about political and social order in the modern world without the appearance of abrupt discontinuity and irrelevance to the biological theory which his book actually offers. But then Monod would have written a different book. We have tried to write that book. Nevertheless we share Monod's aim to build an understanding of life on the foundations of modern biology. His last words before his untimely death in 1976 were: 'Je cherche a comprendre' – I am trying to understand (Judson, 1979, p. 616).

The interpretation of the living in terms of the non-living, which in the recent past has held a stranglehold upon scientific thought, is losing ground on all sides. Alfred North Whitehead (1926*a*, p. 107) long ago pointed out that:

> A thoroughgoing evolutionary philosophy is inconsistent with materialism. The aboriginal stuff, or material, from which a materialistic philosophy starts is incapable of evolution. The material is in itself the ultimate substance. Evolution, on the materialistic theory, is reduced to the role of being another word for the description of the changes of the external relations between portions of matter. There is nothing to evolve, because one set of external

> relations is as good as any other set of external relations. There can merely be change, purposeless and unprogressive ... The doctrine thus cries aloud for a conception of organism as fundamental for nature.

Whitehead also recognised that the conception of organism as fundamental for nature is important for modern physics as well as for biology. This point has been taken up by subsequent physicists, Capra (1975, p. 321) makes it well.

> The mechanistic world view of classical physics is useful for the description of the kind of physical phenomena we encounter in our everyday life and thus appropriate for dealing with our daily environment, and it has also proved extremely useful as a basis for technology. It is inadequate, however, for the description of physical phenomena in the submicroscopic realm ... Beyond the dimensions of our everyday environment, however, the mechanistic concepts lose their validity and have to be replaced by organic concepts which are very similar to those used by the mystics ... Physics of the twentieth century has shown that the concepts of the organic world view, although of little value for science and technology on the human scale, become increasingly useful at the atomic and subatomic level. The organic view, therefore, seems to be more fundamental than the mechanistic. Classical physics, which is based on the latter, can be derived from quantum theory, which implies the former, whereas the reverse is not possible.

We agree with Capra in his evaluation of the relevance of the organic view for modern physics. But the time has now come when this view becomes imperative as well for a proper assessment of what he calls 'science and technology on the human scale'.

That point is also made by another modern prophet, Theodore Roszak (1975, p. 228).

> Here and now, as we restore the orders of reality, we are in the stage of closing up all traditional dichotomies of Western culture, which have served as the bulwarks of the old Reality Principle. Spirit – flesh, reason – passion, mad – sane, objective – subjective, fact – value, natural – supernatural, intellect – intuition, human – nonhuman ... all these familiar dualisms which have divided the spectrum of consciousness vanish as we create the higher sanity. The dichotomies are healing over like old wounds. Even science, in its awkward single-visioned way, has been led to continuities that baffle

traditional assumptions. It can no longer draw hard lines between matter and energy, organic and inorganic, man and lower animals, law and the indeterminate, mind and body. What is this but a final cold reflection of the visionary Whole, the Tao, the One, at last appearing in the alienated mind where it reaches the end of its tether.

The philosopher Skolimowski (1980, p. 4) says:

we are a schizophrenic civilization which deludes itself that it is the greatest that has ever existed, while its people are walking embodiments of misery and anxiety. Our knowledge and philosophy only widen the rift between living and thinking. T. S. Eliot's prophetic cry: 'Where is the life we have lost in living? Where is the wisdom we have lost in knowledge? Where is the knowledge we have lost in information?' rings today truer than ever. [He goes on to say that it is our linear, atomistic and deterministic thinking] ... that chops everything into small bits and subsequently forces the variety of life into abstract pigeon-holes of physical knowledge which I consider diseased, for in the final reckoning it produces diseased consequences. Therefore, when I say that in devising new tactics for living we shall need to rethink our relationships with the world at large, I mean expressly that we shall need to abandon the mechanistic conception of the world, and replace it with a much broader and richer one.

In our day of fragmentation and specialisation, it is rare that any one individual can achieve a holistic vision. Neither of the authors of this book believes that he can do so alone. Together we can come closer. One of us is a biologist, the other a philosophical theologian. We confess that we have dealt with many topics in this book for which specialised knowledge in other fields would have been helpful. Nevertheless, we believe that our collaboration has been fruitful for us both and we hope also for the reader.

Our collaboration has been facilitated by our having much in common. Firstly, despite the fact that we come from different countries (Australia and the United States) and had spent only scattered days together before beginning the project, and that we work in different disciplines, we both encountered the thought of Alfred North Whitehead at formative periods of our thinking, and have both had a long and lively friendship with Charles Hartshorne, who himself was an assistant of Whitehead at Harvard University. It is because of his contribution to our understanding of life that we gratefully dedicate this book to him. Although we do not use much of Whitehead's terminology, this book expresses throughout our indebtedness to his philosophy. Secondly, at least since 1970, both of us have been deeply concerned with global and ecological problems. We have seen the potential relevance to them of Whitehead's thought and have been concerned to relate our disciplines to these problems (e.g. Cobb, 1972; Birch, 1975*a*). Thirdly, we are committed to values and insights which we interpret to be Christian, realising that others may share some of them while rejecting this identification.

Because of our shared convictions we have been able to work together on every part of this book. The scientific information is largely provided by Birch and some of the more specifically philosophical and theological arguments by Cobb. But every chapter reflects the thinking of both of us to a substantial degree. Although some formulations here and there are more natural to one or the other, both affirm what is said throughout.

The book begins with a survey of what is known about life today ordered under the three headings of molecular ecology, organismic ecology and population ecology. It proceeds to an account of evolution which parallels Monod in his treatment of chance and necessity but also introduces purpose as a significant factor through which we may better establish the continuity between biological and cultural evolution. In Chapter 3 we criticise the models in terms of which living things have been predominantly viewed and offer our own, which we call ecological. It is similar to what Whitehead and Capra call organic or organismic, but we believe that 'ecological' better captures our meaning.

The remainder of the book is a clarification of this model and a discussion of its wide ranging implications. Chapter 4 discusses the meaning, both for human self-understanding and the understanding of other animals, of the serious recognition of the continuity between us. Chapter 5 asks for the implications of this model for ethics. Chapter 6 extends this enquiry to theology. Chapter 7 deals with the new capacities of biology to manipulate life itself and struggles to evaluate these. Chapter 8 asks for the implications of the model for the issues, between those who call for unlimited economic growth and those who affirm that there are limits to such growth.

In an article entitled 'Who needs the liberation of nature?', Easlea (1974, p. 89) argues for the importance of spelling out the social goals and practices which different images of nature both reflect and reinforce 'and to propagate those images that at the very least suggest the possibility of liberation'. Our book is an attempt to do this. But in the first eight chapters there is only a preliminary spelling out of the social goals and practices that our image of nature reflects and reinforces. Chapters 9 and 10 are our attempt to advance this process.

In the end we realise how little we human beings really know. For all our scientific knowledge and all our understanding, is the human situation so different from that of Job whom God questioned?

> Who is this obscuring my designs
> with his empty headed words?
> Brace yourself like a fighter;
> now it is my turn to ask questions and you to inform me.
> Where were you when I laid the foundations of the earth?
> Tell me, since you are so well-informed!
> Who decided the dimensions of it, do you know? . . .
> Have you journeyed all the ways to the sources of the sea,
> or walked where the Abyss is deepest?
> Have you been shown the gates of Death
> or met the janitors of Shadowland?
> Have you an inkling of the extent of the earth?
> Tell me all about it if you have! . . .
> Who carves a channel for the downpour,
> and hacks a way for the rolling thunder,
> so that rain may fall on lands where no one lives,
> and the deserts void of human dwelling,
> giving drink to the lonely wastes,
> and making grass spring where everything was dry? . . .
> Who gave the Ibis wisdom
> and endowed the cock with foreknowledge? . . .
> Does the hawk take flight on your advice
> when he spreads his wings to travel south?
> Does the eagle soar at your command
> to make the eyrie in the heights?
>
> The Book of Job, Chapters 38 and 39
> The Jerusalem Bible

The biologist E.O. Wilson in commenting on this passage wrote:

> And yes, we *do* know and we have told. Jehovah's challenges have been met and scientists have pressed on to uncover and to solve even greater puzzles. The physical basis of life is known; we understand

approximately how and when it started on earth. New species have been created in the laboratory and evolution has been traced at the molecular level. Genes can be spliced from one kind of organism into another. Molecular biologists have most of the knowledge needed to create elementary forms of life. Our machines, settled on Mars, have transmitted panoramic views and the results of chemical soil analysis (Wilson, 1978, p. 202).

We, on the other hand, feel more like Charles Darwin when he said that all our knowledge and all our understanding are like a hen's knowledge of a forty-acre field, in one corner of which she happens to be scratching. Yet even this is enough to stagger our imaginations and fill us with awe. As Walt Whitman said:

I believe
the running blackberry would adorn
the parlours of heaven,
And the narrowest hinge in my hand puts to
scorn all machinery,
And the cow crunching with depress'd head
surpasses any statue,
And a mouse is miracle enough to stagger
sextillions of infidels.

From Walt Whitman's *Song of Myself*

We may have all the facts in the world but what are they without understanding, without the vision of a pattern of which they are a part? Poetess Edna St Vincent Millay stated the problem thus:

Upon this gifted age, in its dark hour
Rains from the sky a meteoric shower
Of facts . . . they lie unquestioned, uncombined.
Wisdom enough to leech us of our ill
Is daily spun, but there exists no loom
To weave it into fabric (St Vincent Millay & Ellis, 1967).

Ours is an attempt, albeit only a beginning, to create a loom and to weave a fabric in which we believe we can see a pattern in the maze of threads. In the end, our conceptual environment, not science and technology, will determine the future.

1

Molecular, organismic and population ecology

Every biologist has at some time asked 'What is life?' and none has ever given a satisfactory answer.

Albert Szent-Györgyi (1972, p. 1)

In recent years there has been an explosion in knowledge and understanding about the phenomena of life. Yet, in spite of great advances, there are still vast areas about life which biology has not illumined. Whereas twenty, or even ten, years ago, authors of textbooks on biology showed little reluctance in defining life, now they shy away from doing so. The more we know about life the less we seem able to define it. Sir Peter Medawar, who has written a great deal on this subject, has said 'life is an abstract noun never used in laboratories' (in conversation).

Without a formal definition of life, it is difficult to draw any sharp line between the living and the non-living. Nevertheless, there are major differences between a stone and a jelly-fish. The differences concern the architecture of the atoms and molecules and the relationships among these atoms and molecules. They concern the ecology of atoms and molecules. Atoms and molecules have different relationships depending upon the architecture of the house in which they find themselves. When a jelly-fish dies it is still made of the same atoms as when it was alive. But the relations between these atoms have changed. Their ecology is different. An understanding of these molecular ecological relationships is crucial to an understanding of life. This level of biology is molecular ecology. Second is the level of organismic biology. This is the study of the individual organism. It has to feed, grow, develop, excrete, co-ordinate its functions and reproduce. In the healthy organism there is a proper relationship between these functions and the environment in which the organism lives. This level of biology can be properly referred to as organismic ecology. Thirdly, the individual organism does not live on its own but with other individuals of its own kind and other individuals of different kinds. It is part of a population. Its well-being depends very much upon its ecological relationships with all its neighbours. Biology at this level is population ecology.

In this chapter we consider life at all three levels and how these studies contribute to an understanding of what life is. But first we try to obtain an overall view of life by pretending to look at our planet from the outside.

Looking in from the outside

Imagine a humanoid located on a distant star, far beyond our solar system. He has been scanning the universe for millenia with powerful instruments. Imagine that his instruments are so powerful that they can identify objects, even as small as atoms and molecules, anywhere in the universe. Suppose that his instruments have turned toward our solar system and firstly to the sun. The sun appears to him as an ordinary star with a planetary system similar to some other stars. Even its chemical composition differs little from that of stars elsewhere in the universe. Then about four thousand five hundred million years ago he turns his attention to the planet earth. One feature there is radically different. The temperature on the surface of the planet earth is mild compared with nearly every other celestial object he has looked at. He keeps his instruments turned on earth anticipating that more is to be found there. About half a billion years after the origin of the planet earth he sees a change taking place in the waters of a shallow tropical sea. Very large molecules, which he has not seen anywhere else, become so abundant that they are like a soup in the shallow sea.

Thus began the history of the so-called organic molecules. These are molecules that are built on a framework of carbon atoms. Their appearance is associated with violent electrical storms. Very slowly these large molecules assemble into aggregates which appear to 'feed' on the soup of large molecules in the sea. The soup of large molecules is apparently not replenished after its origin in electrical storms. It is a non-renewable resource for the aggregates of molecules that were assembled from it. These molecules are called organic because they are characteristic of the molecules that constitute living organisms. The largest of the aggregates of organic molecules may have been similar to viruses that we recognise as living when they are inside living cells. However, their existence would probably have been transitory had they not eventually become organised into larger, more complex, assemblages of molecules called cells. Cells have an enclosing membrane that is selective of molecules that can go in and out. This enables them to control, to some extent, their internal environment. The steps from large molecules to cells are still largely a matter of conjecture from what we know about the assemblage of molecules in living cells today. Some of these steps we discuss later.

Cells in turn became aggregated into yet more complex organisms. At first, living organisms seemed utterly dependent upon the finite soup of large molecules on which they 'fed'. As these organisms multiplied, they must have reduced the molecular resource on which they were dependent. They seemed to be headed for a brief heyday of eat and be merry for tomorrow we die! This was not what happened.

Some of these multiplying aggregates of cells hit upon a two-step photochemical process to decompose water, this probably happened first in bacteria then in blue-green algae and later in plants. These photosynthetic organisms used the practically limitless resources of water around them. By means of the sun's energy they split water into hydrogen and oxygen. Oxygen was set free into the atmosphere. So for the first time the earth's atmosphere contained oxygen. Hydrogen was combined with carbon dioxide to build molecules called carbohydrates and fats. These in turn were used as fuels to produce energy to build other molecules such as proteins. The important point is that all these very large molecules were now produced by plants from quite small molecules (carbon dioxide and water) taken from their surroundings. Organisms no longer needed a supply of large molecules from without. So from an unsustainable economy dependent upon a non-renewable resource, life had moved to a sustainable hydrogen economy with a sustainable source of energy from the sun. The umbilical cord with the primeval large molecules was severed.

With an abundant source of energy, plants multiplied and spread. To the humanoid observer from outside, the whole of the surface of the earth was changing. It was becoming greener. Furthermore the atmosphere now contained oxygen and a lot of water vapour giving the earth a heavy cloud cover.

The next important event was the appearance of living organisms that fed on plants. These were animals. They derived their energy, not from sunlight, but from the fuels stored in plants. Using oxygen, now abundant in the atmosphere, they broke down these large molecules into smaller ones, releasing energy in the process. With the released energy the smaller molecules were built up again into larger molecules of their own flesh and being. Hence 'all flesh is grass' (Isaiah 40:6).

The humanoid scanner from outer space would next notice a tremendous proliferation both of numbers and kinds of animals over the face of the earth. From its beginnings in the sea, animal life invaded rivers and then dry land. Plants proliferated in kind at the same time. It may have taken about one billion years from the first macromolecules to the first cells as we know them, and from the first cells to human beings must have taken about three billion years.

The process of multiplication and diversification of living organisms depended upon two distinctive properties: the ability to replicate themselves faithfully and the ability, sometimes, to change during replication. One process conserved existing streams of life, the other changed the direction of the stream.

The thin envelope we call life that covers the earth is earth's most characteristic feature. Having seen it develop from an earth without any such envelope, how would the viewer from outer space describe what it is?

Living organisms are highly ordered macromolecular structures. They maintain this order by means of a complex series of chemical reactions called metabolism (Greek *metabole*, meaning change, which refers to chemical changes). They replicate and thus constitute a spreading centre of order in a less ordered universe. This definition which comes from Grobstein (1964, p. 5) is similar in emphasis to Sir Macfarlane Burnet's (1978, p. 6) statement: 'Probably the most basic definition is that life is the capacity to reconstruct a replica of a complex pattern of organic molecules.'

The second law of thermodynamics maintains that all self-contained systems gradually pass from a state of greater order to one of lesser order. Heat is lost in the process and, with the passage of time, complex arrangements of matter become simple. There is said to be an increase in entropy. Entropy literally means transformation of energy. All the energy of the universe is eventually converted into heat which will be evenly distributed throughout the universe. This means that no more work can be done, such as happens when simple arrangements of matter become more complexly ordered. Maximum entropy is the hypothetical homogenous state when everything will be at the same temperature and all processes will therefore have ceased. Does then the evolution of life run counter to the second law of thermodynamics, resulting as it does in local decreases in entropy? On the contrary, the earth is not a closed system. Energy reaches it from the sun. Increase in order of living matter on earth is gained at the expense of the sun whose order decreases ever so slightly. But in the universe as a whole there is still an increase in entropy. The rule is that in the universe as a whole entropy increases. But there can still be local decreases of entropy as happens in all living organisms while they are still alive.

Decrease in entropy is characteristic of matter in living organisms, but it is not the exclusive property of such organisms. Large organic molecules exist outside living organisms although they never have the complexity of molecules in living organisms. Almost anywhere we care to look in the universe with appropriate equipment it is possible to identify 'organic molecules', that is to say, molecules of several carbon atoms joined with hydrogen and sometimes other atoms. Their formation must have involved a local decrease in entropy. We do not know where they are made or how they are made. Crystallisation also involves a local decrease of entropy, as does a man-made machine that makes bottle tops from sheets of metal.

The existence of organic molecules in the far reaches of the universe, even though those known are simple in structure compared with macromolecules characteristic of living organisms, has led to a modern interest in an alternative scenario of the origin of living things. On earth the simple organic molecules are precursors of the macromolecules characteristic of living organisms. Every day, in every green plant, the syntheses are made from simple organic molecules to macromolecules. The organic molecules in outer space represent early steps in the direction of macromolecules. They may never have got any further anywhere other than on earth. On the other hand, it is conceivable that they have led to further synthesis somewhere in the universe. Living things could have originated beyond the earth and have come here as an invasion from outer space. At present we have no means of falsifying or verifying this hypothesis. That applies also to the alternative scenario of life originating on earth. What we do know is that the large molecules characteristic of living organisms can be synthesised in the laboratory without the aid of living organisms, though as yet these have not been put together to make a living cell.

The humanoid in outer space looking in on planet earth after the emergence of humanity would find amongst the unique features of the earth not only the presence of living organisms but also their machines. Indeed, he would have difficulty in distinguishing between a horse and a tractor moving across a field. Both are large and complex structures. Of both living organisms and machines we can sensibly ask of their parts 'what is this for?' (Thorpe, 1974, p. 17). We cannot sensibly ask this kind of question of natural non-living things, of the solar system or its parts, or of a nebula or of the parts of a complex molecule. The inappropriateness of the question is not due to lack of structure. Even a thunderstorm has a structure as Thorpe (1974, p. 17) has pointed out. It has updrafts and downdrafts and centres of electrical discharge which can be depicted in a complex diagram of the 'structure' of a thunderstorm. Yet of none of these parts can one sensibly ask 'what is it for?' Monod (1974, p. 20) makes exactly the same point when he describes the common property of living organisms and machines and their parts as being 'endowed with a purpose or project'. The legs of a horse are to transport it. The wheels of the tractor are to transport it also. What Thorpe and Monod are saying is that living organisms and machines, as well as parts of both, serve ends which can be identified. How they came to be this way is another matter. The biologist answers that the living organism came to be this way through the operation of natural selection. Those organisms that did not have parts serving appropriate ends for survival and reproduction were eliminated in the struggle for existence. A theologian of a century ago might have answered that living organisms and their parts serve ends because God designed them that way, just as a machine and its parts serve ends because people designed them that way.

Suppose that the humanoid in outer space has successfully distinguished living organisms and their artifacts from everything else in the universe, how does he tell the difference between the two? How does he distinguish a horse from a tractor? Firstly, the horse is made of macromolecules – carbohydrates, fats, proteins and nucleic acids as well as some others. The tractor is not. It is made of small molecules consisting of components of iron, nickel and aluminium in semi-crystalline array. Some parts, or possibly even a whole tractor, could be made of polythene plastics which are surely macromolecules. But it is not the macromolecule as such that distinguishes the tractor from the horse. It is the kind of macromolecule. No machines are made of carbohydrates, fats, nucleic acids and proteins.

Secondly, the horse has a complex metabolism which brings its structure into being from a fertilised egg and this same metabolism maintains its mature structure. Its coming into being from the fertilised egg and the maintenance of that being owe nothing to outside forces, except in so far as components of its environment are supportive of life; the temperature has to be right, food has to be adequate in quality and quantity. No outside force steers the direction of development from egg to horse rather than to oak tree. That designing operation is internal to the organism. The tractor owes its structure solely to forces external to itself, namely the designer and manufacturer. Its structure is maintained by these self-same forces as parts wear out and have to be replaced (Monod, 1974, p. 21).

Thirdly, the horse can reproduce itself. The tractor cannot. The multiplication of tractors is dependent upon the manufacturer who usually makes each tractor *de novo*. Perhaps a manufacturer of tractors might build a tractor that could be programmed to build new ones from the parent tractor as it wore out. Even if this were possible it is the manufacturer outside the machine who makes this possible and who sees that the parental tractor is adequately supplied with spare parts and other materials that can be organised into another tractor.

In summary, the living organism is made of specific sorts of macromolecules, the machine is not; the living organism owes its structure to internal designing forces, the design of the machine is dependent upon forces outside the machine; the living organism reproduces itself, the machine does not.

Aristotle in the fourth century BC was extraordinarily close to the mark when he said 'Life is the power of selfnourishment and of independent growth and decay', a phrase that must surely refer to the coming into being and maintenance of the living organism as being independent of designing influences outside itself.

Molecular ecology

The physicist Erwin Schrödinger gave a series of lectures in Trinity College, Dublin in 1943 which were published in 1944 as a book entitled *What is Life?* Schrödinger selected as the most distinguishing characteristic of life its negative entropy within the general trend of positive entropy of the universe. While the universe as a whole becomes less ordered, the enclave of entities that are living produces order. Life is an ordering process. Negentropy is not, as we have already seen, an exclusive property of living organisms and so is not as unique to life as Schrödinger seemed to suppose. However, Schrödinger was led to ask the question of living organisms: 'How is it that they reverse the general trend of the universe?' He answered – through the receipt of information, or, as he put it, 'life feeds on information'. This idea that the receipt of information is essential to the basic mechanisms of biology has become widely accepted. In all organisms, be they bacteria or people, this information is encoded primarily in the macromolecule deoxyribose nucleic acid (DNA) or ribose nucleic acid (RNA) in the case of plant viruses. Each fertilised human ovum is a repository of information which, in the process of development, is spelt out to produce a human being and not an oak tree and amongst humans a particular and unique one. 'The unfolding of events in the life cycle of an organism', wrote Schrödinger (1962, p. 77), 'exhibits an admirable regularity and orderliness, unrivalled by anything we meet with in inanimate matter. We find it controlled by a supremely well-ordered group of atoms which represent only a small fraction of the sum total in every cell.' A developmental biologist today can hardly take exception to this statement apart from the use of the word 'unfolding' which suggests the stages of development are preformed.

But where did the information that is internal to the organism come from? At one time there was no DNA. The problem is more complex than finding out how a DNA molecule could be formed from its component atoms. The code is meaningless unless it is translated. The DNA code is translated into specific proteins but this is not direct. It involves first a spelling out in the production of complementary RNA. The RNA in its turn guides the synthesis of specific proteins. The proteins then become both builder and building blocks of the new organism.

It is exceedingly difficult to imagine how DNA and protein became linked in the history of molecular ecology. The assembly of DNA into a code in living organisms today requires a protein catalyst and a complex series of proteins for unwinding and repair of the DNA molecules. It is conceivable, though unlikely, that the first DNA synthesis took place in the absence of protein. The origin of the code and its translation mechanism is so baffling that some molecular biologists think that DNA could not have been made on the primitive earth and that life must have started with something else, DNA evolving later.

In either case DNA now codes the specifications of the organism. It is therefore the basis of inheritance. What we call genes are made of DNA. Despite the obscurity of the origin of the genetic code, it is true to say that 'the molecular theory of the genetic code ... does today constitute a general theory of living systems. No such thing existed in scientific knowledge before molecular biology. Until then the "secret of life" seemed to be essentially inaccessible' (Monod, 1974, p. 12). The

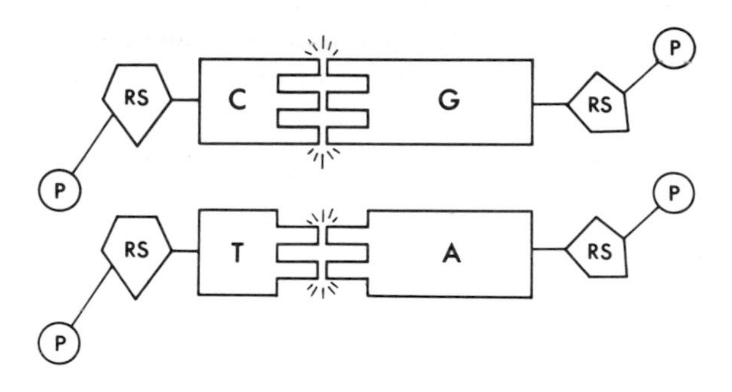

Fig. 1.01. Diagrammatic representation of the nucleotides in the DNA molecule. RS, deoxyribose sugar; P, phosphoric acid molecule; C, cytosine; G, guanine; T, thymine; A, adenine. Each nucleotide consists of P + RS + C (or G, T or A).

secret of life may not yet have been found but it has been made more accessible. For as Monod (1974, p. 94) says: 'It is at this level of chemical organisation that the secret of life (if there is one) is to be found. And if we could not only describe these sequences but pronounce the law by which they assemble, the secret could be declared open, the *ultima ratio* discovered.'

How does the DNA molecule specify the characteristics of living organisms from bacteria to human beings? Its structure is a double helix supporting a sequence of steps, as in a ladder. Each step consists of two nucleotides. Each nucleotide consists of a purine base (adenine or guanine) or a pyrimidine base (cytosine or thymine) linked to a ribose sugar and a phosphoric acid molecule (see Fig. 1.01).

Nucleotides are of four different kinds. They bind together preferentially as two pairs, thymine with adenine and guanine with cytosine, so that the position of one nucleotide in one chain specifies which nucleotide will be on the other chain. This property endows the molecule with its reproductive specificity, whereas the sequence of nucleotides along the molecule spells out the informational code.

The sequence of nucleotides provides 'words' in an alphabet of four letters which are the four nucleotides. So within ten rungs there are 4^{10} (which is well over a million) possible ways of arranging the order of nucleotides. This means that an enormous amount of information can be provided by even ten rungs of a DNA helix. But a typical DNA molecule in the cell of a human being may be composed of millions of nucleotides.

DNA contains the coded information which determines the structure of proteins which play a key role in development and maintenance of the organism. Proteins consist of amino acids. There are 20 different amino acids. So the four letter language of DNA has to be translated into the 20 letter language of proteins. The relation between the DNA language and the protein language is called the genetic code. A triplet of nucleotides spells out an amino acid. With four possible nucleotides there are $4^3 = 64$ possible different triplets that can stand for 20 amino acids. We now know that more than one triplet codes for one amino acid. Some function as a full stop to terminate a code message and others to initiate the message.

In the 1950s it was possible to determine the sequence of amino acids in a protein molecule and 20 years later the sequence of nucleotides in the DNA molecule. This was made possible by the discovery that certain enzymes can cut the DNA molecule into pieces at points along the length of the molecule determined by the occurrence of particular sequences of DNA. The application of these techniques for determining the sequence of nucleotides in DNA shows that genes consist of lengths of DNA carrying, usually at one end, short stretches of nucleotides that act as full stop messages and at the other end short messages initiating the sequence that spells out particular amino acids. Within the length of the gene there are usually stretches of 'silent DNA' which can be as long as the gene itself. Their function is not understood.

Whilst there are only 20 different amino acids there is an enormous number of different possible sorts of proteins. Different combinations of amino acids in different numbers constitute different proteins. For example, an insulin molecule consists of 16 different kinds of amino acid in two long chains of 51 amino acids. The primary importance of proteins is that they play a key role in development and maintenance. The word protein is derived from the Greek *proteios* meaning 'prime' or first. Some of them, the enzyme proteins, are the catalysts for all the chemical reactions in the cell. Some have a structural function and are part of the material of the body.

Which particular protein is made and when it is made is not only a function of the information encoded in the DNA of the nucleus of every cell. It is dependent also, particularly in egg cells, upon a mosaic of information in nucleic acids or nucleotides in the non-nuclear part of the cell in ways that are at present largely unknown. What the DNA does is dependent upon the environment in which it occurs, in other words it depends upon the ecological context.

It is true to say that the structure of the living organism is laid down in information in the fertilised egg. But it is incorrect to suppose that these structures are in any sense preformed in the egg. Development is not a process of unfolding of structures compressed into the egg (the doctrine of preformation). It is a process of spelling out a message from chemical molecules into 'words' and 'sentences' and ultimately structures. This process of development is called epigenesis in contrast to preformation. Monod (1974, p. 87) refers to epigenetic development as a revelation rather than a creation. It is a revealing of something already existing potentially, although concealed.

The system DNA-RNA-protein involves the separation of the functions of reliable storage of information from that of the using of the information as instructions. Presumably there is some great obstacle in combining these two functions in the one sort of molecule.

The two groups of molecules, nucleotides and proteins, are the most distinguishing constituents of living organisms as compared with the non-living, because they are the store and expression of information. There are other large molecules that are also characteristic of life, such as for example adenosinetriphosphoric acid (ATP) which is involved in energy transfer in the cell. When the American Space Administration sent its probes into space to look for life beyond the earth they chose to look for the molecule ATP as the indicator of life, since it is easier to detect than nucleotides and protein.

Although the patents for building the large molecules out of small ones and for organising the larger molecules into living structures are still largely held by nature, biologists have nevertheless gone a long way in synthesising the large molecules characteristic of living organisms. The DNA molecule is what it is by virtue of the special relations of the atoms within this complex molecule. Furthermore the action of the DNA molecule is dependent upon the environment in which it acts, that is to say the particular conformation of molecules in the non-nuclear part of the cell. Molecular biology is ecology at the molecular level.

Information coded in the DNA molecule bequeathed at birth determines a great deal about our lives as of the lives of all living organisms. It determines the difference between an ant and a frog and between a frog and a human being. DNA programmes not only what the fertilised egg will develop into and how it develops but as well it programmes the functioning of the mature organism. For example, our DNA determines the sort of brain we have as compared with say a cat.

Young (1978, p. 7) has said 'the lives of human beings and other animals are governed by sets of programmes written in their genes and brains'. Our genes programme our brains so that a lot of its functions, as with other organs, are automatic. There are daily programmes that have to do with the rhythm of waking and sleeping. There are long term ones, such as the programme that sets the sex hormones to secrete at adolescence, which in their turn convert a child into a sexually mature adult. The hormones themselves have a feedback upon the nervous system determining to some extent its state. There is almost certainly even an inbuilt programme for senescence: genes which function only late in life to reduce the chance of survival (Young, 1978, p. 27).

On the other hand, genetic programming does not totally govern the behaviour of living organisms. This is especially evident in the case of human beings, where cultural differences are so great. Though there certainly is programming of sexual maturation, human sexual behaviour is affected by individual learning and thinking. There are centres for aggression programmed into the brain, but they do not have to be stimulated! Without the necessary programming human language would not be possible, but the particular language we speak and how well we speak it is a matter of culture and individual learning. Just how much of human behaviour is bequeathed by the genes and how much is learned is a subject to which we return in Chapter 4.

Organismic ecology

The life history of a living organism from birth to death can be described as the emergence of order, the maintenance of order and the disappearance of order. The emergence of order is called development. The maintenance of order is called physiology. Scientific understanding of these two sorts of order comes largely from studies of individual organisms. But the individual cannot be studied in isolation. It always has an environment which greatly affects both its development and its physiology. Thus both development and physiology are ecological studies made at the level of the individual organism.

The development of structures from molecules involves an ordering of molecules into cells and cells into organs. Some steps of the assembly process depend upon chemical binding which occurs spontaneously between appropriate molecules. Other steps are controlled by DNA or RNA within the developing cells or by physico-chemical

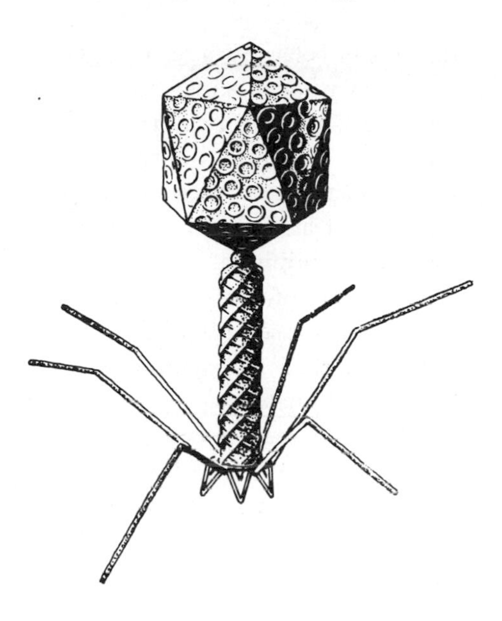

Fig. 1.02. Model of T_4 bacteriophage virus showing the elaborate structure of this self-assembly of molecules. The head, neck, collar, sheath and tail fibres are constructed from many different protein molecules. The head contains a core of DNA.

messages coming from other parts of the organism and the environment.

An example of spontaneous self-assembly is the formation of the fibrous protein called collagen. It forms fibrils that orient themselves helically around each other like a three-stranded rope, forming structures of great strength such as tendons. Collagen molecules will aggregate into helical structures in a test-tube. The particular pattern they form can be changed by altering the components of the solution. The T_4 virus bacteriophage presents a more complex example of self-assembly. It consists of a head, neck, collar and end-plate with tail-fibres (Fig. 1.02). It is constructed from only two types of macromolecule, proteins and (within the head) DNA. In the assembly of the end-plate 17 different proteins are required. The end-plate is formed by a step by step addition of each protein in a specific sequence. Evidently the binding of one protein to the developing end-plate alters its structure enabling it to add the next protein in sequence. The self-assembly of viral proteins yields an extraordinary degree of order at the molecular level. DNA is responsible for the production of the specific proteins. Their arrangement is largely a result of self-assembly, though the host cell in which the virus occurs may make some contribution.

An organelle within a cell such as a chloroplast is far more complex than a virus. It contains elaborate internal stacks of membranes precisely arranged. It is made up of proteins, a special type of DNA, fats and carbohydrates. Self-assembly of molecules and membranes may be important in the development of a chloroplast but this is not the whole story. The assembly of these structures requires an input of information and energy. This assembly cannot be done in a test-tube because the information needed for synthesis and assembly is contained in the DNA of the cell nucleus.

Although the assembly of molecules and organelles is to some extent understood, the processes that underlie the development of the complete cell, particularly its differentiation into a muscle cell or a nerve cell or some other sort of cell, largely remain a mystery. It is not enough to say that the programme is encoded in the DNA because this does not explain how two adjacent cells in an embryo each with identical DNA can follow two completely different patterns of development, one perhaps forming a nerve cell and the other a muscle cell. Apparently the environment of the cell, which includes the other cells around it, is an important component in determining the life history of the cell. Cellular differentiation is thus an ecological process at the cellular level.

Perhaps the greatest pioneer in the ecological approach to the development of the whole organism from egg to adult has been C.H. Waddington. Waddington (1957, 1975) conceived the individual in its development like a ball channelled along the valleys of a complex landscape (Fig. 1.03). The depth of the valleys is largely set by the genes and to that extent the movement is determined. But in other places the landscape is more flattened so the ball could be pushed over from one valley into another. This is essentially the role of environmental stress such as a high or low temperature stress on the developing egg. The movement of the ball is constantly being chivvied by genetic and environmental forces, the interplay of which gives rise to the final adult form or phenotype. So different phenotypes in the one species are a consequence of genetic differences and environmental differences during development.

At each stage in embryonic development certain genes are active and bring about synthesis of their specific proteins. There are two consequences. New structures develop and the presence of the new proteins modifies the cell in such a way as to bring into action another

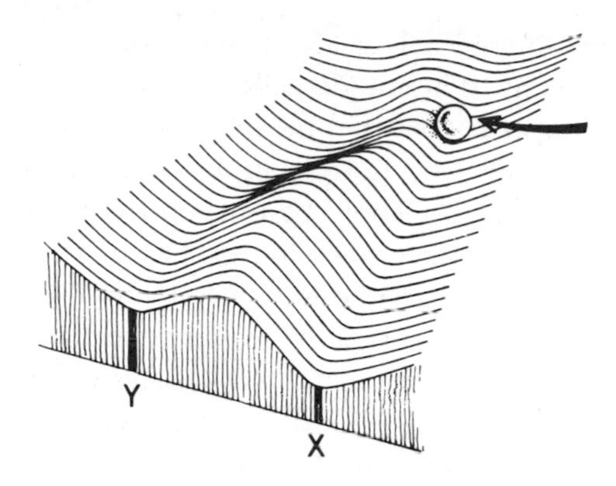

Fig. 1.03. The path followed by the ball as it rolls towards the viewer corresponds to the developmental history of the embryo. The main valley leads to the adult character X. A side path leads to Y but the development does not proceed along the Y path unless an environmental stimulus (arrow) pushes it over the threshold.

set of genes which build in the existing structure of the embryo. The cell changes again in response to these proteins, stimulating again still further genes and so on. Development is thus an ecological succession in which one stage prepares for and initiates the next. The progressively changing gene activities interact with one another in such a way that only certain paths of development are possible, namely those that lead to an integrated whole organism. How that became possible is a matter of evolution.

The process is much more complex than these few paragraphs would suggest. The process of becoming a wing in a fruit fly, for example, is not just a matter of switching on a wing gene at the appropriate time. The development of the wing of a fruit fly is affected by the activities of some forty different genes. The switching on of these genes follows a definite sequence. At each stage the developing wing is 'competent' to respond to a new set of genes. In earlier or later phases it is not competent. Far from being a series of atomistic events the whole process is more like the integration of many players who produce an orchestral symphony. The players not only respond to the score in front of them but also to an outside input from the conductor who determines the pace and many other features of the performance. The score is the genes. The conductor is the environment in which the genes act. The facts of developmental biology lead naturally to the ecological model developed in subsequent chapters.

The maintenance of the order of the mature organism involves a complex of interacting physiological systems which themselves interact with the environment. The single-celled organism that lives in an aquatic environment is bathed in a medium that provides it with all it needs from the outside world. In complex organisms consisting of many cells, most of the cells would be isolated from a watery environment, especially if they happen to be in the body of a terrestrial animal. Yet each cell has to be bathed in a watery medium that brings to it all its needs and takes away from it all its wastes. This is accomplished by means of a circulatory system which in vertebrates ramifies through the body by means of tiny capillaries and fluid spaces. The circulatory system takes from the alimentary or food system the products of digestion the cells need, and it takes from the cells to the excretory system the wastes the cell must get rid of. In addition, there are two co-ordinating systems, the nervous system and the hormonal system, which are closely geared to changes in the external environment. We consider only two examples from the hormonal system to illustrate the close relationship of the whole organism to its environment.

Among seasonal breeders, young are born and raised at a time of the year which is favourable for the new born. Food and warmth are two usual requirements. This regulation of the life cycle to the changing seasons in the outside world requires that mating and pregnancy are co-ordinated with changes in the environment. In many birds and mammals the exciting stimulus is the changing length of days. The increase in day-length as spring approaches is registered by the eyes which communicate to the brain which in its turn causes the pituitary gland to secrete hormones which trigger the maturation of the ovaries and testes. The animals then become ready for breeding at a time which is likely to ensure that food and warmth will be available to the offspring. Where seasons are very irregular, as for example in central Australia, when no rain may fall for years on end, there is no point in having the reproductive cycle geared to changing length of day. Instead, in many inland birds, maturation of the testes and ovaries is triggered by rain.

Another example is the role of the adrenal glands under stress. The adrenal gland, in the human, is a pea-shaped organ attached to each kidney. The central part of the gland, the medulla, secretes adrenalin and noradrenalin. The outer part, the cortex, secretes several other hormones. When danger threatens or in any stressful situation, such as hostility, a narrow escape from peril or cold weather, messages are received by the brain from our sense organs. The brain in turn sends messages to the adrenal glands which then secrete adrenalin and noradrenalin from the medulla. Adrenalin prepares the body for 'fight or flight'. The heart rate and blood-pressure increase, pupils dilate, the concentration of sugar in the blood increases as does the number of certain white blood cells. The blood-flow through the muscles, brain and liver increases by as much as 100 per cent. The digestive and reproductive functions are inhibited and one suffers a feeling of anxiety. Noradrenalin which is secreted about the same time has similar effects. Some cells respond to adrenalin, others to noradrenalin, some to both. This allows for a fine tuning of the system.

Adrenalin, in addition, stimulates the pituitary gland at the base of the brain to secrete a hormone that triggers the adrenal cortex to secrete in turn its hormones. They result in more prolonged adjustment of the body to stress. They also include a hormone that reduces inflammation of damaged tissues should there be any. It is the hormones of the cortex which under prolonged stress maintain the body's emergency mobilisation. Continued and repeated stress can have debilitating effects on the body. The causes of this are not completely known. One theory is that we become 'drunk' on our hormones. As a consequence people get ulcers and other complaints. A system which serves well in special dangers can react back upon the organism if it does not eventually remove itself from these dangers.

Quite independently of stressful situations there is a daily rhythm of secretions of the adrenal cortex. In humans they are produced just before dawn and tone our body for the day's activities. When one experiences a sudden change in the time of dawn, for example, by flying across zones, one's old rhythm continues for a while until it adjusts to the new dawn. This is a contributor to 'jet lag'. In nocturnal animals, such as the rat, the maximum secretion of the adrenal cortex is in the early evening.

These examples are typical of what physiology has to tell about the relationship of the individual organism to its environment. They illustrate the point that there is no such thing as an isolated individual. The individual is what it is in great part through its relationship with the external world around it. Its world is one of physical and chemical and psychological pressures, to which it must adjust appropriately if it is to survive and flourish.

Population ecology

Part of the environment of each individual organism is other members of its own kind and other individuals of different kinds. These influence the individual's chance to survive and to reproduce. They constitute part of the 'struggle for existence' in which every organism is engaged. Darwin used this phrase in a metaphorical sense:

> including dependence of one being on another, and including (which is more important) not only the life of the individual, but success in leaving progeny. Two canine animals in time of dearth, may be truly said to struggle with each other which shall get food and live. But a plant on the edge of a desert is said to struggle for life against the drought ... A plant which annually produces a thousand seeds, of which on the average only one comes to maturity, may be more truly said to struggle with the plants of the same and other kinds which already clothe the ground ... Hence, as more individuals are produced than can possibly survive, there must in every case be a struggle for existence, either one individual with another of the same species, or with the individuals of distinct species, or with the physical conditions of life. It is the doctrine of Malthus applied with manifold force to the whole animal and vegetable kingdoms (Darwin, 1859, pp. 62–3).

In this passage Darwin clearly specified the components of the struggle for existence. Any component of environment of the individual organism qualifies, be it other individuals of the same kind, other species such as predators or disease, shortages of resources such as food or nesting sites or weather. In this passage Darwin lays the foundation of population ecology. This is really the study of what determines the distribution and abundance of living organisms. When and where the struggle is too intense, the species will be rare or may become extinct. Where it is more benign the species may flourish. But in either case there is a struggle in which natural selection of the better adapted individuals is constantly occurring. Evolution and ecology become tied into a single concept – the struggle for existence.

The modern investigation of the struggle for existence is pursued at a number of levels. There are theoretical models in which attempts are made to use mathematical equations to describe the relationship between environment and the chance to survive and reproduce. People also try to reproduce in a simple form the struggle for existence in laboratory experiments. One of the first attempts to do this was described in a rather famous book by G.F. Gause (1934) called *The Struggle for Existence*. Then, most difficult of all, are the attempts to study the struggle for existence in natural populations. There are ways of doing this, and sometimes it is also possible to study evolutionary changes brought about by the struggle for existence (examples of all these operations can be found in Andrewartha & Birch, 1954; Krebs, 1978 and other texts on population biology).

The 'struggle for existence' has been greatly misunderstood. Tennyson described nature as 'red in tooth and claw'. T.H. Huxley, Darwin's great protagonist, referred to nature as 'a great gladiatorial show'. There is cruelty in nature. Predation involves suffering for the prey, at least when they are higher animals such as birds and mammals. The struggle includes predation. But as the passage we have quoted from Darwin indicates it covers much more than that. It is anything that affects life and death and birth. So important is a correct understanding of the concept that Darwin (1859, p. 62) wrote: 'Yet unless it be thoroughly engraved in the mind, I am convinced that the whole economy of nature, with every fact of distribution, rarity, abundance, extinction and variation, will be dimly seen or quite misunderstood.'

What Darwin called the economy of nature and the struggle for existence was given the name ecology by Ernst Haeckel (1870). 'By ecology', wrote Haeckel, 'we mean the body of knowledge concerning the economy of nature – the investigation of the total relations of the animal both to its inorganic and to its organic environment; including, above all, its friendly and inimical relations with those animals and plants with which it comes directly or indirectly into contact' – in a word, ecology is the study of all those complex interrelations referred to by Darwin as the conditions of the 'struggle for existence'.

Ecology is the study not only of what keeps species extant but what makes them extinct. Whereas today on the face of the earth there might be two million species of plants and animals, they are a mere 1 per cent of all species that have existed since the dawn of life (Simpson, 1952). It is the fate of most species, if not all, to become extinct. In the distant past the rate of extinction must have been less than the rate of production of new species. But now, as a result of human activity, it is almost certainly the case that extinction rate is greater than the rate of production of new species. Many species, especially of mammals and birds, have been exterminated by human beings in every continent invaded by human beings since earliest times.

According to estimates of the International Union for the Conservation of Nature, an average of one animal species is lost each year. About 1000 species of birds and mammals are now thought to be in jeopardy. Some ten per cent of the world's flowering plants are said to be 'dangerously rare or under threat'. (Eckholm, 1978, pp. 6–7).

Most of these extinctions are brought about by change in habitat as a result of human activity. The introduction of sheep, cattle and rabbits into Australia less than 200 years ago was probably responsible for more extinctions of native mammals than any other cause. There is now, little doubt that the alteration of the habitat by these exotic animals, including the destruction of food and cover, was the major reason for their extinction. In the mid-nineteenth century the Riverina district of southern New South Wales had a rich marsupial fauna. One expedition recorded 29 species. Today 21 of these species are extinct in the region (Frith, 1973, p. 108). Except for a couple of small marsupial mice, those that survived were either arboreal, aquatic or like the *Echidna* and the large kangaroo extremely adaptable. Ecological studies can help us to understand better the conditions of survival and so help us to prevent the extension of the process of extinction. Other extinctions have been brought about by human hunting, such as for example the passenger pigeon in North America. This may soon be the fate of the blue whale.

The struggle to survive and reproduce is aided by two strategies: individual adaptation and spreading the risks. The modern view of adaptation is that the external world sets certain 'problems' that organisms have to solve if they are to survive. But it is not just a matter of surviving. In adaptation organisms provide better and better 'solutions' to the 'problem'. The end result is the state of being adapted (Lewontin, 1979). Adaptation is thus a dynamic process not a static state. This contrasts with the pre-scientific view of adaptation that led many people to see it as design created from a blue print laid down by a supreme designer.

Two examples of adaptation will illustrate the lengths to which this strategy has been pursued. Both illustrate the special case of co-adaptation in which the life of one organism is adapted to the other and vice versa. *Cryptostylis leptochilla* is a ground orchid in south-eastern Australia. Its flowers look for all the world like a wasp by the name *Lisopimpla semipunctata* that lives in the same region. When the orchid flowers in January the male wasp emerges from its pupal case in the soil and goes off in search of a female wasp. But there are none to be found since they emerge from the soil later. What it does find, however, is the orchid that looks like the female wasp. Indeed, the orchid flower is such a good mimic of the wasp that the male wasp mates with the orchid carrying the process to consummation. It even leaves some sperm behind in the flower. Not only does the flower have a remarkable resemblance to the female wasp but there is some evidence that it gives off a similar odour which attracts the male wasp. Through repeated visits to a number of orchid flowers the wasp successfully pollinates them. It is difficult to know what the wasp gains out of this performance, but it does appear to be essential for the pollination of the orchid.

The Yucca moth of south-western North America has special pollen gathering structures on its mouth-parts into which it packs a ball of sticky pollen from Yucca flowers. The moth subsequently turns to egg-laying. It pierces the wall of the ovary of the Yucca flower with its ovipositor and deposits an egg within. After each egg is deposited the moth climbs to the top of the female part of the flower and rubs part of the sticky pollen into the open end of the stigma tubes. It lays about six eggs. Each egg requires developed fertilised ovules for its growth. But there is a large excess of unmolested ovules that ensure ample seed for the plant. The Yucca flower is constructed to prevent self-fertilisation and the moth is essential to the perpetuation of the species.

A second strategy in the struggle to survive and reproduce is spreading the risks. If I have savings which I wish to invest as a provision for the future I can decide 'to put all my eggs into one basket' and invest my savings in what appears to be the one most profitable enterprise. In doing that I may be taking a big risk. Alternatively, I may spread the risk over many enterprises. This is a logical strategy because the chance of all enterprises becoming bankrupt is much smaller than the chance of one becoming bankrupt.

Nature uses both strategies. A herbivore that feeds on only one species of plant is utterly dependent upon that plant for survival. Yet the plant, if it is an annual, may be around only in the spring and perhaps it lives only in swamps. This hypothetical herbivore is a specialist. This is a good strategy if the chances are that there will always be swamps and the right species of plant in the swamps. It is the wrong strategy if these conditions are not usually fulfilled. Another animal feeds on many species of plants that live in many different sorts of places. It spreads the risk. This is a good strategy if any one of these plants or places has a fair chance of disappearing from time to time.

Many living organisms 'spread the risks' in their lives. They have adaptations that enable them to cope with a diversity of foods and living places and seasons of the year. They do this in two main ways. Their populations are genetically heterogeneous and the habitats they occupy are heterogeneous. They are 'jacks of all trades'. In a changing environment it is a better strategy to be a jack of all trades than to be a specialist.

Genetic heterogeneity means that individuals in the population are genetically diverse, each one being a little different in its adaptations, one better adapted to cold, another better adapted to heat and so on. Some strains of cultivated plants are specialists. They are genetically very uniform. They have a high yield in just the right environment but they are vulnerable to change brought on by changes in weather or arrival of new pests such as fungi and insects against which they are not resistant. Organisms need genetic diversity in their populations if they are to evolve new forms that can cope with these sorts of changes. This is one of the reasons why agriculturalists try to maintain wild species of cultivated plants. These provide a reservoir of genes for cultivated plants that may one day be valuable in the changing environment. So the disappearance of native vegetation from the face of the earth is more than an aesthetic loss. It is a threat to agriculture unless we can maintain special reserves for storing these genes. Living organisms that are to survive cannot be adapted only to the conditions around them today. They must 'anticipate' changed environments. This they do by maintaining a diversity of genes. There is an inbuilt genetic heterogeneity within species.

A second component of spreading the risks is environmental heterogeneity. The great Serengeti National Park in East Africa has one end which has sources of water in the dry season and another end which only has water in the wet season. The game animals that live there can take advantage of the whole area by confining themselves to the wet end in the dry season and spreading out all over the park in the wet season. The story is more complex than that. But in essence the ability of the animals to take advantage of their environmental heterogeneity increases their chance of survival. If they confined themselves just to the area that is wet throughout the year, they would occupy a very much smaller area and have access to a much smaller quantity of food.

In establishing a national park for large mammals it is necessary that the park be large enough to enable movement of animals from unfavourable sites to favourable sites as conditions change in the park. The same sort of principle operates on a small scale, say within a hectare or so, for smaller organisms. Local heterogeneities mean that the risk varies in different places so the chances are that not all places are going to become unfavourable at any one time. By maintaining natural heterogeneities in the environment we help to sustain many species which might otherwise become extinct. A world covered with monocultures of agricultural crops will be a depauperate world even if it proves to be still sustainable.

The web of life and the 'balance of nature'

Species that manage to survive on the earth today are those which have evolved (and continue to evolve) adaptations that enable them to survive and multiply despite checks to increase. Such adaptations may involve appropriate relationships to the environment including adaptations to other species on which they depend. Darwin (1859, p. 73) referred to plants and animals as being 'bound together by a web of complex relations'. Here is Darwin's descriptive account partly based on a paper by a Mr Newman of 1850 and in part upon his own observations.

> From experiments which I have tried, I have found that the visits of bees, if not indispensable, are at least highly beneficial to the fertilisation of our clovers; but humble-bees alone visit the common red clover (*Trifolium pratense*), as other bees cannot reach the nectar. Hence I have very little doubt, that if the whole genus of humble-bees became extinct or very rare in England, the heartsease and red clover would become very rare, or wholly disappear. The number of humble-bees in any district depends in a great degree on the number of field-mice, which destroy their combs and nests; and Mr H. Newman, who has long attended to the habits of humble-bees believes that 'more than two-thirds of them are thus destroyed all over England'. Now the number of mice is largely dependent, as every one knows, on the number of cats; and Mr Newman says: 'Near villages and small towns I have found the nests of humble-bees more numerous than elsewhere, which I attribute to the number of cats that destroy the mice.' Hence it is quite credible that the presence of a feline animal in large numbers in a district might determine, through the intervention first of mice and then of bees, the frequency of certain flowers in that district! (Darwin, 1859, pp. 73–4).

Darwin traced out the web of relations between the humble-bee and other species and other components of its environment. Alter one component and the effects can be far reaching.

It was the fate of Darwin's clear statement about humble-bees that it became elaborated in the telling. The cats were said to belong to spinsters so there must be a correlation between the number of spinsters and the amount of red clover in a district. The women were spinsters because there were so many Englishmen in the Navy. The sailors ate dried beef that came from cattle which grazed on clover fields pollinated by humble-bees. In short, if the spinsters had decided to keep their cats indoors the British Empire might have collapsed long before it did. As the mice increased, the bees, clover, cattle and sailors would have all decreased in numbers. We know now that there is not the simple one to one relationship between the humble bees, clover and mice that Darwin postulated. The real point that Darwin was making concerned the struggle for existence within the complex web of relationships between individuals of different species. Here was an example of 'struggle for existence' which was part of the process of the origin of species by means of natural selection of chance variations that took place within the struggle. Those individuals which were endowed, by chance, with some feature or features that gave them an edge in the struggle would be the ones that would be perpetuated and which would hand on their favourable characteristics to their offspring.

One aspect of the web of life is what ecologists call the food-web. This is usually depicted as a rather complex web of lines on a diagram showing who eats what in a community. At the base of the food-web the organisms are always plants because the sole gateway of entry of energy into the living world (with the minor exception of some bacteria) is the green plant. The green plant traps the energy of sunlight converting it into high energy molecules such as starch. This becomes the source of food for herbivores which in turn are food for carnivores. Then there may be carnivores that feed on carnivores. Food-webs can be typically organised into four such food levels. Very rarely are there more than five food levels but we do not really know why. Food-webs sometimes reveal surprising dependencies. For example, a seemingly insignificant mussel in a Georgian salt marsh is largely responsible for making phosphorus available to the community (Kuenzler, 1961). Pollutants such as DDT tend to accumulate in the higher levels of the food web. For example, in a Long Island estuary in New York the final predators, the tern and osprey, have the highest DDT content of all organisms in the estuarine community (Woodwell, 1971).

The question arises: why don't the herbivores in a food-web eat all the plants or why don't the carnivores eat all the herbivores? Why is the world still green when so many animals feed on green plants?

There are no simple answers to these questions. We suggest some later in this chapter which include the heterogeneity of the natural environment. It is possible that as the environment becomes more homogeneous because huge areas are planted with single species or grazed by a single species, the chance that large sections of the world will no longer be green is increased.

The principle of inter-dependence relates not simply to the food-web but to the relationship of the individual organism to other aspects of its environment. The cycle of elements in nature is an example. The cycle of phosphorus in the Georgia swamp already mentioned is an example of a nutrient cycle on a small scale. The so-called global bio-geo-chemical cycles of carbon, oxygen, nitrogen, sulphur and phosphorus constitute what have been called the life-supporting systems for our planet. These cycles determine the composition of the atmosphere, the composition of sea water and fresh waters, and the fertility of land and water. For example, nitrogen is an essential constituent of living organisms. Very few organisms can 'fix' nitrogen from the atmosphere. Those that can are a few species of bacteria and algae. Although the atmosphere is an enormous reservoir of nitrogen, the continuation of life on earth is almost certainly completely dependent upon the activity of these tiny organisms. Had DDT been toxic to these organisms, who can tell what effect its wide dispersal would have had on life? There is another set of microorganisms that 'unfix' nitrogen and return it to the atmosphere. Unless this was the case nitrogen from the atmosphere would be used up. These organisms are denitrifying bacteria. A breakthrough in modern agriculture came with the invention of industrial processes for fixing nitrogen from the atmosphere. From this enormous reserve nitrogen fertilisers can be produced, although production may be limited by available energy. Nitrogen fixation by the fertiliser industry now amounts to 26 per cent of the present biological terrestrial fixation. The annual growth rate for the fertiliser industry is at present 7–9 per cent. If this high growth rate is maintained, the industrial fixation will be as large as the biological fixation by 1989. The implications of such a change are at present quite unknown (Söderlund & Svensson, 1976).

While nitrogen is taken out of the atmosphere through industrial fixation, it is put back in the form of nitrogen oxide in automobile exhaust gases. Some of this gets into the soil and from there into water supplies which can become overloaded with nitrogen oxides to the extent of making the water toxic for drinking. Atmospheric nitrogen oxides seem to be involved in the destruction of ozone in the stratosphere (Söderlund & Svensson, 1976). It is feared that an increase could affect the ozone shielding the earth from ultraviolet radiation. Without this ozone the sun would kill all green plants on land.

In some communities the cycling of the mineral elements is so complete that the water that runs off is virtually as free of minerals as distilled water. This is the case with some of the mature rain forests in the Amazon basin. Mineral nutrients are taken from the soil by plants, converted by them into plant tissue. Some of this is eaten by animals and so is converted into animal tissue. All these organisms eventually die and fall to the forest floor where they are converted back into mineral nutrients by microorganisms. The cycle is complete. The plants are so efficient in removing these nutrients as they are formed that none is available to be leached out by rain. So the mineral nutrients are recycled, generation after generation, with little loss. All that comes in from the 'outside' is water, carbon dioxide, oxygen and energy from the sun.

By contrast, our human societies are extremely wasteful. We recycle very little of the products we use in our industrial societies. Some of our activities block cycles and convert them into a linear series of reactions with consequent accumulation of products instead of their recycling. This is what happens when inland waters get choked with nutrients from sewage, farms and industry. The cycle stops with the accumulation of undecomposed products in an oxygenless environment. What was a self-sustaining system has been converted into an unsustainable one. Unfortunately human activities seem too often to convert a sustainable system into an unsustainable one. Indeed, a world in which human population keeps on growing at the present rate would become unsustainable soon, if we have not already reached that critical point. Human societies become unsustainable in three ways: when they run out of non-renewable resources such as fossil fuels, when the demand on renewable resources is greater than the environment can sustain and when the pollution absorptive capacity of the environment is exceeded. A critical requirement of our human society is that it be so organised that it is sustainable indefinitely into the future (Chapter 8).

But does this sustainable web of life constitute a balance of nature? The founder of modern animal ecology, Charles Elton, wrote: 'The balance of nature does not exist and perhaps never has existed' (Elton, 1930, p. 17). But isn't it true that we have just given examples of how humans can upset the 'balance of nature'? In a very loose way of speaking this is so, but it is both more honest and more helpful for environmentalists to use more precise language. It is more precise to say that such and such an agricultural programme is unsustainable if that is what it is. When essential interrelationships between organisms and their environments are damaged we should be able to specify in what way rather than simply say the balance of nature has been upset. These problems are better understood in terms of Darwin's web of life.

Sometimes the term 'balance of nature' is used in a way which does not reflect anything real in nature and this is what Elton (above) was complaining about. This is the case when people suppose that in the natural state the abundance of plants and animals in a community does not change and that any change is an upset in the 'balance of nature'. This is myth rather than science. Unfortunately, it is a myth propagated by some scientists who ought to know better. Part of the order of nature is that numbers within a species change, often dramatically, yet the species may continue to exist for millions of years.

Whilst there is an inbuilt tendency for a species to increase in numbers it is obvious that, for any species, numbers cannot increase for ever. Hence the 'struggle for existence' as to who survives to reproduce. In any mature community such as a tropical rain forest we can clearly see the limits to growth. Every rain forest started with seeds that grew into seedlings that grew into shrubs and trees. This was in the early growth phase. There was first an increase in biomass. But eventually the forest reached a mature phase. Energy was then used not primarily for growth at all but for the maintenance of the mature community. Of course, lots of seeds are still produced in a mature forest, but in the struggle for existence most of them never see the light of day. A mature forest has reached the limits to growth.

As far as animal populations are concerned there is much controversy amongst ecologists as to what limits the growth of animal populations. Broadly speaking there are two ways in which the problem is solved, which we can call the 'house cat' solution and the 'alley cat' solution. The population of house cats anywhere is defined completely by the number of territories, that is of cat lovers. The kitten that cannot find a home does not become a house cat. The great tits of Europe and the magpies of Australia are like this. They regulate their numbers through territorial behaviour. Alley cats are different. They are largely non-territorial. Their population growth is limited by many things: shortage of food, accidents, disease and inclement weather.

This is a much more haphazard way of limiting population growth and it would seem to entail more suffering. But it is the way of many species. It is indeed the 'solution' adopted by human populations as the history of human population growth shows. As a global population humans have not yet made the transition from population adolescence to population maturity. In the adolescent phase growth continues until it hits the limits imposed by shortage of food and shelter and by disease. Sooner or later over-population in humans, as in locusts, is followed by a crash, unless deliberate attempts are made toward a smooth transition to no-growth. What we can be sure about is that nothing grows for ever because the environment has a limited carrying capacity for living organisms.

There is an order of nature, but far from being imposed from without it is a consequence of the struggle for existence within nature. This order can be recognised in the web of life and in the wonderful adaptations of living organisms. In natural populations there are forces for stability and forces for change, both of which were first clearly seen by Charles Darwin and A.R. Wallace in their independent elaboration of the theory of evolution by natural selection of chance variation in the struggle for existence.

The existence of an order of nature has for long been an intuition. What is new is the modern scientific understanding of what the nature of this order is. But the phrase 'balance of nature' arose as a pre-scientific answer to real questions about order. It had its origin in Greek thought. Egerton (1972, p. 325) wrote:

> It would be difficult to imagine the rise of Greek philosophy and science unaccompanied by certain basic ideas that were congenial to the development of balance-of-nature concepts ... Greek science was built upon the belief that nature is constant and harmonious ... the Pythagoreans heard musical harmony in the universe ... Greek medicine taught the doctrines of the balance of the humours and the healing powers of nature ... accordingly ecological harmony and balance would have been a compelling expectation.

Herodotus asked why the birds, beasts and men did not eat all the hares? Why did not the whole earth swarm with serpents? His answer was that a superintending Providence had created the different species with different reproductive capacities. Predatory species had fewer offspring than their prey. To illustrate this he assembled reports on reproduction of snakes, hares and lions. His account is the earliest known attempt to obtain biological evidence for the 'balance of nature' idea. His idea that species were specially created with the appropriate capacities to result in a balance was not strongly challenged before Darwin.

Plato's dialogues contain two creation myths that were influential sources for the 'balance of nature' concepts. One is in the Timaeus. In answer to the question: 'In the likeness of what animal did the creator make the world?' Timaeus gives the answer that God did not make the world like any one species but rather as 'one visible animal comprehending within itself all other animals of a kindred nature'. Egerton (1972, p. 327) regards this as the source of the supraorganism concept of nature which still has its modern expounders (e.g. Allee *et al.*, 1949, p. 721 *et seq.*). The second myth in Plato is in the Protagoras in which the god Epimetheus bestowed traits upon each species that would assure it escape from its predators, protect it from the weather and enable it to find the proper food. Both myths contained the idea that providence had endowed each creature with capacities that enabled it to live in permanent association with others and to cope with the perils of the environment, a view which Egerton (1972, p. 327) refers to as 'providential ecology'. This view conflicts with the quasi-natural selection of Empedocles, Democritus and Lucretius. 'Providential ecology' is a view expressed again in the Roman era in Cicero's dialogue *De Natura Deorum* in which the 'balance of nature' became part of the Stoical evidence for the wisdom and benevolence of the creator. The 'balance of nature' was maintained by the differential rate of reproduction of the different species depending upon their place in nature (as a predator or prey) and on the mutual relations between species. Two centuries later Plotinus sought to reconcile the existence of predation with a benevolent creator. Predation, he explained, was one of the evils required for the greatest diversity and quantity of life to come into existence.

It is not until the seventeenth century that we next find an elaboration of the concept of 'providential ecology'. Sir Thomas Browne pointed out an inverse relationship between rate of reproduction and longevity. Species such as human beings and elephants which reproduce slowly tend to be long lived. The long life makes up for the lower rate of reproduction in maintaining the 'balance of nature'. The destructive aspects of 'noxious animals and vipers' are mitigated to some extent, so he thought, by hibernation which puts them out of action for a large part of each year. In 1662 the founder of demography, John Graunt, added still more factors to the list of those supposed to maintain the 'balance of nature'. He discovered the statistical regularity of both the sex ratio in human beings and most of the causes of mortality (apart from epidemics) in human populations. Sir Matthew Hale in 1667 added yet another idea to the 'balance of nature' concept, namely the alternation of hot and cold seasons or what he called 'heat fluctuations'. Most notable of all in the seventeenth century is John Ray the botanist–clergyman who published two widely read works in 1691 and 1693 on what became known as natural theology. For Ray each species had its special place in nature. Each was exquisitely designed to survive best in its special niche. Young birds developed rapidly after hatching which is a vulnerable stage. Plums ripen just when wasps which feed upon them emerge from their cocoons. The fossil evidence that plants and animals had become extinct worried Ray as it was a threat to the 'balance of nature' concept. He remarked that extinction would contradict the 'wisdom of the ages'. However, he argued that since the world was still incompletely explored the species known only as fossils might still live in unexplored regions.

A disciple of Ray, William Derham, published a series of lectures in 1713 entitled *Physico-Theology*. In this publication he recited a long list of examples to defend his thesis of the 'balance of nature'. 'The Balance of the Animal World', wrote Derham (quoted by Egerton, 1972, p. 333), 'is, throughout all Ages, kept even, and by a curious Harmony and just Proportion between the increase of all Animals and the length of their Lives, the World is through all Ages well, but not over-stored.' The existence of plagues was an awkward fact to contend with in this view. He concluded that they were to chastise us and to excite our wisdom and industry. In the same century, Alexander Pope in his *Essay on Man* suggested that all species were so closely interdependent that the extinction of one would lead to the destruction of all living nature.

In 1749 Linnaeus, the great Swedish botanist, wrote an essay entitled *Oeconomia Naturae*. In it he wrote: 'the divine wisdom has thought fit, that all living creatures should constantly be employed in producing individuals, that all natural things should contribute and lend a helping hand towards preserving every species, and lastly that the death and destruction of one thing should always be subservient to the restitution of another' (quoted by Egerton, 1972, p. 336). Here we find once again a resurrection of the 'balance of nature' idea. Propagation, preservation and destruction maintain the economy of nature. If it were not, said Linnaeus, for the selective feeding of insects, some species of plants would get so abundant as to crowd others out of existence.

The conflict between the fact that species do become extinct, which was generally admitted by 1800, and the traditional view of the 'balance of nature' seems to have been appreciated by only a few naturalists. One of them was Alfred Russel Wallace who wrote in his 'species notebook' around 1855 (Egerton, 1972, p. 339) as follows.

> Some species exclude all others in particular tracts. Where is the balance? When the locust devastates vast regions and causes the death of animals and man, what is the meaning of saying the balance is preserved? Are the devastations of Sugar Ants in the West Indies and the locust which Mr Lycll says have destroyed 800 000 men an instance of the balance of species? To human apprehension there is no balance but a struggle in which one often exterminates another.

Here we have the first critical questioning of the age old 'balance of nature' concept, a questioning which was continued by Charles Darwin in *The Origin of Species*. The theory of evolution which Wallace and Darwin independently developed involved a complete acceptance of the fact of extinction. Indeed it seemed to be the fate of all species to become extinct. Their ideas of the struggle for existence gave a new emphasis, namely the instability of nature.

Darwin's revolutionary ideas had considerably less influence on ecological theorists of the 'balance of nature' school than they should have had. Ecology today needs to be liberated as much as any other part of biological science. Its besetting sin is to become constrained within deterministic notions that treat plants and animals as so many symbols in complex mathematical equations, which, however complex, bear little resemblance to the complex web of nature and the multifarious forces of the struggle for existence within which every individual finds itself. Mathematical models have their place, but they cannot replace observed facts and sound thinking, both of which were crucial to Darwin's approach.

Sound ecology will depend less upon glib phrases such as 'balance of nature' and more upon real observations and experimental manipulation in the field. Experiments are needed to find out just how much DDT will produce what effects on non-target organisms before we release such materials into the environment. That is done by means of controlled experiments. It is not science to say that DDT is a foreign agent and will therefore upset the 'balance of nature'. Before scientists can intelligently plan such experiments with DDT or anything else they need to know much more about Darwin's 'web of complex relationships' that exist between an organism and all components of its environment, be they other organisms or food or weather. Sound ecology is built upon an analysis of environment and how its components affect the chance to survive and multiply. For every species, in every part of its environment, there will be complex relationships that have to be understood by observation and experiment in order to learn what accounts for its distribution and abundance and how our activities may be changing these relationships.

Conclusions

The modern way of looking at biology as a whole is in terms of levels. There are three such levels: the cell, the organism and the population. At each level ecology is the most appropriate category for considering the phenomena. The life of the cell is best understood in terms of ecological relationships among molecules. The living organism is best seen in terms of its ecological relationships to its environment. The interdependence of each living organism with other living organisms and with other components of its environment is the principle of population ecology. The importance of ecology for the understanding of life at these three levels points forward to the proposal of an ecological model of living things in Chapter 3. But already this chapter has begun to clarify what life is.

The process of building complex molecules, cells, tissues, organs and organisms that goes on every day in all living creatures is a process of decreasing entropy or increasing order. Living organisms are amongst the few components of the universe where this occurs. The second law of thermodynamics is the law of increasing entropy in the universe as a whole. Life is a special sort of organisation of atoms and molecules that decreases entropy. The organisation of life owes its particular form to internal designing forces in the DNA molecules. Life is characterised by the capacity for self-reproduction. It thus constitutes a spreading centre of order in a less well-ordered universe. All along the line, from the first large molecules to cells to organisms, the properties of life are to be understood in terms of the ecological relationships of the parts to one another and to their environment.

The beginnings of life centre on two very large molecules: deoxyribosenucleic acid (DNA) and protein. Both molecules can exist in an infinite array of forms. Which came first and how they became conjoined as the basic molecules of life is largely unknown. What is clear, is that life builds on information which is provided by the DNA molecules which is translated eventually into the language of specific proteins. Proteins are both structural as in muscle and catalysts as in enzymes that promote the multifarious chemical reactions of the cell.

The life of the whole organism can be summarised as the emergence of order, the maintenance of order and the disappearance of order. Development is the progressive appearance of order. It is best understood in an ecological context. The developing individual is a product of internal forces that stem from the genes and of external forces imposed by the environment. It is the interplay of these two sets of influences which determines the final adult form or phenotype. Physiology is the maintenance of order in the individual organism. But it is an order that is geared to an ever changing environment of day and night, seasons of the year and longer term changes. So physiology too is best understood within an ecological framework.

There is an order of nature in the life of the population of living organisms that clothe the earth. Although much less 'fixed' than the order of the cell or of the individual organism, it can nevertheless be understood in ecological terms. The key idea here is 'struggle for existence' which includes among its strategies individual adaptation and spreading the risks. From these dynamic processes emerges the principle of interdependence of living organisms and their environment. This is exemplified in the web of life and in the bio-geo-chemical cycle of the elements. These are both parts of the life-supporting system of the earth. If essential pathways in this system are broken, the life veins of earth could be cut. The sustaining of life on earth is now very much in human hands and has been, to an increasing extent, ever since human beings ceased to be hunters and gatherers.

2

Evolution

> How have all these exquisite adaptations of one part of the organisation to another part, and to the conditions of life, and one distinct organic being to another being been perfected?
> Darwin (1859, p. 60)

Following the quotation that heads this chapter Charles Darwin wrote about a parasite that clings to the feathers of a bird for transport, the special structure of a beetle that dives in water for food and the plumed seed that is wafted in the wind. In the first chapter we discussed other examples of adaptations of plants and animals. Their numbers could be extended endlessly.

The modern view of the origin of adaptation is the answer Darwin provided, but understood now in terms of genetics. Evolution by means of natural selection is the mechanism of adaptation. Darwin knew that 'organs of extreme perfection and complication' were a critical test case for his theory, and he took these up in a chapter on 'difficulties of the theory'. He wrote:

> To suppose that the eye, with all its inimitable contrivances for adjusting the focus of different distances, for admitting different amounts of light, and for the correction of spherical and chromatic aberration, could have been formed by natural selection, seems, I freely confess, absurd in the highest possible degree. Yet reason tells me, that if numerous gradations from a perfect and complex eye to one very imperfect and simple, each grade being useful to its possessor, can be shown to exist; if further, the eye does vary ever so slightly, and the variations be inherited, which is certainly the case; and if any variation or modification in the organ be ever useful to an animal under changing conditions of life, then the difficulty of believing that a perfect and complex eye could be formed by natural selection, though insuperable by our imagination, can hardly be considered real (Darwin, 1859, pp. 186–7).

The function of this chapter is to outline the main elements in the currently emerging form of Darwin's theory.

The first section explains that chance constitutes the foundation of evolution. It is the fact that DNA does not always replicate itself exactly that makes possible the emergence of new forms of life.

The second section explains how natural selection works in the preservation and multiplication of some of the chance mutations described in the first section. This occurs in such a way that it is possible to think of chance and necessity as two great principles operative in evolution.

However, in fact the selection of animals depends in part on their own intelligent and purposeful response to their situations. The third section shows that purpose is indeed a part of the explanation of the course of evolution even though there is no pre-established goal for this process.

Especially when animal purpose is included among the explanatory factors in evolution, human evolution can be seen as continuous with biological evolution in general. The fourth section presents human evolution both in its continuity and in its distinctiveness.

Chance and mutation

The evolution of a living cell from organic molecules may have happened more than once on the earth. But probably only one original cell gave rise to all the rest of life on earth. This seems to be the only possible explanation of the basic similarity of the cells of all living organisms. All use the same DNA code and similar amino acids. The doctrine of evolution holds that from one beginning all the diversity of life on earth, its two billions of species (of which two million known species are alive today) and the many varieties within these species have arisen. Life is like a great branching tree with one central stem. The specifications of the first cell were contained in its particular DNA. In so far as this cell produced subsequent generations of cells with precise replications of the original DNA (which is what usually happens when cells replicate), there would have been no evolution. The original information in the first cell would have been faithfully transmitted to subsequent generations, each generation spelling out the same message for ever. But that is not what happened. On occasions, rare occasions to be sure, the DNA was not faithfully replicated. It replicated with a difference. The sequence of nucleotides changed from the original order; some were dropped out, others were added. This is mutation. It is the basic ingredient of evolution. Without it there could be no diversity of life. Eventually billions of species arose with their different sorts of DNA providing them each with different information.

What is known about mutation comes from studying it in living organisms today. In the growing organism, such as a human being developing from a fertilised egg, with each multiplication of cells there is faithful copying of the DNA of the parent cell. This is true also in the mature human being when the sex cells are produced (excepting that each cell receives only half the information). Once in a few million or a few billion times the replication of multiplying cells is not exact. The mistake or accident in replication in the sex cells is the beginning of something new. It is a mutation. Usually the mutation spells out a message which is disastrous for the organism by making it less well adapted. But it may, by chance, spell out a message that confers characteristics that enhance the chance for survival and reproduction. Mutation that changes the DNA molecules such that it confers upon an insect resistance to DDT is a lucky accident. Before DDT was invented by human beings this particular sort of DNA (that confers resistance to DDT) was produced but not perpetuated, because it conferred no advantage on the organism possessing it. There is something profligate about nature producing a vast array of different mutations as though in anticipation of the day when their time will come. Many are called but few are chosen. The mutation that changes the DNA in a human being such that it causes its carrier to have the disease haemophilia is a highly deleterious one. Unfortunately there are over 1600 human diseases now identified as being caused by specific mutations (Birch, 1975*b*, p. 10). This is one of the costs of creation by mutation.

About the chance aspect of mutation Monod (1974, p. 11) wrote: 'We say that these are accidents, due to chance. And since they constitute the *only* possible source of modifications in the genetic text, itself the *sole* repository of the organism's hereditary structures, it necessarily follows that chance *alone* is at the source of every innovation, of all creation in the biosphere.' Whilst agreeing with Monod that chance plays a critical role in evolution, we argue in this and later chapters that chance events do not exclude a role for purpose which becomes particularly crucial in human evolution.

When mutations are referred to as spontaneous or chance in occurrence, this does not mean that they are not caused. Accidents have causes. Some of the causes of mutation are known. Cosmic rays, X-rays, other sorts of radiation, as well as certain chemicals, are known to be mutagenic. But the mutation induced bears no relation to the needs of the organism at the time it is produced, except by chance. It more often harms than helps survival just as random interference with the mechanism of a watch is more likely to harm the mechanism than to improve it.

Theoretically it is possible to imagine a system in which mutations are produced by the direct effect of the environment in which organisms live and in such a way as to enhance their chance to survive and reproduce in that environment. It might be imagined that extreme cold could induce a mutation conferring cold resistance or that DDT might induce a mutation that conferred DDT resistance. There is nothing wrong with this as an idea. But it does not happen. Some environmental stresses such as radiation can cause mutation but their effects are rarely, if ever, appropriate in the sense of conferring an advantage such as radiation resistance. It would perhaps be an efficient system if it happened. The difficulty, presumably, is that the genetic material DNA, which stores the genetical information, must be inert and unreactive if it is to be a reliable store. If it were capable of being changed by all sorts of environmental influences, such as changes in temperature and changes in any other component of environment, it would soon be reduced to jibbering nonsense. Living organisms have to make do with the very indirect process of relying on purely random mishaps to produce changes in their genes. On the face of it one might think this is an extraordinarily inefficient method of creation. In fact it is not. Two special genetic mechanisms seem to extract the maximum value out of this way of producing variation. They are recombination of genes through sexual reproduction and storage of genetical variability.

The molecular events involved in mutation, which we have discussed, were discovered quite recently. The role of mutation in evolution has been known much longer. Darwin knew that there had to be a source of genetical variability if there was to be any evolution. He did not know what that source was. In desperation he fell back on Lamarck's thesis that genetical variation originated from the inheritance of acquired characters. According to Lamarck the duck got its webbed feet as follows. Birds on dry land do not have webbed feet. If the climate changed and the countryside became flooded, some birds would have attempted to paddle. This new form of exercise would lead to a bodily change, the development of webbing between the toes. This is what Lamarck called an acquired character. The change in body led directly to a change in genes. The offspring would then be born with webbed feet. The sequence of Lamarck's argument is that change in environment produces change in habit, which produces change in body form, which produces change in genes. It is the complete reverse of the modern view that change in gene comes first and that leads to change in body form. Lamarck's argument is a perfectly logical hypothesis. The trouble about it is that experiments have falsified it.

Mutation was identified for the first time in the laboratory by T.H. Morgan in 1910 when white-eyed fruit flies appeared in his culture of normally red-eyed flies. Morgan was able to show that the gene for white-eye had mutated from the gene for red-eye in these flies. It happened only rarely, but when it did happen it was transmitted to the offspring according to Mendel's rules. Neither Morgan nor his contemporaries for a decade or more considered mutation to be of any importance in evolution. All it seemed to do was to produce misfits: non-pigmented eyes, missing eye elements, incomplete wings, malformed legs and so on. Secondly, they did not see mutation happening in natural populations. Indeed, they were more impressed with the uniformity of individuals in natural populations than with any evident diversity.

It is quite natural, when observing a population that is well adapted to its environment, to come to the conclusion that all mutations are misfits. All the 'good' ones will already have been incorporated into the genetic composition of the population. However, when the observations are made on populations which are deliberately put into a stressful environment, we can observe mutations that increase the chance for survival and reproduction in the stressful situation. This happens every time a culture of bacteria is subjected to an antibiotic. Most of the bacteria are killed. Those few that survive will be those that have a mutant gene or genes for resistance to the antibiotic. Similarly with development of resistance of an insect population which finds itself in the stressful environment of DDT or some other insecticide. It is important to appreciate that in these sorts of examples it is not the toxin that induces the mutation. All the toxin does is to provide an environment in which the mutant can flourish. The mutation occurs at a low rate all the time, whether or not the toxin is present.

The second problem about mutation as provider of the source of variation necessary for evolution was the apparent lack of variation in natural populations. This problem was resolved in 1927 when the Russian geneticist Tschetwerikoff ventured into the steppes of Russia, chasing down fruit flies in the wild. The flies he collected appeared uniform enough on the surface. However, by appropriate crossings he was able to reveal a great deal of 'concealed' genetic variability beneath the surface. Flies which looked the same had different genes. These different genes originated from mutations which had happened, perhaps, many generations back. This led to another idea, namely, the 'storage' of mutant genes. A mutant gene that is deleterious today may be advantageous in a different environment tomorrow. Not all deleterious genes are wiped out at the time they occur. Some become stored in various ways, which render them innocuous in the current environment. A very simple way is as a recessive gene. Its deleterious effects are only evident when present in a double dose. So it is preserved when present only in a single dose. There are several other ways too. Stored genes await the day when they may be useful to the organism.

Mutation is the ultimate source of differences in genes of different organisms. There is a secondary source of variability and that is genetic recombination which is achieved by means of sexual reproduction. Unlike asexual reproduction, sexual reproduction produces variety in the offspring. In the special cell divisions in the ovary and testis that result in the production of ova and sperm, not all sex cells get the same complement of genes (DNA). Probably none get the same complement. Each one has a selection from what is called the gene pool. Furthermore, when a sperm cell fertilises an ovum there is a further source of variety since sex cells combine at random. In sexually reproducing organisms such as human beings the number of possible combinations of different genes is so great that no two human beings are genetically identical except they be identical twins. Each of us owes his or her genetic identity to at least 10 000 and probably many more pairs of genes given to us in the fertilised egg. Each of these pairs of genes is a selection of two out of about ten possible different genes. What then is the number of different combinations of ten thousand pairs? The answer is $10^{10\,000}$ which is more by far than the number of electrons in the universe, which is of the order 10^{80}.

Genetics teaches the uniqueness of the individual. Each human being is genetically unique. Which complement of genes she or he gets is a matter of chance, depending upon which collection of genes each sex cell got and on which sperm fertilised which ovum. Chance at this second level is thus responsible for producing a tremendous pool of genetic variability in a sexually reproducing population.

Natural selection

It is from the pool of genetical variability derived from mutation and the recombination of genes through sexual reproduction that natural selection operates. Natural selection is not a force so much as a necessary outcome. Hence Monod's (1974, p. 114) statement: 'drawn from the realm of pure chance, the accident enters into that of necessity, of the most implacable certainties'. The necessity of which Monod speaks so emphatically is that individuals endowed with a higher capacity to survive and reproduce will necessarily contribute more of their kind to successive generations than those individuals less favourably endowed. However, what follows from chance variation is not the sort of deterministic necessity which statements such as Monod's imply. The operations of natural selection are more subtle than that. We shall argue later in this section that for creatures having the power to choose and learn about different environments 'necessity need no longer be spelt with a capital N' (Thorpe, 1978, p. 33). In the later stages of evolution 'behaviour is always a jump ahead of structure', so that 'habits, traditions, and behavioural inventions must have played an ever-increasing role in the evolutionary story as the animals mounted the ladder of complexity' (Thorpe, 1978, p. 75). Secondly, the naïve selectionist approach too often gives us an evolutionary biology of organs and genes, but not of organisms. 'It assume(s) that all transitions could occur step by step and understate(s) the importance of integrated developmental blocks and pervasive constraints of history and architecture' (Gould & Lewontin, 1979, p. 597). For example, humans are not optimally designed for upright posture. This is because so much of our structural design, particularly the skeleton, evolved from quadruped life. Natural selection is a fact of nature but we misunderstand its mode of operation when we divide organisms into parts and try to explain each as a direct adaptive response to natural selection. As is explained in more detail later, it is the whole developing organism, from the egg to the mature form, which is subject to selection and this within the constraints imposed by architectural and evolutionary inheritance.

The theory of natural selection has often been described as a tautology. It is a tautology to say that the 'fit' survive. To be 'fit' in the evolutionary sense is to have a high capacity to survive and reproduce. So individuals are less fit or more fit than one another. They are not simply fit or not fit. It is no tautology to say that individuals which are endowed by their genes with adaptive features such as efficient wings will increase in numbers in the population compared with those with less efficient wings. An analysis in which problems of design are posed and characters are understood as being solutions to the problems of design, breaks through the tautology by predicting in advance which individuals will be fitter. Scientists can identify characteristics which are adaptive by all sorts of experiments, both physiological and ecological. Then they can independently see how their possessors fare in the 'struggle for existence'.

A simple and accurate definition of natural selection is that it is the differential survival and reproduction of individuals in a population. It is an ordering process because it moves a population in the direction of greater adaptiveness. When an environment changes, most individuals may be unadapted to the new situation. If the population has a rich store of genetic variability at its disposal, natural selection will move the population toward one that is adapted to the new environment. As the environment changes, so will the direction of natural selection in each successive generation. This elemental process of change would eventually lead, so argued Darwin, to the origin of new species. The diversity in nature today is to be explained on the basis of natural selection of random variation over vast periods of time. Darwin did not witness natural selection in operation. He knew about artificial selection in which the animal or plant breeder did the selecting. From that he inferred the possibility of natural selection. Provided we have the appropriate techniques, we can today demonstrate natural selection in action in almost any population we choose to study. The evolution of resistance of bacteria to antibiotics and of insects to insecticides are examples. We can also study the way in which weather in successive seasons is a selecting agent. So is a selective predator. So is any component of environment when we define environment precisely (as is necessary in evolution and ecology) as anything that affects the chance to survive and reproduce.

It usually takes a long time, thousands of generations, for populations to so diverge from one another that we would call them different species. Populations that are so different as to be called different species will normally differ not just in a few but in a very large number of genes, let us say 500 mutant genes. What then is the chance of incorporating 500 mutant genes into a population when the rate of mutation is so slow and the proportion of mutations that is beneficial is also very small? Let us suppose that 10 000 generations are needed for the transformation of one species into another. Secondly, let us suppose that the mutation frequency per gene is one in every 100 000. Thirdly, suppose that only one in every 100 000 mutations is beneficial. It is a matter of simple arithmetic to show that only one in 200 000 of the beneficial mutations that occur needs to be established to get the 500 mutations needed to transform one species into another. So despite the low rate of occurrence of beneficial mutations, plenty of them occur, given enough time, to fulfil the requirements of the theory. The time is so long that we cannot expect to witness this transformation in a human life-time or even many human life-times. Does that mean that we shall never witness the transformation of one species into another? No – there are more direct ways in which this can happen in plants in a matter of a generation or two. A case in point is the now famous plant *Raphanobrassica*, which is a man-made hybrid between a radish and a cabbage and which, by any definition, is a new species. It can produce fertile offspring but not with either parent, the cabbage or the radish. Unfortunately, it has the root of the cabbage and the leaves of the radish! A large number of man-made cultivated plants come into this sort of category. Human beings have witnessed the origin of species.

There is plenty of evidence that evolution is a continuous process and is going on today. Because that is the case we would expect to find in nature all stages in the evolutionary process, from little or no divergence between populations of a single species right through to divergence so great that we call the populations different species. The modern study of natural populations of plants and animals reveals just that.

Imagine the evolution of life as a great river that branches out across a huge delta, like the Ganges at its mouth. The student of evolution who studies living organisms on the face of the earth today is like a person making a journey that cuts across the branching river system. As he crosses these branches he will have no doubt that in some places one branch is quite distinct from another. It has branched off from its neighbour a long way back. But at another point in the journey he will be at the fork where two streams begin to branch and make their independent journeys. At another place he may find himself a little down from the fork, they are two streams but only just. So in nature today we find some groups of organisms that are so different we call them different species. There are others that are distinct, but not so distinct. They may even be reproducing one with the other, perhaps only occasionally. The dividing line between other groups may be so indistinct that no one would call the different groups different species, yet differences are recognisable. The old idea of Linnaeus that nature is divided up into totally distinct types which never change is no longer tenable. Darwin saw this over a hundred years ago when he concluded the fourth chapter of *On the Origin of Species* with these words: 'As buds give rise by growth to fresh buds, and these, if vigorous, branch out and overtop on all sides many a feebler branch, so by generation I believe it has been with the great Tree of Life, which fills with its dead and broken branches the crust of the earth, and covers the surface with its ever-branching and beautiful ramifications' (Darwin, 1859, p. 130).

There were three phases in the development of the theory of evolution by natural selection. The first was the bold outlines in Darwin's *On the Origin of Species* in 1859. The second had to wait until 1930 when quite independently three brilliant geneticists developed the genetical theory of natural selection. These were R.A. Fisher and J.B.S. Haldane in England and Sewall Wright in the USA. This phase was first highly theoretical. It came to earth in the studies of Theodosius Dobzhansky and his students on natural populations, especially of fruit flies. The third phase had its origin in the unravelling of the structure of the DNA molecule by Watson and Crick in 1953. This opened the door to the interpretation of mutation and selection in molecular terms.

Some have charged that the evolution of life by means of natural selection of chance variations was as credible as to suppose that a million monkeys banging at random on a million typewriters would chance to produce one of Shakespeare's plays. For evolution to have proceeded that way is as incredible as the production of a great picture by a blind painter sprinkling a canvas at random with a brush dipped in unseen colours. But such analogies are misleading for several reasons. Anyone with an understanding of the genetical theory of natural selection would see that to be so. The metaphor has to be changed in some such way as this. Not one, but a million – nay billions – of blind painters each sprinkle a few splashes of colour on millions of canvases. Of these, only the few that show the first feeble suggestion of a meaningful picture are preserved. The rest are destroyed. The selected rudimentary pictures are reproduced a millionfold, and again millions of blind painters add a few random touches of paint to them; again the best are selected and reproduced, and so on millions of times corresponding to the number of generations that have elapsed since life appeared on earth. This metaphor corresponds a bit closer to reality than the monkey typewriters because it provides for reproduction and selection. But of course it is no more than a metaphor and is therefore quite inadequate to convey the subtleties of the genetical process of natural selection.

Natural selection has been compared to a sieve which retains the beneficial materials and lets the deleterious ones be lost. That analogy too is a poor one because it leaves out of account sexual recombination of genes and the interaction between genes. The usefulness or harmfulness of a genetic variant often depends upon the other genes in whose company it finds itself. A variant beneficial in one company can be detrimental in another. Dobzhansky (1967, p. 42) has said that to make the sieve analogy valid, one would have to imagine an extraordinary sort of sieve. It must be designed so as to retain or to discard particles, not on account of their size alone, but in consideration of the qualities of all the other particles present. 'We have then', he says (p. 42), 'not a sieve but a cybernetic device which transfers to the living species "information" about the state of its environments. This device also makes the evolutionary changes that follow dependent upon those that preceded them. The genetic endowment of a living species is an integrated system, the parts of which must fit together to be fit to survive.' Some idea of the subtle workings of natural selection can be gathered from the following example of what Waddington (1957) called genetic assimilation of environmental effects.

The thickening on the soles of our feet is an obvious adaptation to walking on rough surfaces with bare feet. As Darwin pointed out, the thickening already appears in the embryo before the foot has borne any weight. The structure therefore cannot be a direct response to external pressure, but must be produced by the hereditary constitution independently of the specific external influence to which it is an adaptation. Similar thickenings are found in other parts of the body in other species. For example, the ostrich squats down in such a way that the under surface of the body comes in contact with the ground at its two ends, fore and aft. In just these places considerable callosities develop in the skin. Furthermore, these make their appearances in the embryo before hatching. Let us consider how these callosities might have become genetically determined. Consider the ostrich before it ever had callosities. Presumably its skin, like that of other animals, could react directly to external pressure and rubbing by becoming thicker. Let us now suppose that the capacity to react in this way is dependent upon genes. Since the individuals in a population of animals are never quite uniform in any character, we must expect that the ostrich ancestors varied in the genes that conferred different capacities to produce callosities. There would then be natural selection of those which had the genetic propensity to react effectively to friction. Friction on the ground is an environmental stress. A race would eventually be developed by selection which had the genes that enabled its members to develop the most appropriate adaptive thickenings of the skin. At this stage the thickenings would still not be genetically fixed and independent of the pressure and rubbing. They would still be acquired characters in the conventional sense. We have to find a hypothesis which could explain how they could become genetically fixed.

Waddington found such an hypothesis which worked in experiments with flies. He was able to demonstrate that after several generations of selection of those individuals which had the genetic constitution to react in the appropriate way, such as enlarging the anal papillae of the larvae to cope with a saline environment, genes were accumulated such that the adaptation eventually appeared even in the absence of the environmental stress. Natural selection not merely ensures the survival of those animals that have something near the optimum characteristics, but favours as well those whose genetic constitution is such as to produce such animals under any conditions, stressful or otherwise. Thus a stage was reached in which the evolving ostriches nearly always developed callosities of just the right sort, even in those individuals which sit down very seldom or those which loll around a lot of the time.

Waddington referred to this process as genetical assimilation of an environmental effect. It is not Lamarckism but a thoroughly Darwinian process. The important point is that genetical assimilation reduces the hit and miss element of simple models of natural selection. As a matter of empirical observation we know that stressful environments modify the organism in structure or physiology in a way that is more often than not adaptive. Organisms have this strong tendency to adapt their bodies to the environmental stresses they live with. If there are genetic mechanisms that tend to fix these changes genetically then we have a much closer relationship between environment and genetic constitution than a simple reading of Darwinism would give. The theory becomes that much more credible. Natural selection is not, as is so often stated, the selection of the genotype. It is not even the selection of the phenotype. It is the selection of the developing phenotype.

In Chapter 1 we discussed the interplay of genes and environment in the development of the adult organism from a fertilised egg. At each stage of development, environment plays a role in determining the phenotype. We expressed this by means of the analogy of a ball moving along the valley of a landscape (Fig. 1.03, p. 25). Which particular valley the ball moves down depends in part upon environmental influences that may move it from one valley to another. Those individual organisms that develop in the most adaptive way in response to their inherited genes and the environment in which they develop will have the highest chance of surviving and reproducing. The capacity to respond adaptively is what gives survival value. It is the ostrich which developed callosities on its thighs in response to abrasion on hot dry soil which had a phenotype better adapted than those which did not have the propensity to develop callosities.

Enter purpose

Monod entitled his important book '*Chance and Necessity*'. He correctly showed that mutation, which is the basis for all evolution, is a matter of chance and that once variety is introduced there is a necessity of selection. But in his passionate commitment to the exclusion of purpose from evolution he underestimated, although he did not ignore, the extent to which purposive behaviour does have an extensive effect upon the process of evolution. This is a field to which biologists have as yet given insufficient attention.

Waddington has shown that a thoroughly Darwinian account of the development of callosities by ostriches can be provided by his concept of genetic assimilation of acquired characters. Once this concept is accepted it becomes understandable how acquired behaviour can also have an effect upon the process of selection.

Considerable work has been done on insects which suggests the importance of genetical assimilation of acquired behaviour. For example, the codlin moth had not been recorded as a pest of walnuts in the USA. before 1918. A few years later walnuts were consistently infested under circumstances which suggested the appearance of a new genetic race which preferred walnuts to apples. How the original change in habit could be reinforced is indicated by experiments with other insects on what is called 'host conditioning'. Larvae of the fruit fly *Drosophila* which are fed on peppermint flavoured food produce adult flies which, when given a choice, have a preference for laying their eggs on peppermint flavoured food. The fly is conditioned to choose the smell of the food it was raised on. We may suppose that in the case of the walnut race of codlin moth there were genetic differences amongst the original individuals of apple-bred flies in their flexibility of behaviour. Some no doubt never made a mistake. They always chose apples. But some were genetically more flexible in their choice of hosts and some of these found walnuts. Having started their lives on walnuts, 'host conditioning' would help to keep them there. Mutations which had no survival value in the old environment of apples could be advantageous in the new environment of walnuts (Andrewartha & Birch, 1954, p. 692 *et seq.*). Amongst insects there are many examples of 'host' races or of closely related species which differ in the hosts they feed on. There are a number of races of the European fruit fly *Rhagoletis cerasi* which no longer interbreed. In two closely related species of a gall-forming fly, each of which forms galls on different host species, a single gene controls the ability to discriminate between the two hosts (Bush, 1974).

Whether this acquired behaviour in insects is in any significant sense purposive may be open to dispute. But it is adaptive behaviour which was not initially programmed by the genes. Sir Alister Hardy, who has been one of the few biologists to emphasise the importance of these ideas, remarked 'it must often be the restless, exploring and perceiving animal that discovers the new ways of living, new sources of food. . . . The restless, exploratory type of behaviour has no doubt been fostered and developed by selection just because it pays dividends' (Hardy, 1965, p. 172). Even in the case of insects we cannot exclude that there may be restless, exploratory behaviour.

A clearer example can be found in the famous case of the tits. It was in England that tits, by their exploration, learned to open milk bottles to feed on the milk, first the cardboard tops and then the metal tops. The habit was learned by subsequent generations and spread throughout the tit populations of Europe. There is reason to suppose that what started as a change in behaviour which was advantageous to tits will eventually become a genetically determined characteristic, though we do not know whether this has yet happened with the tits. Indeed, tits may be expected to evolve a beak more suited to opening milk bottles than they have at present. When Sir Alister Hardy gave a lecture on this subject to the Linnean Society of London, someone half-jokingly said: 'And eventually I suppose we shall have a race of tits with beaks like tin-openers.' To which Hardy replied: 'Exactly!' (Hardy, 1965, p. 177).

This instance introduces purposive behaviour much more clearly, because tits learned this behaviour from other tits. This learning must be an individual matter. A tit learns behaviour that achieves goals it finds desirable. This is not random imitation or random exploration. It is because of the widespread intelligent adaptation of tits to their new environment that genetic changes are also likely to occur.

It is unlikely that this is a rare phenomenon in evolution. On the contrary, it is probable that environmental changes have frequently led individual members of species to adopt new behavioural patterns which prove adaptive. The following quotation from Monod shows that he too appreciated such developments, though their importance is little recognised in his general theory of evolution:

> It is also evident that the initial choice of this or that kind of behaviour can often have very long-range consequences, not only for the species in which it first appears in rudimentary form, but for all its descendants, even if these should constitute an entire evolutionary subgroup. As we all know, the important turning points in evolution have coincided with the invasion of new ecological spaces. If terrestrial vertebrates appeared and were able to initiate that wonderful line from which amphibians, reptiles, birds, and mammals later developed, it was originally because a primitive fish 'chose' to do some exploring on land, where it was however ill-provided with means for getting about. This fish thereby created, as a consequence of a change in behaviour, the selective pressure which was to engender the powerful limbs of the quadrupeds. Among the descendants of this daring explorer, this Magellan of evolution, are some that can run at speeds of fifty miles an hour; others climb trees with astonishing agility, while yet others have conquered the air, in a fantastic manner fulfilling, extending, and amplifying the 'dream' of the ancestral fish (Monod, 1974, p. 121).

The process of exploratory behaviour and learning, which is assumed in the case of the tits and recognised to have some evolutionary importance by Monod, has been directly observed in many areas. For example, in the case of the Japanese macaque the young are the innovators. In a natural group on Koshima island a young female called Imo was seen to wash sand from sweet potatoes in the sea. Her playmates were the first to imitate her, followed by their mothers. Subsequently the infants of these monkeys learned the custom from their mothers. Later different styles of potato washing developed, roughly along kinship lines (Kawai, 1965).

What was observed in the monkeys is unlikely to have any effect on biological evolution. The needed flexibility for this behaviour is already fully present in the species. Macaques individually and as troops are capable of a wide range of behavioural habits. That a particular troop takes to washing potatoes in the ocean will differentiate it from others culturally rather than biologically. That different families within the troop develop different styles of potato washing will further enhance the cultural diversity.

As human beings with increasing patience study the behaviour of other species in greater detail, such cultural diversities within species are becoming more apparent. For example, Marais (1969, p. 68) demonstrated that the behavioural differences between the arboreal baboon and the mountain troops in South Africa are in fact cultural. An infant from the former brought up by the latter group grows up 'with the complete knowledge necessary for it to exist in its new environment'.

Human evolution

Through chance, necessity and purposive behaviour an ape-like creature became transformed into the Australopithecines of Africa some five to six million years ago. These in turn transformed through stages into the human like creatures *Homo erectus* who roamed the earth from a million years ago until a little over half a million years ago. *Homo erectus* in turn transformed into *Homo sapiens* who came on the scene probably about a quarter of a million years ago. The transition involved many anatomical changes, particularly in the brain and the limbs. Hands became more manipulative enabling Australopithecines to use crude tools and *Homo erectus* to make and manipulate sharp instruments. *Homo erectus* was the first to discover the use of fire. The evidence for this is with the remains of *Homo erectus* in Peking.

But most important of all were two other features; the development of a brain which conferred intelligence and particularly the capacity of abstract thought and, secondly, the capacity to communicate by means of language. Human beings are the only biological species with a highly developed capacity for symbolic thought and the use of language.

These made invention, be it a wheel or a computer, possible. So we see in the human the development of a body of learned information which we call culture. Culture is what is learned and transmitted to subsequent generations through teaching. In addition to a genetical inheritance humans acquired a rich cultural inheritance

The evolution of a brain which made intelligent action and culture possible was gradual. Evolution toward greater intelligence involved enlargement of the cerebrum and especially its outer convoluted part or cortex. The cortex is the site of activities associated with intelligence, learning and other complex functions of the brain. It is quite tiny in the brain of a fish. It becomes progressively larger in proportion to the rest of the brain in reptiles and mammals and specially in humans (see Eccles, 1979, pp. 84–5).

Behaviour has evolved from highly programmed stereotyped responses to a much more flexible adaptation to the environment by means of learning from experience. An animal with a small brain and a small cortex is more like a programmed automaton keyed to certain principal features of its environment, the sign stimuli, which it must respond to if it is to thrive. It is guided home by one of them, it recognises a mate by another, it seizes food by two or more and so on through its life. The large brained animal with the large cortex is also programmed but it can modify its behaviour greatly by learning. It can learn to select among the stimuli in its environment. It moulds itself to the local environment and responds with greater precision and safety than the animal with the small brain and small cortex (see Pulliam & Dunford, 1980). The evidence suggests that all animals have some ability to learn, even single celled ones (see Chapter 4). But this ability becomes very much greater in the higher animals, particularly in birds and mammals and notably in the human.

Elemental forms of distinctively human cultural activity existed with Australopithecines two million or more years ago. There must have been an overlap of over a million years from the beginning of human culture to the appearance of human beings like ourselves. Instead of culture being added onto a finished animal, it is an ingredient in the production. As culture accumulated and developed, a selective advantage accrued to those individuals in the population most able to take advantage of it – the effective hunter, the efficient tool-maker, the resourceful leader. A remarkable insight into the operation of natural selection for such qualities in humans is to be found in a paper by Alfred Russel Wallace published in 1864 some seven years before Darwin published *The Descent of Man*.

> From the time, therefore, when the social and sympathetic feelings came into active operation, and the intellectual and moral faculties became fairly developed, man would cease to be influenced by 'natural selection' in his physical form and structure; as an animal he would remain almost stationary; the changes of the surrounding universe would cease to have upon him that powerful modifying effect which it exercises over other parts of the organic world. But from the moment that his body became stationary, his mind would become subject to those very influences from which his body had escaped; every slight variation in his mental and moral nature which should enable him better to guard against adverse circumstances, and combine for mutual comfort and protection, would be preserved and accumulated; the better and higher specimens of our race would therefore increase and spread, the lower and more brutal would give way and successively die out, and that rapid advancement of mental organization would occur, which has raised the very lowest races of man so far above the brutes (quoted by Hardy, 1966, p. 36).

Between the cultural pattern that developed on the one hand and the brain and body that were selected on the other, a positive feedback was created in which each shaped the advance of the other. A graphic example of this is correlation between the changing anatomy of the hand, increased use of tools, and the expanding representation of the thumb in the cortex of the brain.

Nevertheless, in the last 50 000 years there has been a shift to a predominance of cultural evolution. Geertz (1965, p. 112) says:

> What happened to us in the Ice Age, is that we were obliged to abandon the regularity and precision of detailed genetic control over our conduct for the flexibility and adaptability of a more generalized, though of course no less real cultural control over it. To supply the additional information necessary to be able to act, we were forced, in turn to rely more and more heavily on cultural sources.

Still cultural evolution was a slow process. Only 35 000 years ago human beings learned to cook. A mere 10 000 years ago we discovered agriculture and how to weave clothes. Only with the discovery of writing was the process greatly accelerated.

Karl Popper (1972) has suggested a simple mental experiment to demonstrate the role of culture in human development. Culture he calls 'world three', world one being the physical world, and world two, the world of consciousness. Imagine that tools and machines are destroyed and with them all subjective learning, including the knowledge of how to use tools and knowledge. All libraries are also destroyed so that the human capacity to learn from books becomes useless. There would be no re-emergence of human culture for millenia. Humanity would be transported far back into pre-history and would have to begin again the long upward climb. Had libraries survived in this scenario the world would have got going again, though no doubt after much suffering.

We are not genetically identical with our ancestors of 50 000 years ago. But Popper's thought experiment helps us to see that the main differences between human beings now and then are not genetic but cultural. 'Man is a most extraordinary product of evolution', wrote Dobzhansky (1956, p. 6). 'He is so much unlike any other biological species that his evolution cannot be adequately understood in terms of only those causative factors which are operating in the biological world outside the human kind.'

Although cultural evolution is dominant in human evolution, genetic selection has not ceased in the human. It is not now the force it must have been in our early history, nevertheless it continues. Its direction has changed. We no longer find it adaptive to have a physique effective in warding off wild animals or that would be advantageous for a hunter and food gatherer. There is no particular advantage for most men to become Tarzans. Some types of natural selection have been lessened by the practice of modern medicine. Many people now survive and have children, thanks to insulin and other forms of therapy, who in former times would not have survived to hand their genes on to their children. In our society higher proportions of people probably suffer from genetic diseases than in former days. But they may not physically suffer much if therapy is effective. On the other hand, we may be encouraging the selection of traits that had little selective value in times past. People with genetic constitutions such that they might better be able to withstand the stresses of living together in crowded cities would be at a selective advantage today. It is now possible to manipulate in a more direct way the genetic constitution of human beings by eugenic procedures. The extent to which this might ever mean that human beings could directly control their genetic destiny is discussed in Chapter 7.

The biological understanding of human history is that the decisive qualities that made the difference between the human and the pre-human did not appear first in one individual at a specific moment in history. No such moment and no such individual could have existed. A threshold had to be crossed. But it was no mere hop, step and jump but two or more million years wide. Secondly, a population made the journey. Mutation, recombination of genes and natural selection are phenomena that happen to populations. The human gene pool does not belong to any one individual. It belongs to the total population of human beings alive at any one time. The most elemental fact of evolution is change in the proportion or frequency of different sorts of genes in the population as a whole. We cannot measure evolution by comparing one individual today with a forebear sometime back.

In the crossing of the human threshold the far shore was reached in a photofinish with no one individual getting there first. Hence the telling expression of Teilhard de Chardin that 'the first man is a crowd and only can be a crowd, and his infancy is made up of thousands of years', or again, 'man came silently into the world. As a matter of fact he trod so softly that, when we catch the first glimpse of him as revealed by those indestructible stone instruments, we find him sprawling all over the world from the Cape of Good Hope to Peking' (Teilhard de Chardin, 1959, p. 186). Or as Geertz (1965, p. 110) says: 'Men have birthdays, but man does not.'

What is the evidence that 'the first man is a crowd'? Evidence for this view comes from population genetics and palaeontology. In the pre-Darwinian view the species was a type. It was 'typified' by a single specimen called the 'type specimen' placed in a museum as representative of the species. Typological thinking is still unfortunately imprinted into the minds of many people. The type is supposed to reproduce its 'kind'. There is no room for differences between individuals. Population genetics shows us that species cannot be typed by a single specimen. The species can only be characterised by the means of characters and their variability. What happens in evolution is not that the type suddenly changes but the mean of the population in many of its characters gradually moves. The species takes a trip. Typically that happens when environment changes.

The fossil evidence supports this picture of the evolution of human beings. Information from this source is, of course, severely limited to anatomical changes and to changes in the artifacts used by human beings. Every year brings forth new finds of this sort especially from Africa. Despite the incompleteness of the fossil record of human origins there is a continuum of transitions between early ape-like creatures, Australopithecines, *Homo erectus* and *Homo sapiens*. The 'missing link' is no longer missing.

'There is grandeur in this view of life,' wrote Darwin (1859, p. 490), 'with its several powers, having been originally breathed into a few forms or into one; and that, whilst this planet has gone cycling on according to the fixed law of gravity, from so simple a beginning endless forms most beautiful and most wonderful have been, and are being, evolved.'

Conclusions

The purpose of this chapter has been to take a fresh look at evolution. It began with the living cell, the mystery of whose origin was discussed in Chapter 1. Jacques Monod is right that the whole evolutionary process is absolutely grounded in chance. Random mutation is the foundation of all the variety of living things.

Next, given the multiplication of different kinds of living organisms, Monod is right that there is an inexorable necessity that there will be a natural selection of types. We agree with Monod as to the fundamental importance of the two principles of chance and necessity which he has emphasised.

Nevertheless, the picture is more complex than Monod allows. For example, C. H. Waddington has shown the role of the developing phenotype in the process of natural selection. It is the developing phenotype as moulded by genes and environmental influences that is selected. The actual shape that an individual organism will take is greatly affected by environment. This in turn plays a direct role in natural selection.

Furthermore, the actual process of evolution cannot be understood apart from the purposive behaviour of the animals that are evolving. The restless, exploratory behaviour of animals, especially of increasingly intelligent animals, leads to new patterns of behaviour which in turn affect the processes of natural selection. With the emergence of human beings the role of purposive behaviour in evolution becomes greatly magnified. Any effort to exclude such behaviour from the principles explanatory of the actual course of evolutionary development must lead to a profoundly inadequate account.

There are two clear implications of this account. The first is that evolution, with all of its discontinuities, is still fundamentally a continuous process. There is no great gulf, for example, between human beings and other living things. Human beings are fully a part of nature, even though a unique part. This assumption will pervade the remainder of the book, and its meaning will be particularly discussed in Chapters 4 and 5. The second implication is that ecology and evolution belong together. This implication is important because of the strong tendency to separate them in human vision and imagination. The social and spiritual effects of the doctrine of evolution, on the one hand, and of ecology on the other, have been almost opposite. The image of evolution has encouraged the idea of movement forward to new heights. It has focused on individual species and their superiority to others as shown by their competitive success. It has encouraged a certain human indifference to the fate of lesser species. In the form of social Darwinism it has even supported economic and political competition in the belief that this will lead to the survival and increase of the best and strongest.

The image of ecology has downplayed the separate development of species and any evaluation of their diversities of value. Because all are interdependent, all are necessary. Humans should recognise that their arrogance and their efforts to manipulate their environment are destructive of the web of life. In this perspective the need is for a deep spiritual transformation that will lead human beings to experience themselves simply as a part of the whole web and not as agents of purposive change.

If, however, ecology and evolution belong together, both of these images are to be rejected. Evolution is not a process of ruthless competition directed to some goal of ever-increasing power or complexity. Such an attitude, by failing to be adaptive, is in fact not conducive to evolutionary success. A species co-evolves with its environment. But equally, there is no stable, harmonious nature to whose wisdom humanity should simply submit. Intelligent purpose plays a role in adaptive behaviour, and as environments change its role is increased. Human culture is an immensely important factor in ecology and one which necessarily introduces profoundly new elements into the web of life.

There will be no single attempt to formulate the implications for our lives of the views of an ecological evolution and an evolutionary ecology which have been put forward. But the later chapters of this book, which work out some of the implications of the view of life developed in these first chapters, reflect what is entailed.

Every view of evolution is at the same time a view of life. Living organisms must be the sort of things that can evolve in the way the theory asserts. The chapter that follows turns to the construction of a model of living things. It must be a model which is consistent not only with the foundational roles of chance and necessity but also with the fact that animals behave purposively.

3

Models of the living

> In my hunt for the secret of life, I started my research in histology. Unsatisfied by the information that cellular morphology could give me about life, I turned to physiology. Finding physiology too complex I took up pharmacology. Still finding the situation too complicated, I turned to bacteriology. But bacteria were even too complex, so, I descended to the molecular level, studying chemistry and physical chemistry. After twenty years work, I was led to conclude that to understand life we have to descend to the electronic level, and to the world of wave mechanics. But electrons are just electrons, and have no life at all. Evidently on the way I lost life; it had run out between my fingers.
>
> Albert Szent-Györgyi (1972, p. 7)

In our hunt for the secret of life we too have started with the facts of life as understood in contemporary biology. Chapters 1 and 2 discussed concepts of biology at the three levels of the molecule in the cell, the organism and the population. They showed that biological concepts are more than collections of facts just as a house is more than a collection of bricks. Facts are put together in models such as the DNA model of the gene, the epigenetic model of development, the feedback model in physiology, the 'web of life' model of populations and the natural selection model of evolution. The question now is: do the models that have dominated biology tell us the complete story of what life is, or is it the case, as Szent-Györgyi suggests in the quotation above, that in the process of studying life it has run out between our fingers?

This chapter begins with some reflections on what models are and how they work. It then considers in turn the three most prominent models that have been used to account for living organisms and their behaviour: those of mechanism, vitalism and emergent evolution. These are considered critically and their adequacy is denied.

The fifth and sixth sections develop a different model, an ecological one, which can show the limited, but very real, values of the other models, and especially the mechanical one. And the chapter concludes with further reflections on the definition of life and in particular the arbitrariness of drawing any definite line between what is alive and what is not.

The function of models

There are a number of reasons why the biologist's models of life do not tell us the complete story. There is the very obvious one that every day new experiments produce new information. Then there are less obvious reasons why the picture is incomplete. Human beings get pictures of the outside world through their sense organs. These sense impressions are so constructed by the nervous system that they automatically carry with them an interpretation of what is seen, felt and heard. A dog with a different assemblage of sense organs and nervous system from the human has a different picture of the world. For one thing it is a black and white world. There is yet a third reason why scientific models are incomplete representations of what is there. This has to do with the way scientific information is obtained. Science deals with only a part of nature.

Any scientific experiment presupposes the division of the world into what is regarded as relevant and what is regarded as irrelevant. Jack is put in a box. But there is a lot outside the box and there are connections between what is outside and what is inside. The answers obtained are necessarily limited to what is put inside the box. This basic assumption, that some things are relevant and others irrelevant for our purposes, is a lie. The universe is not constructed in a series of disconnected boxes. Maybe no picture of the whole can be obtained, but at least there should be the awareness that science carves it into bits. Moreover, the very process of putting Jack in the box alters the nature of Jack. Hence Wordsworth's retort to the biologists: 'We murder to dissect'. To suppose that what is found is a complete picture, when it is an abstraction from the complete picture, is to commit what Whitehead (1926*a*, p. 58) called 'the fallacy of misplaced concreteness'.

Implicit in any model is a metaphysical framework. The implicit framework in most biological models is mechanism. An electron or a cell or a brain or the universe is represented as being like a machine. Biologists have not always chosen the mechanistic model, as is indicated by the famous historical antitheses that have developed in biology: mind and body, vitalism and mechanism, preformation and epigenesis, nature and nurture, reductionism and holism. Some of these antitheses seem to be resolved in one generation only to reappear in the next in a new dress. Nature and nurture return as sociobiology and environmentalism, mechanism and vitalism return as chance and purpose.

Scientific models are like the shell of a crab which contains the contents of the crab for a time. As the crab grows the shell splits, and the crab grows a larger one more appropriate for the larger animal. The process of casting off an old shell and growing a new and larger one is slow and painful. The shell of biology today is still an old one from the past. It has matured and hardened after some 300 years of good service. There are indications that it is now cracking. It is bursting at the seams and bits are falling off. It is time to heave off the old shell and set about making a new one. The old shell can no longer hold the facts. The conventional framework within which we construct our biological models is at best inadequate and at worst false.

The precise form and texture of the future shell need not be known in advance. But some good growing points from which to develop a new shell need to be established. Without these the view of life will remain confined in a cramped structure.

To break down the old shell and to begin to grow a new one is to begin the process of liberating the concept of life. Liberation is called for at every level of biology, the molecular level, the organismic level and the population level. If we can liberate the concept of life we might better be able to liberate life itself. In this chapter we discuss some conceptual models of living things that are inadequate or false, yet are widely accepted, and then move on to the ecological model which is a liberated and liberating alternative.

The mechanistic model

A mechanistic view of the living organism may be invoked on two main grounds. One is the conviction that the organism is a machine. Alternately, one may avoid making any metaphysical assertion about the organism and simply say that, whatever its nature may be, we can only study it scientifically by treating it *as if* it were a machine. We recall a meeting of a group of distinguished biologists including three Nobel Laureates whose discussions resulted in a book by Ayala & Dobzhansky (1974). Most of the participants were mechanists in the metaphysical sense. A few were not. However, all agreed that in their experimental studies they treated the organism as if it were a machine. They were all methodological mechanists.

A metaphysical mechanist would say that the experimental method is designed to reveal the mechanisms of organisms and that that is all there is to be revealed. A methodological mechanist might say that science reveals the mechanical aspects of the organism but that there may be other aspects which the scientific method does not reveal. The theoretical justification for methodological mechanism as the only valid methodology, even in science, is not as obvious as is sometimes supposed. Woodger (1929, p. 237) quotes this passage from P.W. Bridgeman's *Logic of Modern Physics*:

> It is difficult to conceive anything more scientifically bigoted than to postulate that all possible experience conforms to the same type as that with which we are familiar, and therefore to demand that explanation use only elements familiar in everyday experience. Such an attitude bespeaks an unimaginativeness, a mental obtuseness and obstinacy, which might be expected to have exhausted their pragmatic justification at a lower plane of mental activity . . . I believe many will discover in themselves a longing for mechanical explanation which has all the tenacity of original sin. The discovery of such a desire need not occasion any particular alarm, because it is easy to see how the demand for this sort of explanation has had its origin in the enormous preponderence of the mechanical in our experience.

A modern mechanistic biologist (Calow, 1976, p. 4) says that:

> in seeking machine analogies in biology, scientists have really been asking 'can the rules governing the operation of machines be used to explain the operation of living things?' or more deeply, 'can the theory of physical things be used to explain the behaviour of biological things?' This is an important question, because if we answer it in the affirmative we are really claiming that there is nothing special about biological entities and perhaps there is no need for a subject called 'biology'.

We gave our answer to these questions in Chapter 1 when we said that organisms are like machines in some respects but differ in others. Organisms possess pumps and levers and circuits for transmitting messages. In these and many other respects they are like machines. It is one thing to say that living organisms are like machines in some respects. It is another to say they *are* machines, which is what the metaphysical mechanist says. The metaphysical mechanist regards these mechanisms – pumps, levers and so on – as consisting of smaller mechanisms (the cells) tied together into a large mechanism, the organism. The ultimate mechanical model involves 'dissecting' the organism down to its constituent controlling mechanisms and building it up from these building blocks. These building blocks for the biologist have often been the cells. With the rise of molecular biology they are now usually molecules. Typical of this newer view is Jacques Monod's doctrine that the controlling building blocks are the DNA molecules. In the preface of his book it is claimed that the molecular model of the genetic code (DNA) 'today constitute(s) a general theory of living systems' (Monod, 1974, p. 12). In principle, of course, there is no reason to stop at the DNA molecule. After all, more ultimate constituents are electrons, protons and the other so-called fundamental particles.

The mechanistic model is correlative with the assumption that these ultimate constituents are indeed particles and that these particles are correctly understood as bits of matter. The principles which govern material particles are the laws of mechanics, and the behaviour of larger entities composed exclusively of material particles must be machine-like. Further, such machine-like behaviour is fully deterministic. Hence mechanism, materialism and determinism belong together.

The mechanistic model has a long history. About the year 400 BC Democritus declared: 'There is nothing but atoms and space, all else is an impression of the senses.' His was a universe of multi-shaped little billiard balls moving in empty space, colliding with one another, grouping together and separating from one another. His atomistic theory was elaborated by the successive speculations of Epicurus (342–271 BC) and Lucretius (99–55 BC). Their ideas became part of the background of physical thought in the renaissance of science in the fifteenth and sixteen centuries through the works of Copernicus, Bruno, Galileo and later Newton.

Modern biology began in this mechanistic climate with the anatomical investigations of Vesalius in the sixteenth century and with William Harvey's publication on the circulation of the blood in 1628. The lives of Harvey and Newton overlapped for a quarter of a century. But whereas Newton had inherited a great legacy of precise and detailed physical science, Harvey had inherited almost nothing from the biology of his predecessors. He was a great innovator.

Sitting in the wings, or more precisely by his stove, was René Descartes (1596–1650). He, more than anyone else, made the doctrine of mechanism in physics and biology explicit. Trained as a mathematician and engineer, he later became a philosopher. 'Give me matter and motion', wrote Descartes, 'and I will construct a universe'. A similar claim was made later by Laplace when he said that if he were given the original positions and motions of every particle in the universe he could predict the entire course of subsequent events.

The success of Newton's system in explaining the motion of bodies led to the demand of the seventeenth century, spearheaded by Descartes, that all explanation in science should be in terms of mass particles or atoms moving in space. This is what Whitehead (1926*a*, Chapter 3) called the 'billiard ball' universe in which the course of nature is conceived as being merely the fortunes of matter in its adventure through space. When Descartes set out to explain the phenomena of life by means of the two agencies, matter and motion, the works of Vesalius and Harvey fell into his hands. Here he found levers, pumps, valves and a host of other machines by which to explain the living organism. In 1664 Descartes wrote the first scientific treatise of mechanistic physiology, *Traité de l'homme*. In it he pointed out that one had only to consider the ingenious machines worked by water to be found in the royal gardens, some of which play musical instruments and others utter words, to be convinced that the activities of the human organs may be similarly caused.

But Descartes, like anyone seeing for the first time the marvellous structures of the body, was overwhelmed by the intricacy of design. The body, he said, was far more marvellously constructed than any machine made by human beings. That, he said, is because it is made by God.

There was one exception that did not fit into Descartes' mechanical explanation. This was human thought. To account for that Descartes required the existence of a mind. A human being was a machine but with a difference. It had a mind attached.

This is the origin of the Cartesian doctrine of the bifurcation of nature into mind and matter. Some have claimed that Descartes has influenced the background thought of scientists more than any other philosopher. In addition he had some effect on stimulating scientific investigations. Something is owed to Descartes for the biochemical movement led by Franciscus Sylvius of Leyden, the discovery of the muscular nature of the heart by Nicholas Steno and the application of mathematics and physics to biology by Giovanni Borelli, Sanctorius and others. Mechanism became the creed of most biologists in the seventeenth century. The knowledge of physical happenings was so vastly in advance of the knowledge of biology in the seventeenth and eighteenth centuries that mechanistic interpretations of life were almost inevitable.

Mechanism held sway also in the nineteenth century with the rise of organic chemistry and in the entirely mechanistic interpretation of evolution in Darwin's *On the Origin of Species*. The twentieth century opened with the rediscovery of Mendel's work on the laws of inheritance, which were to be understood on the basis of genetic factors, later called genes, which obeyed mathematical rules. Indeed, it seemed pretty clear that Mendel had, on the basis of his mathematical hypothesis, predicted the results of his experiments before he did them. Mendel's work eventually led others to elaborate the complex mechanical model of gene action in the DNA molecule, which Monod (1974) regards as the ultimate triumph of mechanism.

Also early in the twentieth century the elusive problems of animal behaviour were being interpreted in entirely mechanistic terms by Jacques Loeb in Chicago, who was thinly veiled as Martin Arrowsmith in the novel of that name by Sinclair Lewis. Loeb constructed a mechanical 'insect' that followed him with his torch around a dark room. Each 'eye' was a photoelectric cell which was connected with a wheel on the opposite side of the body. When the light was head-on each photoelectric cell received the same illumination, and the two wheels were geared to move at the same rate. So the 'insect' moved straight ahead. When Loeb moved with his torch to the left the right eye received less illumination. This triggered the wheel on the left side to slow down. The body then veered toward the light until illumination on both cells was the same again. This simple model incorporated the mechanical principle of negative feedback. The machine responded to differences of illumination on either side by moving to reduce this difference to a minimum. What looked like purposive behaviour on the part of the 'insect' was in fact due to negative feedback.

Negative feedback is a principle incorporated in governors on engines, goal-seeking missiles and goal-seeking anti-aircraft guns. Hence the rise of cybernetics in the development of complex goal-seeking machines and in interpreting 'purposive' behaviour in animals. Chapter 1 showed the difficulty an observer from outer space would have in distinguishing a horse from a tractor. Both serve ends that can be identified. They both appear to be goal-directed. It does not, of course, follow that they do it by the same means. This is a subject we return to later. However, Jacques Loeb declared his faith that all behaviour could eventually be understood in mechanical terms. The behaviourists who followed him had a faith that the human brain would also yield up all its secrets in completely mechanistic terms.

Despite the great heuristic success of mechanistic models in biological research, it is now clear that mechanism by itself cannot be an adequate way to understand organisms. As we noted in Chapter 1 there are differences between organisms and machines. Organisms consist of complex molecules and have complex metabolisms; machines have neither. Organisms reproduce; machines do not.

Mechanists recognise these differences. Their expectation is that they can explain these differences in terms of operations at the sub-cellular level that are machine-like. They have hailed the discovery of DNA as a great step in their programme. But if their programme is taken seriously as mechanism, it is no step in that direction at all. A DNA molecule is no more like a tractor than is a horse. This has nothing to do with the question of whether a DNA molecule can be synthesised in a laboratory outside of a cell. This has been done. But it is instructive to see how different this act of synthesising was from the way a machine is constructed. In the first synthesis of DNA in the laboratory Arthur Kornberg did not put 200 000 pieces of the DNA molecule of a bacterium each in its correct position. What he did was to provide the appropriate constituents whereby, step by step, the sequences were linked together by virtue of their own combining properties.

Most mechanists in fact do not intend to push the model of the machine to the limit. They are not really seeking a little machine in the cell. What they seek are physical entities which on the one hand behave according to physical and chemical principles and on the other hand explain the distinctive properties of organisms. The physical structure of the DNA molecule is thought to meet these requirements. But, of course, it does not. Its capacity for self-replication and its counter-entropic function do not derive from anything known about its physical constituents as they behave in other contexts. Scientists know what type of ordering of physical elements is necessary for cellular life to exist. But they are no closer than before to deriving biological laws from physical and chemical ones. And, ironically, if they did so, the laws in question would be statistical rather than deterministic ones.

It turns out, however, that a deterministic understanding of DNA was not of central concern to many mechanists either. Their real interest was to show that the evolutionary process and all the apparently free and purposeful phenomena of animal life are in fact fully determined. They would like to show that the emergence of the DNA molecule itself was fully determined by antecedent occurrences, but for the moment they rejoice to discover how chance mutations caused by external forces can explain the evolutionary process and how even very complex and apparently intelligent behaviour can be programmed.

Even if the mechanical model is treated as nothing but a name for any form of determinism whatever, its applicability is still limited. In the first place, it is still a matter of metaphysical faith that an animal's behaviour is programmed in absolute detail. Such programming must, of course, be a programming of reactions to the environment, but the environment is infinitely complex and constantly changing. Nothing now known about DNA shows that animals are wholly lacking in flexibility in adapting to their varied situations. Indeed the account of evolution developed in Chapter 2 emphasises the importance of such flexibility.

In the second place, if biologists assert absolute determinism with respect to animals, then they must either assert that they themselves are absolutely determined or else posit a radically unique and supernatural element in human beings. There is often a curious inconsistency in their thinking.

The process of evolution is conceived by biological determinists, such as Jacques Monod, as an entirely determined set of operations that reduces itself to the organisation of atoms and molecules into complex forms. In a totally programmed organism with a computerised brain there is no room for freedom. Thoroughgoing determinists hold that human beings are such organisms. No one can deny that there is a great deal about human beings that is determined, if not by genes then by the environment they did not choose to be born into. But strict determinism means that all is determined. Yet these self-same scientists who speak of people as evolved machines speak with another voice about the future of people. Human beings, we are told, are unlike their forebears in that they are now in control of their own evolution. Which way that evolution goes is for us to determine. People can choose the future. That is the meaning of cultural evolution. What people learn and pass on in the form of knowledge and invention shapes the future in a way which renders genetical evolution almost superfluous. They now change the environment instead of being changed by it. The human species is now in a position to destroy itself or to save itself. So it is popular for modern biologists to speak of the human responsibility to be rational in choosing a future. But to talk about human beings as responsible and rational and moral and making choices about the future presupposes something very different from determinism.

There is then a self-contradiction between determinism and responsibility for the future in much discourse about the human. Science appears here in two roles, as a body of conclusions about the nature of things, including human beings, and as a means through which human beings can exercise control over the world, including themselves. In the first of these roles it is presented as teaching strict determinism. In the second, it is offered as providing human beings with a radical control over human destiny that implies radical freedom. In this discussion the two doctrines are unreconciled.

More plausible in the light of the scientific evidence and immediate experience than this combination of absolute determinism and absolute control is the belief that human beings are partly determined and yet partly able to determine their own destiny. What scientists have learned about DNA enriches knowledge about one crucial factor in the determinism. Nothing human beings ever become will be incompatible with the instructions given by their DNA. But it does not follow that a study of the DNA code would make possible prediction of all that they do and say and think and feel. If there is any distinctiveness in cultural evolution, and surely there is, then fresh responses to the physical environment must be acknowledged as having led to changes even in the genetic code. There is interaction between learning and choice on the one side and the genetic constitution on the other. Neither the mechanistic model as such nor any deterministic model can do justice to this situation.

The vitalistic model

There is nothing new in rejecting the mechanical model of life. The difference between organisms and machines has been recognised from the beginning of history. The major theory through which this recognition has gained expression is called vitalism. Vitalists have sought to solve the problem by arguing that in addition to the physical components of the living organism there is also an additional principle or force. In recent times vitalists have been those who, in opposition to mechanism, asserted that the living organism consists of physical atoms and molecules plus another entity of a totally different nature variously called vital spirit, life force, *élan vital* and entelechy.

Vitalism is an ancient doctrine. The vitalist ideas of the first experimenter in ancient biology, Galen, of the second century, were accepted by physicians right throughout the middle ages and lingered on into the seventeenth and eighteenth centuries. Galen was a physician. At the age of twenty-eight he became surgeon to the gladiators in Pergamum, where he got considerable practice in studying human internal anatomy. He, like so many of the ancients, could not conceive that the exquisitely designed living organism could be explained in terms of a mindless interplay of atoms. He asked: when in not a thousand or ten thousand people but only one in ten thousand times ten thousand we observe malformation of the sixth finger can you consider this accomplishment by nature accidental? Like Erasistratus before him, Galen conceived that a vital principle was absorbed through the lungs from the air. He thus regarded the blood, which reached the left side of the heart from the lungs and which then surged through the arteries as rich red arterial blood, to be charged with a vital principle or vital spirit. If he had substituted the word oxygen for vital principle he would have been absolutely correct. But Galen's vital spirit was no mere chemical element. He invented a second sort of vital spirit, associated with the nervous system, which he called the animal spirit. For all his spirits, Galen was the last outstanding physiologist for a millenium.

With the rise of mechanism in the sixteenth century and its triumphs in the seventeenth century, there were only a few dissident voices amongst biologists. Such were Francis Glisson, an English anatomist (1597–1677), Marcello Malpighi of Bologna (1628–1694) and George Ernst Stahl of Halle (1660–1743). In the eighteenth century the claims of the vitalists came to the fore again. Stahl's followers were active as were others, such as the physician genius Francis Xavier Bichat of the Hotel Dieu. The physician John Hunter recognised a 'living principle' in addition to mechanics.

A resounding blow was struck against vitalism with Wöhler's synthesis of urea in the laboratory in 1828. Although the vitalists had been forced to admit that the living body could be shown to be composed of the same elements that constituted the non-living world, they had held it as a victory over mechanism that certain substances elaborated by living bodies could not come into existence in the absence of vital principles. This belief was shown to be false when Wöhler produced urea from nothing more mysterious than non-living chemical reactions.

Vitalism or neo-vitalism, as some of its proponents preferred to call it, returned in the nineteenth century in the writings of the embryologist Driesch. He had done experiments with embryos that were critical in the controversy over the role of preformation and epigenesis in development. The preformationists thought that development simply involved the unfolding of pre-existing structures in tiny form in the egg or sperm. The epigenesists believed that development involved the creation of new structures from primordial parts. Driesch had been able to obtain quite normal embryos of sea urchins from single cells taken from embryos in the 2, 4, 8 and even the 32 cell stage. Driesch was quick to point out that it was difficult to imagine a machine capable of being cut up into an indefinite number of pieces, each of which could restore the whole machine again. Driesch performed many other experiments on embryos, transplanting parts from one place to another and always ending up with complete organisms. He did at first try to produce mechanistic explanations of what he found. In 1881 he published a book entitled the *Mathematico-Mechanistic Investigations of the Problem of Morphology in Biology*. But he had second thoughts. It began to seem inconceivable to him that any mechanism could account for what he had observed. In his Gifford Lectures of 1907–1908, entitled *The Science and Philosophy of Organism*, he proved to his own satisfaction that no conceivable mechanism could account for the phenomenon of development. Living organisms were endowed with something no inanimate object possessed. He termed this entelechy. The concept was left vague and was to have an early death in the history of embryology. Many, but by no means all, of the problems that puzzled Driesch are now resolved by our understanding of the way in which the genetic instructions coded in the DNA of the egg are communicated by chemical means during the course of development of the embryo.

The most thoroughgoing vitalist of the twentieth century was probably Bergson (1911) with his concept of the *élan vital* which gave direction and purpose to life. The English zoologist E.S. Russell (1945) also invoked such a concept to account for development and behaviour. But as Julian Huxley remarked, the vitalists who ascribe development and evolution to an *élan vital* no more explain the history of life than would ascribing an *élan locomotif* to a steam engine explain its motion. Two American biologists of this century, E.W. Sinnott (1950) and R.S. Lillie (1937, 1945), regenerated a brief interest in vitalism with their rather more philosophical arguments on the subject, but the issues they raise can be much more satisfactorily dealt with in the ecological model which we propose later in this chapter. Vitalists of today are mostly non-biologists such, for example, as Elsasser (1966).

The emergent evolution model

Vitalism amounted to the assertion that living beings are composed of some material things which operate mechanistically plus some additional life quality lacking in non-living things. Mechanism is the assertion that the living organism is eventually to be understood in terms of the relationships between its material components, which is to say in terms of the physics and chemistry of its material components. Early in the twentieth century a half-way house between the conflicting doctrines of mechanism and vitalism was provided by the theory propounded by Lloyd Morgan in his book *Emergent Evolution* (Morgan, 1923). According to Morgan, in the course of evolution there were a number of miracles that were interposed into the stream of evolutionary events. Two were of special importance, the emergence of life and the emergence of mind. Their appearances were miracles in the sense that they were not understood and could not be understood in terms of physics and chemistry. Morgan (1923, p. 204) explicitly said 'that life cannot be interpreted in physico-chemical relatedness only; that human affairs, which depend on the quality of mind, require something more than biological interpretation'. Morgan believed with other emergent evolutionists such as J.C. Smuts, a Prime Minister of South Africa, that when these properties emerged in evolution new laws besides those of physics and chemistry came into existence.

This doctrine would hardly be of more than historic interest now except that the doctrine of emergence, shorn of the miracles posited by Morgan, is part of the framework of thought implicit, if not explicit, in the writings of many biologists who use the concept of emergence as though it explains something. Dobzhansky (1967, p. 32), for example, has said that 'the origin of life and the origin of man were evolutionary crises, turning points, actualizations of novel forms of being. These radical innovations can be described as emergences or transcendences, in the evolutionary process.' In the sentence immediately following this quotation, Dobzhansky made it clear that he believed evolution produced novelty in the sense of something completely new. Life emerged from the lifeless. Mind emerged from the mindless. He puts the emergence of these two qualities into the same category as the emergence of new anatomical features such as the five-toed leg from a fin-like structure or the feathers of a bird from the scale of a reptile. People who think this way also say that when sodium and chlorine atoms come together to form common salt new qualities emerge in the compound sodium chloride which were not contained previously in the isolated atoms.

But to say this or that property 'emerges' is to say nothing more than from A comes B. It explains nothing. Rather the term emergence signifies a problem requiring solution. How one anatomical structure such as a wing emerges from another sort of anatomical structure such as a leg can be explained by normal evolutionary theory. But how qualities such as livingness and mentality can be derived from something which totally lacks these qualities cannot. These are two different sorts of problems. If, as Dobzhansky (1967, p. 32) says, life and mind were totally new qualities when they first appeared in evolution, he is not explaining anything by saying that they emerged. He is only using the word emergence to point to what is left totally mysterious.

Toward an ecological model

The basic objection to vitalism and to theories of emergent evolution is that they leave the mechanical model unchallenged in its sphere. They point out correctly that this model cannot explain everything, and accordingly they introduce additional concepts to name these elements that are not explicable mechanically. But they have been of little scientific use, since these concepts point to a problem without explaining the phenomena. So far as explanation is concerned, they leave the field to mechanism. They show that livingness and mentality are not properties of machines, but they do not explain how they come to characterise organisms.

We propose a quite different response to the shared perception of the inadequacy of the long-reigning mechanistic model. We propose substitution of a different model through which both the machine-like traits of living things and the traits that are not machine-like can be considered together.

Even scientists who think of themselves as mechanists have already modified their approach far in the direction indicated by this model. To recognise this, to affirm it and to develop its implications will prove liberating both of the concept of life and of life itself.

The image of the machine is of an entity whose internal structure is given and whose behaviour follows from this structure. Of course, the machine cannot be entirely abstracted from some environment. In order to function it may require an external source of electrical power, for example. But the relevant features of the environment are conceived as limited and controllable. Given the presence of a few necessary conditions, the functioning of the mechanism can be understood as constant in a wide variety of environments. A machine is expected to perform in the same way whether it is in the factory where it is made or in the kitchen in which it is installed. If it fails to do so, one looks for dislocations of the mechanism caused in transportation and installation.

To view animals as machines is to approach them with similar assumptions. One seeks explanations of their functioning in terms of the physical structure of their bodies. This is assumed to be the same whether they are in their native habitats or in laboratories. Of course, it is recognised that some 'inputs' are necessary, especially food and water. But on the whole it is assumed that the animal can be appropriately studied apart from specific attention to the environment.

Of course, much has been learned by investigations of animals in which consideration of the environment has been minimal. But today few would question that animal behaviour is deeply altered by environmental conditions. Serious efforts are made to study the behaviour of animals in their native habitat. Even laboratory study gives extensive attention to differential behaviour as the environment is altered. In short the actual experience of scientists has led them to an increasing emphasis on the importance of environment. An ecological model of living things, in distinction from a mechanistic model, is one which pictures the organism as inseparably interconnected with its environment. To some extent students of animal behaviour have already largely shifted to this model.

However, those who for practical reasons employ an ecological model may still believe that at a deeper level explanatory categories must be mechanistic. One may suppose that although the environmental influences on the behaviour of a living organism are far more complex than are those on the machine, so that more attention must be paid to them, nevertheless, all these influences can be mechanically explained when we probe to the cellular level. Hence, even if an animal as a whole does not have the relatively self-contained character of a machine, if its behaviour is better studied as relative to environment rather than as simply inwardly programmed, the question must still be asked about a cell. Can a cell be appropriately studied on the analogy with a machine?

Once again it seems that the effort to view a cell as like a machine leads to distortion and one-sidedness. The behaviour of a cell is profoundly affected by its environment, as the experiments of Driesch and many others have abundantly shown. One and the same cell will function very differently when placed in a different environment.

It may still be held, however, that this is analogous to a complex machine designed to respond differentially to diverse inputs. The interconnection of cell behaviour with environment, which conforms to the ecological model rather than the mechanical one, may be interpreted as the product of mechanism within the cell. For example, it may be held that the mode of action of DNA is closely analogous to that of a machine. This view seems to be the final refuge of the committed mechanist; so it is necessary to examine the issue more closely.

The DNA in the nucleus of the fertilised egg contains all the instructions necessary to make all the different proteins and all the different sorts of structures in all the different sorts of cells in the body; muscle, nerve, skin, glandular cells and so on. But not all the instructions are needed by every cell. The liver cell needs some, a brain cell needs another lot. Developmental biologists are only just beginning to find out how the selection takes place as each different cell is made and carries out its particular function. More is known about this in the way in which a bacterium makes different enzymes to meet different situations. The bacterium *Escherichia coli* normally resides in our intestines. If a culture of these, living in glucose solution, is given a new sort of sugar, lactose, then within a few minutes the bacteria begin to produce the enzyme β-galactosidase, which was not there before. This enzyme is necessary if the bacteria are to get energy from the lactose sugar. In their normal life in our intestines these bacteria must be ready to change their enzymes quickly to suit the sort of sugar we send down to them. They are selecting just one from the several that their DNA allows them to produce. The part of their DNA that is not used at any time is held suppressed. When the new sugar arrives it must first be detected by some sensory system on the surface of the bacterium. Then a signal is passed through the cell, a process of de-repression is set in action and the DNA is activated to set in sequence a spelling out of its message to produce β-galactosidase.

Does this indicate that the functioning of DNA is machine-like? Monod (1974) believes that it does. According to him the exclusive choice of a substrate is determined by its steric structure, i.e. the bacterium recognises which sugar is present and sets up a chain of chemical reactions which can then be described fully. In commenting on Monod's statement, Young (1978, p. 21) asks:

But is this really true? What one tends to forget is that describing the chemistry of DNA and its reactions does *not* completely describe

> the process. The bacterial cell can produce the appropriate enzymes, say β-galactosidase, only because it is endowed by its past history not only with DNA but with a very particular composition and organization. The cell must have receptors at the surface to detect the new sugar and a communication system to transmit the information to the DNA. Then there is an elaborate system of polymerase enzymes, which allow the selection of the new set of instructions and their 'transcription' and 'translation' to make a new enzyme. It is the presence of this elaborate *individual*, with a history, acting as a chooser with an 'aim', that makes the whole process so *unlike* all known chemical reactions except those of other living things.

Young thus makes the critically important point that the cell is not just a bag of enzymes that are turned on and off by mechanical forces. When the bacterium recognises a particular sugar the cell then selects one of a series of chemical pathways already established in the non-nuclear part (cytoplasm) of the cell. The cell does not set up this pathway *de novo*. It has already been established in the cytoplasm by virtue of the cell's evolutionary history.

It is significant that in all experiments so far that have been successful with cloning, the donor nucleus which may come from the intestine or skin or elsewhere, must always be transplanted into the cytoplasm of an ovum, not just any cell. The ovum by its evolutionary history has within its cytoplasm the necessary means (and we have almost no idea what they are) of turning on and off particular DNA at the appropriate times to produce a complete organism such as a frog. This capacity of the ovum's cytoplasm continues to exist in the daughter cells up to about the 16 cell stage of the embryo. Up to this stage any one of the 16 cells can, if isolated, produce a whole animal. But after that, say at the 32 cell stage, none of these cells when isolated can produce a complete animal. They develop into an incomplete and rather structureless embryo. Something is lost in the cytoplasm of individual cells in the process of differentiation after the 16 cell stage. What that is scientists do not know. The whole process, so far as is known, is very different from the way a machine, or even a series of machines in tandem, functions.

The contrast of the mechanistic and ecological models can be restated at this molecular level. The mechanistic model entails that the constitutive elements in the cell behave like the constitutive elements in a machine, a behaviour relatively independent of the environment of these elements and subject only to the laws of mechanics. The alternative view is that the elements in the cell relate to one another and to the cell as a whole more like the way the animal as a whole relates to its environment. Most research on the inner functioning of the cell has been carried out by persons chiefly influenced by the mechanical model. But what is learned appears to fit the ecological model better.

One might, of course, press for a mechanical explanation at a still more fundamental level, that of the electromagnetic field. But this is surely doomed to frustration. Physicists have long recognised that mechanical models do not work in the subatomic realm. There is a field of events no one of which can be abstracted from the remainder. Each event expresses the whole field in that spatio-temporal locus. It cannot be viewed first as a self-contained event which then secondarily has relations to others. On the contrary, it is constituted by its relation to the field and has no other existence. The ecological model is far more appropriate than the mechanical model for expressing what is known at this level.

The ecological model holds that living things behave as they do only in interaction with the other things which constitute their environments. It does not deny, of course, that many features of their behaviour are determined by structures of the organism itself or that much has been learned when this structure has been examined by scientists with the mechanistic model in mind. But the ecological model proposes that on closer examination the constituent elements of the structure at each level operate in patterns of interconnectedness which are not mechanical. Each element behaves as it does because of the relations it has to other elements in the whole, and these relations are not well understood in terms of the laws of mechanics. The true character of these relations is discussed in the following section as 'internal' relations.

Thus far we have presented the ecological model simply as one which views each living thing, and each component of a living thing, as part of a system. There is no way to go below the level of the systematic interconnections of events to another level at which there are self-contained entities whose properties in isolation from one another explain the system. There are no atoms in the sense of Democritus.

To a large extent the implications of this way of viewing living things have already been appropriated by scientists, even by those who continue to describe themselves as mechanists. Hence we do not wish to claim that the consequences of frankly surrendering the mechanistic model in favour of an ecological one would be enormous. Nevertheless, paradigm shifts do have their importance. Instead of looking for mechanistic explanations of phenomena and viewing structural and ecological explanations as provisional and unsatisfactory, scientists can clarify the conceptuality and methodology appropriate to a world which is far more complexly interconnected than mechanical models have suggested. The direction of research will be affected by the types of questions thereby suggested. Just as in physics Newtonian science is not rejected but included in a broader context; so also in biology, what has been learned under the influence of the Newtonian mechanistic paradigm can be subsumed within what can be learned with the use of an ecological paradigm.

From substance thinking to event thinking

Although a general recognition of the importance of structure and interconnectedness is already pervading the sciences, and although the summary of the present state of biology in Chapter I under three kinds of ecology is unlikely to appear forced or shocking to scientists, nevertheless, more is entailed in a shift of paradigms than this. The continuing hold of the mechanistic model, despite its manifest difficulties and limitations, indicates that it is closely connected with widespread habits of thought and basic modes of perception. When Immanuel Kant saw that David Hume had shown the lack of empirical evidence for the mechanical view of causality, he postulated that it is the necessary nature of the human mind itself to organise the world in mechanical terms. At the time he wrote he understood himself to be providing the needed grounding of science. In this century physics has struggled to free itself from mechanistic categories, but with only partial success. The difficulty of doing so gives some support to Kant's contention, but the effort to do so indicates that Kant was wrong in supposing that physics totally presupposes this mode of thinking.

The effort of physics to find more appropriate paradigms leads naturally to raising philosophical questions about the nature of reality. Unfortunately a philosophy that, since Kant, has grown unaccustomed to discussing such questions is not well-equipped to assist scientists in the framing of new models and paradigms. Nevertheless, science and philosophy need to work together at this task.

The paradigms that have dominated Western thinking since Aristotle have all thought of objective reality primarily in terms of substantial and enduring objects. Many thinkers tried, scientifically and philosophically, to understand rocks, mountains, stars, trees, dogs and human beings. In analysing entities like these, they have sought in their parts other entities which were even more substantial and enduring than these. The goal has been to explain the qualitative changes in the world experienced through the senses in terms of movements of entities which in their own nature do not change. The atomism of Democritus is one of the clearest expressions of this ideal. Descartes' understanding of the world as composed of material and mental substances goes far in the same direction.

The word substance is a convenient one to apply to this family of models. They take substantial entities as the objects to be explained and they seek the explanation in terms of the substances that underlie the changing forms and relations of these entities. When the notion of substance is fully refined a substance is usually understood to be in itself unaffected by the changes which occur at secondary levels. In the mechanistic model, which is so natural an expression of thinking in terms of a substance, the parts of the machine are not thought of as themselves affected by the changing operations of the machine. A particular wheel, for example, is viewed as moving, but it is not conceived as being itself affected by its motion. More broadly, qualitative changes are understood to be the result of changes in relative position of entities whose own properties are not affected by this motion.

The difficulties of substance metaphysics are now well known. They were exposed radically by David Hume and the philosophy of Immanuel Kant took account of them. Hegel freed his thought from the remaining elements of substantialism in Kant's philosophy and represented reality as a process. He called it a dialectical process. However, the reality of which he spoke was human reality. German idealism left the objective world of science to the categories of Newtonian science. This reflects substance thinking even when the metaphysical ground of such thinking was explicitly rejected.

The banishment of metaphysics had ambiguous consequences. On the positive side, it freed scientists to explore alternative models of reality. On the negative side, it discouraged radical questioning of the inherited habits of thought. The new models, for the most part, are still not free from the influences of substance thinking. Nevertheless, progress has been made.

The conceptual progress that has occurred has depended on shifting from substance to event as the basic way of identifying what is to be explained and the principles of its explanation. Instead of taking for granted that the world is composed of the substantial objects of sense experience or of the substances which underlie them, one may think instead of a world composed of events and the smaller events which in turn constitute them. What is to be explained, then, is why things happen as they do. And the explanation will consist in analysis of the causal relations among events and of the component occurrences which make up the larger ones.

Of course science has always been interested in events. Biology has concerned itself with animal behaviour and the functioning of cells. The difference is that the explanation of behaviour and functioning was sought in something unlike behaviour and functioning, that is, in substances and their spatial movements. Our argument is that it is now time to seek the explanation of behaviour at one level in terms of behaviour at other levels and to recognise that behaviour at any level is to be accounted for in terms of complex interacting. This complex interacting is an event, not a substance.

Substance thinking recognised the occurrence of events and undertook to explain them in terms of substances. Event thinking must recognise the existence of relatively enduring 'substantial objects' and undertake to explain them in terms of patterns of interconnectedness among events. From this point of view an atom is not a substantial entity but a multiplicity of events interconnected with each other and with other events in a describable pattern. A mouse is a far more extensive society of events, electronic, cellular and organismic, interconnected in far more complex patterns.

Field theory, relativity physics and quantum mechanics all point in the direction of event thinking instead of substance thinking. We believe the same is true of biology and that this is witnessed to in much that is now being said and done. But too little effort has been made to construct new models and categories out of sustained attention to what is entailed by the priority of events in relation to substantial objects. Alfred North Whitehead is a remarkable exception to this generalisation, and we are profoundly indebted to him in our present proposals and exposition.

Even when one thinks of an event one is still too likely to bring from substantialist habits of mind the notion of something self-contained and self-sufficient. If so, one will have failed to attend sufficiently to the evidence, whether from physics or biology. An electromagnetic event, for example, cannot be viewed as taking place independently of the electromagnetic field as a whole. It both participates in constituting that field as the environment for all the events and also is constituted by its participation in that field. In abstraction from that field it is nothing at all. It does not have independent existence and then relate to the field. It is constituted by the complex interconnections which its place in the field gives to it. The same is true when the event in question is the functioning of a gene, a cell or a rabbit. This functioning does not exist in itself apart from its total environment and then relate to the environment. It *is* a mode of interacting, of being affected and affecting.

The point of this philosophical excursus is to say that the shift from the mechanistic to the ecological model cannot fully occur apart from the more difficult shift from substance thinking to event thinking. As long as substances are supposed to be more basic than events, explanations of the interrelatedness to which the ecological model points will be sought in some deeper level. It will be consciously or unconsciously supposed that self-contained entities change only in outward relations. That will mean also that explanations of living things are still sought in terms of mechanisms.

We are not opposed, of course, to learning as much as possible about why animals behave as they do through the study of the behaviour of cells, or explaining as much as possible about the behaviour of cells by examining the behaviour of the elements which compose them. But the ecological model makes impossible the reductionism that is inevitable with the substantialist prejudice. In the ecological model an event at a higher level can be explained partly by events at a lower level, but the event at the lower level cannot be explained fully without reference to the event at the higher level. The behaviour of molecules in a cell is partly a function of their environment in the cell as a whole. Cellular behaviour can be partly explained by the behaviour of the molecular components. But the molecular behaviour, in terms of which cellular behaviour is explained, cannot be explained fully without reference to its cellular environment.

This language still expresses too much the inherited prejudice in favour of substances. We should speak of molecular events or events at the molecular level rather than molecular behaviour, since the latter language still implies that there are first and primarily substantial molecules which then incidentally behave in certain ways. But it will take time to develop a language which is truly appropriate to the recognition of the primacy of interconnected events rather than of substantial objects. Until that can be done, the ecological model will have to struggle against much of the language in which it must continue to express itself.

The distinction between substance thinking and event thinking is closely connected with the traditional philosophical distinction between external and internal relations. Philosophically understood, 'internal relation' does not refer to a relation among the parts of a machine in distinction from the relation of the machine to its outer environment. As long as substance thinking is dominant, the relations among the parts are viewed as just as external to the parts as the relation of the machine to what is spatially outside it.

External relations are incidental or accidental to the entity. Their occurrence or non-occurrence does not affect the being or the character of the entity. A stone is a typical example for substance thinking. A stone on top of a desk appears externally related to the desk in that its own constitution seems to be unaffected by its spatial relation to the desk. It would be the same stone and have the same properties if set on the floor.

The idea of an internal relation is of a relation which is constitutive of the character and even the existence of something. Such relations cannot characterise substances, for substance is defined in terms of independent existence. Internal relations characterise events. For example, field theory in physics shows that the events which make up the field have their existence only as parts of the field. These events cannot exist apart from the field. They are internally related to one another.

Substance thinking sees relations as external to substances. The substance is what it is independently of relations and then enters into relations. These relations do not affect its fundamental nature or existence. Event thinking sees relations as internal to events. Events are constituted by their interconnections with other events. There cannot first be an event which then is superficially characterised by its pattern of spatio-temporal relations. There is no event that does not have its very occurrence specified by spatio-temporal relations to all other events. The character of that four-dimensional event must be a character that could only occur in that spatio-temporal locus through a pattern of relations to other events.

This close connection between the ecological model and event-thinking indicates not only how difficult it is fully to appropriate the model for reflecting about living things but also that the model is not restricted in its application to living things. Events in an electromagnetic field exemplify the ecological model as much as does animal behaviour in the wild. Although this book focuses on living things, it is important to recognise that the ecological model counts against any dualism between the inorganic and the organic, the non-living and the living. We are emphatically not proposing an ecological model for living things while supposing that the mechanistic model is adequate for non-living things.

Of course the mechanistic model is relatively adequate for discussing much of what takes place in the non-living world and even in the sphere of life. It is remarkably appropriate to machines! But even in the case of machines, a full physical analysis will require that we view the apparently substantial and enduring wheels and levers as complex societies of events at the molecular level, and ultimately at the electromagnetic level. Models derived from the gross functioning of the wheels and levers are poor guides to what takes place there. The most complete account will explain the functioning of the machine in terms of what is occurring at the molecular and electronic levels where mechanistic explanation is inadequate. The mechanistic model is a very useful way of describing much that takes place among inorganic entities at the levels of the world which our sense organs are adapted to identify for us. It is not the clue to the ultimate nature of the physical world. At that level the ecological model is far more appropriate.

This does not mean that the ecological model should totally displace the mechanistic one. When events at the molecular level attain the kind of stable structured order they have in a stone, the relevance of the ecological model to the stone as a whole becomes trivial. Of course even a stone is affected by its environment, and some of the influence of the environment upon it must, ultimately, be understood in ecological rather than mechanistic terms. But for most practical purposes the behaviour of the stone can be discussed adequately in the simpler terms provided by the science of mechanics. The mechanistic model works well.

The mechanical model is valid and illuminating to whatever extent the structures studied are relatively independent of environment, such as the stone or a metal lever. Since no structure at any level is totally independent of all environmental influences, the mechanistic model always involves some abstraction or qualification. But since many phenomena can be studied quite satisfactorily on the basis of such abstractions and qualifications, the mechanical model has a wide range of practical usefulness. The ecological model allows for differing degrees of importance of environmental factors, and the mechanical model points to the limiting case where, with carefully stated qualifications, the influence of the environment is negligible.

Our quarrel is not with the practical use of the mechanical model but with the prejudice introduced into research and explanation when its final adequacy is assumed. This can be illustrated in the case of the atom, and since atoms have been taken as a special stronghold of materialism, this illustration has some importance. What happens when it is assumed that atomic behaviour can be studied adequately in abstraction from the environment? Are important features of atomic events thereby obscured? Can atomic activity be better understood through the ecological model?

It would be nonsensical to propose an ecological model of the atom if we took our modern atoms to be what the Greeks philosophically defined as atoms. These were essentially material and could be nothing else. But the Greeks deduced correctly from their doctrine of atoms that atoms could not be destroyed, and we know that this is not true of what we, under a misapprehension, named atoms. If we are to continue to believe that an atom is unaffected in its internal constitution and properties by its environment, this would have to be on the basis of strong empirical evidence. But such evidence is lacking. It seems instead that atoms exhibit different properties in different environments.

The significance of this view can be illustrated in the interpretation of common salt, which is a compound formed of sodium and chlorine atoms. According to the classical materialism that informs the mechanical model, the sodium and chlorine atoms are unaffected by their combination. Hence in principle all the properties of salt should be discoverable in sodium and chlorine atoms investigated in isolation. But in fact this proves impossible. Hence many scientists speak of emergent properties, and we agree that some properties 'emerge' when sodium and chlorine atoms are combined. But as ordinarily used this doctrine of emergence explains nothing. It assumes that the atoms remain unchanged, having only those characteristics they had in isolation, and that the salt is nothing but the combination of these atoms. Yet it recognises in the salt properties not derivative from those of the atoms.

With the ecological model in mind it is possible to do better. The events that are occurring at the atomic level are internally related to one another. The events that make up the sodium atoms and chlorine atoms are affected by their environments, and when these environments include each other in appropriate ratios, the atoms exhibit properties they do not exhibit in other environments. What these properties are cannot be discovered by examining the atoms in these other environments.

When certain arrangements of carbon, nitrogen and hydrogen atoms, together with a few others, exhibit properties that we recognise by the name of enzyme, and other arrangements of the same atoms result in cells that conduct nerve impulses, we have discovered something new about the nature of those societies of events we call atoms with their remarkably stable structures. When they are organised in these particular ways, the resultant events have characteristics they do not have when this organisation is lacking.

Life without boundaries

We have written a great deal about life in the first three chapters of this book, but the discerning reader will notice that we have not defined it. We turn now in concluding this chapter to the question of definition. What do we mean by 'life'? Can we draw a line between the animate and the inanimate?

Until 1935 there seemed to be a sharp discontinuity between the inorganic and the living, such that several different criteria could be used equally well to indicate the presence of life. The biologist could provide a range of living creatures stretching in size all the way from a whale down to a bacterium. The chemist's range extended in the other direction from the electron-type particle through atoms to molecules, but it included nothing that approached the biologist's bacterium. Between the chemist's largest molecule and the smallest organisms that the biologist regarded as living there was a wide gap. That gap was filled in 1935 when W.M. Stanley isolated for the first time a virus in crystalline form. Here was something that exhibited properties of both the 'living' and the 'non-living'. Here was something the size of the chemist's largest molecules yet with some of the properties of living organisms. The part of the virus that replicates in living.cells turned out to be one large molecule of DNA in animal viruses and of RNA in plant viruses. The virus consists of one such molecule surrounded by a coat of protein. On infecting an organism, the protein coat is squeezed off and the virus simply becomes a naked nucleic acid molecule. It cannot reproduce on its own, but it has this capacity when in the cell of a living organism, its host. In a test tube the purified virus is a white powder. Put into a host it exhibits some of the properties of life.

Scientists now know that cells contain small entities called organelles, such as plastids (which contain the chlorophyll in the green plant), mitochondria and many others. They too reproduce and transform simpler combinations of molecules into complex ones. But they do not have an independent life outside the environment of the living cell. No one has found a free living organelle. However, it is sometimes difficult to distinguish a plastid from a blue-green algal cell or a mitochondrion from a mycoplasma. Mycoplasmas are the simplest cells we know. Their architecture is much simpler than any other sort of cell. They have not, for example, any membrane-lined internal organelles, though they have some naked ones. Among them are the smallest cells known, which contain probably not more than about 600 genes, about one-fifth the number in bacteria. Most of them are parasites on other cells or in other ways take in ready-made products of considerable complexity. They are possibly a result of degeneration of more complex cells. There is a hierarchy in complexity from viruses to organelles to mycoplasma type cells to more complex cells, but there is no clear dividing line between the inanimate and the animate. To be sure, the fully fledged living cell has properties that we do not find in individual molecules in isolation, but some of these properties are exhibited by individual molecules or groups of molecules, as is the case with viruses, when in the environment of a cell.

Given this complexity, scientists could only draw a clear line between the animate and inanimate by means of a very precise and arbitrary definition. If a definition of life would imply that there exists objectively to us some self-contained thing called life, then no definition is appropriate. Although we have used the word life frequently because of its convenience, we do not believe that life as such exists as a discrete property or substance. We have rejected the vitalist doctrine. Our interest is in living things and the characteristics of those things that lead us to think of them as living. We have said something about that.

But we have not made unequivocally clear where the line should be drawn between the living and the non-living. We have not specified absolutely what characteristic or group of characteristics is the necessary and sufficient ground for the attribution of life to any entity. In short we have left the boundary between the living and the non-living blurred.

This blurring is compounded in the proposal of an ecological model. 'Ecological' is used to point to the internal relation of *every* entity to its environment. This means that the model proposed for the study of living things is actually appropriate to the study of non-living things as well. Instead of interpreting living organisms in terms of materialistic mechanism, the objects studied by physics as well as those studied by biology can now better be understood by using an ecological model. As mechanists affirmed, and vitalists tried to deny, there is continuity between what we have conventionally called inorganic and the organic and between non-living and living, but it is a continuity now better studied with models first developed for the understanding of the clearly living.

This does not mean that there are no lines to be drawn at all. Three criteria for drawing lines have been emphasised. One can be drawn in terms of entropy. There are things which simply participate in the increase of entropy. There are others that locally reverse this. This distinction can be treated as the line that separates the living from the non-living. In that case organic molecules and crystals should probably be regarded as forms of life. Or one can draw the line in terms of ability to replicate or reproduce. In this case most organic molecules would not be classified as living. But without more careful specification of our meaning, the case of viruses and organelles could not be settled. These can reproduce within the living cell but not otherwise. Shall they be viewed as alive when in the living cell, or shall we regard the cell as a whole as the lowest unit of self-reproduction?

Once it is clearly recognised that there is no such *thing* as life, little hinges on the answer to such questions. For different purposes the line may be drawn at different places. Or, better still, if people can get used to the idea, all such line drawing can cease. We have a widely varied set of entities which manifest to varying degrees and under varying circumstances different features which are associated with living things. Any line is arbitrary.

When biologists say that the unit of life is the cell, they mean that you have to have at least a minimal cell organisation if you are to find a combination of properties that includes reproduction and the transfer of chemical information from nucleic acids to build the complex molecules found in living organisms. A collection of DNA molecules, protein molecules and all the rest is not enough. They must be organised within an architecture of membranes. It is the presence of membranes that makes possible the co-ordination of the activities of the large molecules. Biologists have very little information on the steps that led from complex molecules to their organisation within membranes in the cell. Because it is the cell alone that reproduces and metabolises as an independent unit, biologists have regarded the cell as the unit of life and the origin of the cell as virtually synonymous with the origin of life while at the same time conceding that the cell was preceded by entities such as organelles that some have described as sub-vital. As long as we recognise the element of arbitrariness in this language, we can continue to use it.

Conclusions

Science cannot function without models or paradigms. Successful models tend to become deeply entrenched in habits of mind, in the structure of scientific theory, and in the very methodology employed. In biology as in physics the mechanical model has been brilliantly successful and has become deeply entrenched.

The recognition of the limitations of the mechanistic model has gone further in physics than in biology, although it has a longer history in the latter discipline. Vitalism has opposed mechanism for centuries, but neither vitalism nor the more recent theory of emergence has been able to make much headway against mechanism. A major reason is that both leave the main field to mechanism and that neither actually provides a mode of explaining those phenomena which they see that mechanism cannot account for. They rightly point to the limitations of mechanism, but they do not in fact provide an alternative model which is of use to the scientist.

This chapter identifies an alternative model. We call it ecological. It has affinities with what has sometimes been called an organismic model. Indeed, Whitehead, to whom we are so greatly indebted, called his position 'the philosophy of organism'. We have no objection to this language, but we fear that some have used it less carefully than Whitehead. Instead of emphasising the internal relatedness of living things to their environment, some have used the idea of organism to support questionable doctrines of the whole as greater than the sum of its parts.

The ecological model is of entities which are what they are because of the environment in which they are found. The objection to the mechanical model is that it leads to the study of living entities in abstraction from their environments. As described in Chapter 1 much, if not most, of current biology has in fact recognised the importance of the environment of what it studies. Hence the ecological model in this sense is already widely operative in the discipline.

The question is whether what is described ecologically at one level can be explained mechanistically at a more fundamental one. We believe it cannot. The elements by which such explanation is sought turn out again and again to be understood better ecologically than mechanistically. This is true of the living cell and of its parts including DNA. Pressing the analysis to the level of subatomic events contributes nothing to the mechanist's cause. What happens at every level is a function of a structure of interacting events.

One major reason why the evidence against the adequacy of the mechanical model fails to overthrow its dominance in practice is that unconscious metaphysics, rooted in Indo-European languages, favours it. People in the West are accustomed to think of substantial things as the primary realities and of events as the result of the interaction of these substances. Hence they seek explanations of events in substances. The ecological model can only be truly understood and appropriated when this process is reversed. Events are primary, and substantial objects are to be viewed as enduring patterns among changing events. It is these events which are constituted by their relations to one another in the way the ecological model indicates.

Relations that are constitutive of events are properly called internal to those events. The ecological model is a model of internal relations. No event first occurs and then relates to its world. The event is a synthesis of relations to other events.

The shift to the ecological model has the least importance for the study of machines and such inanimate objects as stones. To study these in some abstraction from their environments, as the mechanical model does, introduces only minimal distortions which, for most purposes, can be ignored. But the ecological model is important for the study of the constituent elements of machines and stones as well as of living things. After all, both machines and organisms are composed of molecules and of atoms. The ecological model does not propose a dualism between the organic and the inorganic. It proposes that both are composed of events which are best understood ecologically, but it recognises that these events may achieve such stable structures that the important properties of the resulting substantial objects are, within limits, independent of their environments. The Newtonian science of mechanics works well here.

The ecological model overcomes the need to draw a sharp line between the living and the non-living. Among events which are internally related to other events, diverse features which are prominent in complex living organisms may appear to different degrees. Living things 'emerge' from an inorganic world through the attainment of structures which give scope for these features to function and become prominent. The living cell is a marvel of such structure.

The ecological model is pursued further in Chapter 4 in the context of the human and its relation to the non-human.

4

The human and the natural

> It seemed to me sufficient to indicate, in the first edition of my 'Origin of Species' that by this work light would be thrown on the origin of man and his history; and this implies that man must be included with other organic beings in any general conclusion respecting his appearance on earth.
>
> Darwin (1871, p. 389)

Chapter 2 sketched a contemporary version of a Darwinian account of evolution. It showed that human beings are among those things that have evolved in this way. Purposive behaviour has played a role in animal evolution in general, but it has played a much larger role in human evolution. It has given rise to human culture which is today of enormous importance for the entire planetary biosphere.

Human beings play a very distinctive role, and they have separated themselves a long way from all the other products of the evolutionary process. Nevertheless, no point in the evolutionary sequence can be specified at which human beings first came into being. Biologically understood, humanity constitutes one animal species among others. *Homo sapiens* is part of nature. Doctrines affirming that human beings are to be understood in other terms and categories, that there is a sharp division between the human species and the rest of life, inevitably find themselves in tension with the biological evidence for continuity. Most such doctrines came into existence well before biology was a science. This helps to explain how the doctrines arose, but it does not justify continuing to hold views which are now obsolete.

When humanists view the non-human world in terms of the mechanical model, they are justified in positing a radical division between humanity and this world even when they cannot specify just where to draw the line. They are rightly confident that human feeling, thinking and willing are not merely mechanical processes. When the rest of the world could be viewed only as a machine, dualistic thinking had profound justification despite its difficulties. However, Chapter 3 has shown that the mechanistic model, which dualists have rightly resisted in its application to human beings, is inadequate to other living things as well. It is not adequate to atoms and subatomic particles either.

The superiority of the ecological model in interpreting the non-human world reopens the question of dualism. Granted that a mechanistic model cannot account for human feeling, thinking and willing, what about an ecological one? Can an ecological model be used for understanding human experience as well as rabbits and cells? Or does the ecological model, like the mechanistic one, prove applicable only to the non-human world, thus requiring continuation of dualistic modes of thought?

This chapter begins by testing the adequacy of the ecological model to human experience against the deep-seated objections in Western philosophy and religion to considering human beings as essentially part of a continuous natural process. In this testing, features of the ecological model will be brought out that were not immediately evident in the treatment of the model in the preceding chapter.

The second section of this chapter surveys the recent debates on the understanding of present human existence and its future possibilities arising from a view of the human species as continuous with the rest of nature.

The third section draws an evaluative picture of the development of human beings from their pre-human ancestry through the emergence of civilisation.

Having developed the ecological model more fully in the first section to display its applicability to the human, we turn in the fourth section of this chapter to the applicability of this model to other organisms. If human beings with their subjective experience are continuous with the rest of nature, this implies something about other living organisms as well. If human beings resemble them, they must resemble us. The fourth section asks whether the features of the ecological model that appear through its illustration by human experience are found in other animals.

On the assumption that there has been a continuity and development of experience along with that of visible structures, the chapter concludes with a conjectural reconstruction of the evolution of experience.

The ecological model of human existence

Human beings, especially in the West, have been reluctant to consider themselves as part of nature. Aristotle recognised that we are animals, but he insisted that we are rational animals, meaning thereby that our essence is our rationality. Accordingly the influence of Greek philosophy has favoured identification of the human with reason. Descartes intensified and systematised the distinction between us and the rest of the world. He divided everything into two fundamentally different orders of reality. On the one hand was the rational mentality of human beings and on the other the spatially-extended physical substances that comprised everything else, including human bodies and other animals in their totality.

The Bible recognised human beings as part of God's creation. But it distinguished humans as those who alone participate in the image of God. In Christian theology this image is usually conceived as spirit, that which transcends nature, which involves self-consciousness and freedom and which makes fellowship with God possible. The Christian tradition has emphasised how this image was distorted and shattered in 'the fall'. But Christians have been convinced that the destiny of all human beings is of a different order from that of all other creatures. Given this background, Christian theology found it all too easy to accommodate itself to Cartesian dualism, including a mechanical concept of the world. For many Christians the earth was no more than a material stage on which the drama of human history was performed. Nature and human history were distinct domains. There was no sense in which one arose from the other. Mechanism's most famous metaphor of the universe as clockwork (with God as the clockmaker) was provided by a bishop, the Frenchman Nicole Oresme in the fourteenth century (Lynn White, 1975). According to Lynn White (1975) the schism between our understanding of nature and our understanding of human nature as detached from it, began 650 years ago in a theologically motivated effort to salvage revelation as essential to the Christian life. Most of the steps were taken by good Christians, not least by Sir Isaac Newton.

In the eighteenth century nature was increasingly interpreted through the eyes and methods of physics. The physics of that time viewed its objects as passive matter obedient to deterministic laws. Thus nature appeared as a great machine. Most people were impressed that human beings, as those who could grasp the principles of nature, were above nature. And so the view of the human as rational mind was increasingly the common sense view of the age.

Despite the widespread acceptance of a dualism between mind and matter across the centuries, this dualism has always been difficult to state intelligently. Does it mean that mind and matter have no effect on each other? If so, then the human mind is not affected by the natural occurrences it seems to know. Then what is science all about? Or does dualism imply that mind has no effect upon the material world? If so, then the apparent responsiveness of the human body to rational human intentions is an illusion.

Some bold thinkers began to draw the conclusion that human beings, too, are simply part of the great machine! The brain, they argued, is as much a part of nature as is any other part of the body, and thinking is the activity of the brain. If this thinking is 'rational', it is because the machine happens to be so constituted as to produce rational thinking through the human brain.

While materialists were erecting the mechanistic model into a complete cosmology or metaphysics, other philosophers, the empiricists, were pressing the question of knowledge. On the whole, they were as impressed by physics as were the materialists, and they wished to make all thinking scientific. This was true even of David Hume. But the unintended consequence of his analysis was to threaten scientific claims to knowledge and, even more clearly, to undercut any metaphysical extrapolation from those claims. All that is really experienced, he held, was a flux of sensory impressions or phenomena. Carefully examined, this flux justifies neither the idea of substantial material objects nor the idea of laws establishing objectively necessary connections between events. Scientific laws can be nothing more than generalisations from repeated conjunctions of impressions. Since, according to this position there is nothing but a flux of phenomena, it is sometimes called phenomenalism.

The human mind fared no better in Hume's examination of experience. He could find no mental self underlying the impressions or acting upon them. Whereas dualists posited a rational mind alongside a material nature, and materialists saw human beings as a part of the machine, Hume's phenomenalist view locates human beings fully within the flux of sensory impressions or phenomena.

The double attack on human pretensions by materialism and phenomenalism elicited a brilliant reply from Immanuel Kant. To this day his influence underlies the thought of most continental intellectuals as well as many others. It has provided the main defence of human uniqueness in the modern world.

Kant's central insight was that the human mind actively shapes the world it knows. He pressed Hume's analysis of what is given through impression still further and argued that even time and space are not derivable from these impressions. Space, time and causal relations must all be contributions of the structure of the human mind rather than of an objective material world. Instead of viewing the mind as a mere part of either a mechanical or a phenomenal nature, Kant saw the nature investigated by science as a construct of the human mind. He acknowledged an extra-mental reality as source of the phenomena ordered by mind, but he held that nothing more could be said about it. Instead of viewing mind as a mere part of either a mechnical or a phenomenal nature, Kant assigned it the prior and primary role in all reality. In this way he founded German idealism.

Next to Kant the greatest thinker of the idealist tradition is Hegel. Hegel accepted the priority and primacy of mind, but instead of examining its constant and universal structure, he stressed its historical development. Human hstory came to be seen through Hegel's influence as the unfolding of reality itself. The natural world is understood as playing its role in, and for, the movement of mind.

Although most subsequent thinkers have not identified themselves closely with the details of the doctrines of Kant and Hegel, the thought of these two figures has had vast influence. Most of nineteenth- and twentieth-century philosophy has focused attention on the way human knowing shapes the human world. Science is viewed as one of those means of knowing which has its own quite appropriate way of shaping its world. But most intellectuals have been sceptical of claims that any way of knowing grasps an extra-human reality as it is in itself. For them the nature which is known, is nature as it is ordered and even constituted by human knowing. What is of central interest is the human knower and the ways of knowing, not an extra-human reality which humans purport to know. For those who adhere to this idealist tradition, the knower certainly cannot be explained in terms of the objects of knowledge.

Most of the twentieth-century defences of human dignity against materialist and phenomenalist reductionism have stemmed from this idealist tradition. For example, although the existentialists criticised some features of idealism, shifting the focus of attention from thought to decision, they continued its general view of the separation of humanity from nature. Ordinary language analysts sometimes reflect the influence of this tradition when they treat as a category mistake the effort to understand human beings in the language appropriate to the understanding of nature.

The mode of thinking that distinguished the human sphere from nature is built deeply into our academic disciplines. In Germany a fundamental distinction is still made between the *Geisteswissenschaften*, the sciences of mind or spirit, and the *Naturwissenschaften*, the sciences of nature. In the USA the distinction between the humanities and the natural sciences is similar. In both cultures a deep-seated impression is communicated that the two sides of the educational curriculum have little bearing on each other, that human meaning is not bound up with the world view suggested by the natural sciences.

Despite the hold of idealistic dualism, it is being eroded. Its first great opponent was Marxism. Marx was deeply affected by Hegel's concern to understand history as the fundamental reality, but he turned Hegel upside down. Whereas Hegel believed that history was the dialectical advance of mind or thought, Marx insisted that history is fundamentally the dialectical advance of the material grounds of human existence. The sharp division between humanity and the natural world is blurred, if not entirely overcome, in Marxist theory.

A second blow against idealism was Darwin's evolutionary theory. If human beings emerged gradually as a development from other natural forms, how can we believe that nature is the construction of the human mind? Are we to dismiss the whole Darwinian theory as an appropriate construct for science but one which tells us nothing about our own human reality?

A third blow was struck by the development of psychology and the social sciences. Here scientific study of nature was extended to the study of human nature. The possibility of this was recognised by Kant, who could yet deny that the scientific study of human beings deals with their true nature. Only the observable phenomena can be scientifically treated, he taught. Such denial continues to be theoretically possible, but as psychology develops it becomes more difficult to sustain. Psychology has provided therapies through which human experience is profoundly changed.

The hegemony of idealistic ways of thinking is further challenged by the extension of evolutionary theory to anthropology and ethology and the emergence of increasingly detailed and plausible hypotheses about millions of years of proto-human and early human development. Here the scientific study of human experience and behaviour and the evolutionary study of organisms merge. The results are far more difficult to distance from our fundamental human self-understanding than was evolution viewed as simply a biological theory.

The deepening contact with Asian modes of thought has been a fifth factor in weakening the hold of idealistic dualism. In Asia we find a highly sophisticated understanding of the human and the natural that emphasises continuity rather than discontinuity. Of course Western dualists can continue to affirm their philosophical analyses against these non-dualistic views. But in global perspective dualistic idealism increasingly appears as a response to problems that arose at a particular stage of a particular culture. It becomes more difficult to understand it as the final truth.

Finally, quite recently we have become acutely aware that the problems with viewing ourselves as outside of nature are not just intellectual ones. They are urgently practical. Our disregard of the extra-human and extra-mental world has distracted our attention from its systematic exploitation and destruction. Even now many Western thinkers, shaped consciously or unconsciously by German idealism, are uncomfortable with direct attention to questions about the natural world and prefer to focus on the human world, hoping that by dealing with the problems of the human world those of the natural world will fall into place. Those who realise the importance of dealing with environmental and ecological problems often find themselves handicapped for lack of categories within which to understand these in their bearing upon human existence.

Surely these blows would long ago have destroyed the power of idealistic dualism had it not been that the alternative appeared even more unacceptable. Many Western thinkers have believed that the rejection of idealism must lead to seeing human beings as mere objects, mere parts of that mechanistic world which deterministic science posits and studies. They rightly insist that for scientists to view themselves exclusively as simply part of the objective world makes no sense. The knower cannot be part of the known as long as the known excludes such activities as knowing.

But there is no need to suppose that deterministic materialism is the only alternative to idealism. The previous chapter offered a quite different model. Indeed, deterministic materialism is inadequate for the understanding of all living things and even of atoms! An ecological model will account for all that the mechanistic model has illumined and more. It allows us to see all living things as taking account of their environments in an active way. This is clearly a more promising model for understanding ourselves than is the mechanistic one.

Nevertheless, there is always danger in interpreting human beings through models developed for the understanding of other things. It is precisely against this practice that humanists have insisted on the separation of the human from the world studied by science. We must consider the ecological model more carefully to determine its appropriateness to human beings.

The model is one of internal relations. This was explained in Chapter 3, but there is more to be said. The task now is to explain internal relations in the context of human experience. This will at the same time display the applicability of the ecological model to human beings.

Consider a simple situation. One woman (A) is observing two other people (B and C) conversing. The normal account of the conversation by A would present B and C as questioning and answering one another, responding to challenges and otherwise interacting. However, A may be eager to be purely objective and 'scientific' in her account. In this case A will only describe what she observes in the strictest sense. She can then describe the sequence of gestures and sentences. She can even speak of correlations between certain gestures on the part of B and others on the part of C. But she cannot speak of C as answering B or of C's statements as exciting or upsetting B. To speak in that way would imply that B and C were taking account of one another's behaviour, although she cannot directly observe such taking into account. What she can observe are spatial and temporal relationships between the behavioural patterns, and about these she can make certain generalisations, such as, perhaps, that there are correlations between certain linguistic expressions by B and increased animation on the part of C.

What A observes are external relations between B and C. Yet A in fact believes that more is going on. A knows that B and C are taking account of one another's gestures and speech. B sees C's gestures, hears what C says, and responds more or less appropriately; and the same is true for C's relation to B. B's relation to C is an important part of B's experience; and C's relation to B is an important part of C's experience. In other words, in addition to the external relations between B and C, that is, the relations as observed by A, there are also internal relations between them. These are not observed by A, but they explain what A does directly observe. It is because C is excited by some of the things B says that there is a correlation between some of the linguistic expressions by B and increased animation on the part of C.

The internal relations are composed of all sorts of ways of taking into account: sensing, feeling, hearing, seeing and experiencing in every other way. B's seeing of C is a relation to C which is internal to B. To an important extent B's relation to C is constitutive of B's experience.

Indeed, one cannot describe B's experience during the conversation apart from this relationship. This distinguishes internal relations from external relations. When A limits her account to what she directly observes she cannot include internal relations. She cannot assert that the external relations she observes are constitutive of B and C. In order to speak of constitutive relations one must allow the standpoints of B and C to be considered. In other words, B and C must be recognised as subjects experiencing each other and not only as objects being experienced by A.

Newtonian science attempted to deal only with external relations. The kinds of things that are conceived to be related to each other only externally are substances. The ways material substances acted upon one another could best be understood on the model of a machine. The assumption that living things should also be viewed as having only external relations to each other and to the remainder of their environments led to the mechanical model for the understanding of organisms. The extension of that model to human beings is certainly dehumanising.

But the ecological model is based on internal relations. It asserts that living organisms take account of their environments, that is, that their relations to their environments are constitutive of what they are. The full meaning of internal relatedness is clearest to us in the case of human experience. Experience is the way in which human beings take account of the world. Human relations to this world are constitutive of human experience. A human experience is the way one takes account of one's world and responds to it. In short, human experience is the ideal exemplification of internal relations. To apply to human beings a model based on internal relations, then, is not at all to understand human beings in terms of a model that can be adequately derived from the objective study of other organisms.

The actual movement is in the other direction. Truly to understand the internal relations that are so important to the ecological model of events is to understand them as illustrated in human experience. If the observer A had been watching two dogs instead of two people, she could have described their interaction, also, either purely objectively or in terms of their mutual experience. If she chose the second course, the ecological model would come into play. It is dog B's experience of dog C which relates B internally to C. Our proposal is that something important can be learned about all living things by taking human experience as the prime illustration of the ecological model.

What then can be learned about the meaning of being alive from the consideration of the human experience of aliveness? This is an inexhaustible topic, for the depths of aliveness of immediate experience can never be fully plumbed. But a beginning can be made on the expansion and clarification of the ecological model. Quite apart from reflection, even quite unconsciously much of the time, we enjoy life in the present, we want to continue to live, and we want our life to be as alive as possible. In short, the sheer givenness of life is accompanied by a positive valuation of life that is deeper than reflection about values, meanings and purposes. There are some experiences in life that are so precious in themselves as to prove that not everything is a means to some end other than itself. The positive valuation can indeed endure through great frustration and suffering. William Faulkner ended his novel, *The Wild Palms*, with this sentence: 'Between grief and nothing I will take grief.' The survivors of concentration camps were often those who, through sheer will to live, were able to endure the apparent hopelessness and meaninglessness of life. They were able even to bear the incapacity to grasp life's unconditional meaningfulness in rational terms (Frankl, 1964, p. 75 *et seq.*). But this positive valuation is not absolute. There are circumstances in which death is preferable to the only kind of life – a living death – that seems possible. And with the passing of the years and the completion of the life-cycle some are prepared to welcome death as a friend.

From such experience of how it is to be alive, it is reasonable to generalise to other living things. Whitehead (1929*b*, p. 8) proposed that all living things have a threefold urge '(i) to live, (ii) to live well, (iii) to live better. In fact the art of life is first to be alive, secondly, to be alive in a satisfactory way, and thirdly, to acquire an increase in satisfaction.' That is, life is bound up with an urge to live. It is not a mere fact; it is a value. That is, being alive is valuable in itself. If life were not prized by those who live, death would soon triumph. In short, if life were not a value, it would cease to be fact. For of all physical structures, those which produce life are the most fragile. They are built against the tendency of all things, when left to themselves, to decay. To maintain itself, the living organism must be constantly active. If all its constituents remained unchanged for even one second it would quickly return to a bundle of inanimate physical elements.

But human experience shows that, as Whitehead says, people want not only to stay alive but to live well and then to live better. To live well is to be more alive. We are most alive when we are most attuned, most in harmony, most stimulated, most integrated, most responsive, most loving, most accepting, most spontaneous, most honest and most innocent. A common consequence of this aliveness is a feeling of gratitude. In that context human beings may experience an all embracing love for everybody and everything, leading to an impulse to do some good in the world, an eagerness to repay, even a sense of dedication and obligation. There comes a fundamental trust with respect to the meaningfulness of life.

The narcissist, by contrast, has lost any fullness of meaning. Because he has few inner resources he looks to others to validate his sense of self. He needs to be admired for his beauty, charm, celebrity or power; attributes that usually fade with time. Unable to achieve active involvement in the world in love and work he has little to sustain him when youth passes him by.

In general, aliveness can be correlated with two facts: how rich is the world to which one is attuned and how fresh is the response of feeling, thought and action to that world. Both are matters of novelty. To stay alive requires new stimuli and newness of response. It is this novelty which affluence seems to dull. It throws into the clearest, coldest light the spiritual, ethical and philosophical hunger of people. 'Striving for something one lacks inevitably makes one feel that life has a meaning and that life is worthwhile,' says Maslow (1970, p. 38). 'But when one lacks nothing, and has nothing to strive for, then ...?' Hence the plethora of new words and phrases to describe the modern state; rootlessness, anomie, meaninglessness, existential boredom, emptiness, estrangement. In *The Farther Reaches of Human Nature*, Maslow (1973, p. 26) recalls the frontispiece of a book on abnormal psychology which he studied as an undergraduate. The lower half of the frontispiece was a picture of a line of babies, pink, sweet, innocent, lovable. Above that was a picture of a lot of passengers on a subway train, glum, gray, sullen, sour. The caption underneath was very simply 'What happened?'

To live well people must be stretched, even to the limits of their strength. Popper (Popper & Eccles, 1977, p. 558) agrees: 'Life is a struggle for something; not just for self-assertion but for the realization of certain values in our life. I think it is essential for life that there should be obstacles to be overcome. A life without obstacles to overcome would be almost as bad as a life with only obstacles which could not be overcome.' Our best moments and our best work may be on the rim of hell. Contentedness and discontentedness belong together.

Without new stimuli, which often arise our of a discontentedness, people relapse into dull habit. What is needed is not a tensionless state but rather the striving and struggling for some goal worthy of human life. Without novel response to integrate the new stimulus people will either blunt the new stimulus or impoverish the possibilities of inheritance that could lead to greater fulfilment. Thus life is bound up with novelty.

This novelty is not mere change. When there is no novelty there is decay. That, too, is change. That change expresses the ever present entropic tendency. This tendency can only be countered in two ways. It can be resisted by the very stable structure of many inanimate things. These endure by endless repetition of unchanging patterns. It can be overcome locally and temporarily by creative novelty which rises above external determination. Hence, as Whitehead said: 'Life is an offensive against the repetitious mechanism of the universe ... Life can only be understood as an aim at that perfection which the conditions of its environment allow. But the aim is always beyond the attained fact. The goal is some type of perfected things, however lowly and basically sensual. Inorganic matter is characterized by its acceptance of matter of fact' (Whitehead, 1933, p. 102).

The incredible fact is that this changing thing called life endured through the ages. Had we been present at its misty birth in some shallow sea with electrical storms raging above and volcanic eruptions below, who would have predicted that a tiny delicate jelly-like blob with no physical power to its credit could have endured through change while so much around perished!

Each human life is not merely a product of its genetic endowment and its environment. It is also a creative response to these given conditions. This is the element of transcending. This element is the urge of life to meet its unrealized possibilities. To live is to relate to existence in terms of the possible as well as the actual; to be awake to all the possibilities of any occasion and eager to comprehend the meaning of these possibilities; to rise above the situation-bound here – now; to experience the joy of attempting to do what we think we cannot do and then doing it.

During Hitler's Germany, newspapers carried a picture of an old Jewish man with a beard being paraded before the jeering crowd in Berlin in a garbage truck. He had compassion written all over his face for the crowd whom he seemed to look upon with pity and forgiveness. To live fully is to 'forgive them for they know not what they do'; to become so absorbed in something or someone as to lose all sense of time and space. It is the travelling toward rather than the arriving; the state of being something rather than having. It is gratuitous grace.

If this transcending were wholly absent from the rest of nature, it would introduce such a fundamental cleavage between human beings and other living things that the acknowledged physical continuities would appear unimportant. But, according to the ecological model, all living things share in this experience of transcending. In Whitehead's words (1933, p. 266), 'The essence of life is the teleological introduction of novelty, with some conformation of objectives. Thus novelty of circumstances is met with novelty of functioning adapted to steadiness of purpose.' Life is thus a transcendence of physical causality. It is 'the origination of conceptual novelty' (Whitehead,* 1978, pp. 102/156) and 'a bid for freedom' (pp. 104/159). Living things inherit from their physical environment, but their response introduces something new.

Our intelligence, our language, our culture enable us to transcend far more radically than can any other living things of which we know. It is very unlikely that the sense of responsibility for society as a whole, possessed by some human beings, could be even remotely approximated in other creatures. But the act of transcending which expresses itself in these rich and distinctive ways in us establishes, not our discontinuity with other living things, but our continuity.

Supporters of human uniqueness are correct in calling self-contradictory the view of human existence as simply a part of the mechanistic world. Even to think about the question is an activity quite unlike that of machines. For scientists to assert that human beings are machines would imply that machines are making such assertions about themselves, and since machines can be programmed to assert the contrary just as well, there would be no reason to take them seriously. Those who say human beings are machines deserve to be taken seriously only because they are not in fact machines.

On the other hand, there is something disparate about the expedient of denying the continuity of human life with other natural things in the face of so much cumulative evidence. We are fortunate that another option is available, one that takes seriously the evidence both of human distinctiveness and of the continuity of life. This we explore in the next section.

* The two sets of page numbers in Whitehead, 1978 refer to revised edition 1978/original edition 1929.

The human condition in dispute

For the most part social scientists, as students of the human condition, accept the conclusions of biologists on biology. Yet most of them make little use of evolutionary ideas in their own studies. They may ridicule the Biblicist arguments against Darwinism, but for practical purposes their position is often more on the Biblicist side. Barash (1977, p. 7) complains that 'like religious hypocrites, most behavioural scientists have been content to profess the faith, while functioning on a daily basis as though the organizing constructs are somehow irrelevant'.

One reason for the modest role of evolution in the social sciences is that those who have made use of it have given it a bad name by their errors and excesses. Critics of evolutionism remember that in 1851 Herbert Spencer argued in *Social Statics* that poverty and starvation were natural agents cleansing society of the unfit. According to him it was wrong to help them. Later in the century other social Darwinists gave comfort to the captains of industry in their ruthless competition for wealth and exploitation of colonized people. In Hitler's Germany Konrad Lorenz (1943), who was then Professor of Psychology at the University of Königsberg in East Prussia, evoked his ethological theories to recommend a 'self-conscious scientifically based race policy' to eliminate 'the degenerates' of society. The arbiters of this scheme, who were to decide who were the degenerates, were to be 'our best individuals' (*führerindividuen*) (translation by Chatwin, 1979, p. 8).

Others have concluded from their understanding of how the human species is shaped by evolution that it is ill-made and, hence, that brain engineering is required. Koestler (1967) in *The Ghost in the Machine* proposed that the human species could be converted from maniac into man by the use of some as yet undiscovered drug. In *The Call Girls* he describes a series of ways in which scientists might deal with basic human problems, for example, by electronic control through implanted electrodes (Koestler, 1971). More recently, Carl Sagan (1978) has taken up this proposal quite seriously in *The Dragons of Eden*. Sir Macfarlane Burnet (1978) urges a genetic purgation process that would last for generations. His proposal is for a programme of selective breeding to avoid genetic deterioration and to reduce aggressive propensities.

Marshall Sahlins (1977) brilliantly exposes the ways in which ideas about evolution and about *laissez-faire* capitalism have gone hand in hand. He quotes (p. 71) from Ghiselin's (1974) book, *The Economy of*

Nature and the Evolution of Sex, to show that drawing morally destructive conclusions from evolutionary ideas is not a thing of the past. Ghiselin wrote (p. 247):

> The evolution of society fits the Darwinian paradigm in its most individualistic form. Nothing in it cries out to be otherwise explained. The economy of nature is competitive from beginning to end. Understand that economy, and how it works, and the underlying reasons for social phenomena are manifest. They are the means by which one organism gains some advantage to the detriment of another. No hint of genuine charity ameliorates our vision of society, once sentimentalism has been laid aside. What passes for co-operation turns out to be a mixture of opportunism and exploitation. The impulses that lead one animal to sacrifice himself for another turn out to have their ultimate rationale in gaining advantage over a third; and acts 'for the good' of one society turn out to be performed to the detriment of the rest. Where it is in his own interest, every organism may reasonably be expected to aid his fellows. Where he has no alternative, he submits to the yoke of communal servitude. Yet given a full chance to act in his own interest, nothing but expediency will restrain him from brutalizing, from maiming, from murdering – his brother, his mate, his parent, or his child. Scratch an 'altruist', and watch a 'hypocrite' bleed.

Given the way in which the evolutionary interpretation of human and social phenomena has been dominantly expressed, it is not surprising that most social scientists prefer to keep their distance. Indeed, seeing that their theories have been misunderstood and applied inappropriately, many biologists are equally eager that their theories not be used in other disciplines. Both are generally agreed that the safest approach to the study of the human is to formulate categories geared to human distinctiveness. While affirming that *Homo sapiens* is a biological species, they prefer that this be ignored so far as the social sciences are concerned.

Nevertheless, human beings are part of nature. They are products of an evolutionary development. Even if this is ignored in the social sciences, the fact is entering more and more deeply into the collective consciousness. Those who emphasise it and make proposals derived from it usually get a wide hearing. Often their proposals are rightly rejected, but the questions they raise cannot be ignored. Their critics insist that human problems are cultural and not biological. But as humanity faces the future, many people want to know from where they have come. Religions used to answer these questions of origin and destiny, but the growth of scientific knowledge has made their traditional answers unconvincing. People want to learn from science what is to take their place.

Precisely because of this religious interest, the debates are often intense. Today they are more frequently cast in political than in explicitly religious terms, but one can see the same concerns at work, for example, in the response to the sociobiology of E. O. Wilson (Sociobiology Study Group, 1976). Merely factual and scientific questions are subordinated to questions of ultimate meaning.

The heat of the debate is all the greater because the question of the human future has an urgency today hardly matched before. The end of the human venture may be near. Scientists have already manufactured the weapons that could annihilate us, and they continue to perfect our destructive capacities. The global society is pressing the limits of the planetary carrying capacity and must either change its direction drastically or face the imminent possibility of collapse. Solutions to one problem generate others, so it is difficult even to conceive a happy scenario into the future. There is a decay in social order that does not respond to our efforts at healing.

The irrationality of collective and individual behaviour is more evident than ever before. The question is why? Is there something in human nature that drives people to destruction? If so, what is it and how did it come about? Is our species programmed in ways that now prevent adaptive response?

The answers of biologists, and writers on biology, to these questions are various. Morris (1967, p. 241) refers to aggressive and territorial feelings in the human as biological urges as basic as our sexual and parental impulses. There are, he thinks, severe limits to the extent to which these urges can be controlled by cultural constraints. Lorenz (1966) differs little from Morris. The violent response of our ancestors to enemies has not disappeared from the human constitution. However, Lorenz suggests that it can be given new outlets such as competitive sports and space exploration. In contrast to these authors Lewis & Towers (1969) consider that stereotyped behaviour is not characteristic of human beings. Human reactions are 'infinitely malleable' (p. 55). They are not determined by a set of programmed drives. There is very little substantial evidence in the writings of any of these authors that could lead to anything more than speculation on the subject.

Recently, however, this debate received a new name and moved nearer the centre of the academic community with the publication of Wilson's (1975*a*) *Sociobiology: The new synthesis*. Most of this book is concerned with the behaviour of ants and other animals, but the last chapter brings in the human. Wilson proposed (p. 547) that in the 'macroscopic view the humanities and social sciences shrink to specialized branches of biology; history, biography, and fiction are the research protocols of human ethology; and anthropology and sociology together constitute the sociobiology of a single primate species'. The implication of such a claim appears to be sweeping indeed! It seems to imply that the methods of all the human sciences would be no more than specialisations of the methods of biology.

In fact Wilson's position provides no basis for this conclusion. The function of sociobiology will be to throw as much light as possible on the ways in which the human genetic constitution evolved and how it affects all forms of social behaviour. Wilson's probes in this direction are far-reaching and often suggest that a very great deal can be explained in this way. But when he confronts the issue directly, his claims are much more modest. He recognises that human social behaviour is largely determined by culture. If the choice is between environmental and genetic conditioning, Wilson affirms that on the basis of objective evidence the truth appears to be somewhere in between, closer to the environmentalist than to the genetic pole.

In his most modest formulation of the genetic influence on human behaviour Wilson pictures this as the genetic provision of a large array of possible behavioural patterns which none the less exclude others that characterise other species. Within this array different cultures will encourage different possibilities. This could be interpreted as meaning only that some behavioural patterns possible to other animals are impossible for human beings. Whether the limitations are strictly physiological is not asserted.

But if this were all that Wilson was asserting his book would have aroused little note. In fact he seems to be saying a great deal more. He makes many generalisations about human behaviour in which he sees parallels to animal behaviour and leaves the reader with the distinct impression that the genetic make-up of the species not only allows such behaviour as one pattern among others, but favours it. One example will suffice. 'The members of human society sometimes co-operate closely in insectan fashion, but more frequently they compete for the limited resources allocated to their role-sector. The best and the most entrepreneurial of the role-actors usually gain a disproportionate share of the rewards' (Wilson, 1975*a*, p. 554). Yet he provides no scientific evidence favouring the genetic explanation of these phenomena.

That our genetic programming exercises a strong influence on our behaviour rather than merely making many patterns possible is also suggested by Wilson's repeated insistence that human beings are not infinitely malleable and that this fact must be taken into consideration in reference to proposals for the future. He asserts that there are ineradicable drives for survival and self-esteem, and that recognising the depth of our genetic inheritance, people should be reserved about Utopian proposals. Yet the restraints involved in that inheritance are, apparently, not absolute. Though they constitute human nature, and though Wilson himself seems positively disposed toward these restraints, he writes:

> Although they are the source of our deepest and most compelling feelings, their genetic constraints evolved during the millions of years of prehistory, under conditions that to a large extent no longer exist. At some time in the future it will be necessary to decide how human we wish to remain, in this the ultimate biological sense, and to pick and choose consciously among the emotional guides we have inherited (Wilson, 1976, p. 189).

He thinks that male dominance, which he believes has a considerable genetic component, is one of those traits of human nature about which human beings must soon decide.

Wilson (1978) followed up *Sociobiology: the new synthesis* with *On Human Nature*. He recognises that this is not a scientific work, although it includes a great deal of scientific information. 'Its core is a speculative essay about the profound consequences that will follow as social theory at long last meets that part of the natural sciences most relevant to it' (p. x). Much of the presentation shows how human cultures function in ways that are adaptive in their diverse contexts just as other animals are adapted genetically. The implication of the way this is stated is that such adaptive behaviour is at least in part genetic. But the argument against it being purely cultural is only rarely developed.

The best clarifications are in the chapter on 'Aggression'. To the question whether human beings are innately aggressive Wilson's answer is an unequivocal 'Yes!' (p. 99). But Wilson proceeds to a careful statement of what is entailed. Lorenz wrongly conceived aggression as a drive which would require some outlet and accordingly proposed that to reduce the pressure for war, society should emphasise aggressive sports. The evidence, Wilson states (p. 101), is against this view. Aggression is a tendency which culture can either encourage or discourage rather than a drive which must find release. Nevertheless,

> human beings are strongly predisposed to respond with unreasoning hatred to external threats and to escalate their hostility sufficiently to overwhelm the source of the threat by a respectably wide margin of safety. Our brains do appear to be programmed to the following extent: we are inclined to partition other people into friends and aliens, in the same sense that birds are inclined to learn territorial songs and navigate by the polar constellations. We tend to fear deeply the actions of strangers and to solve conflict by aggression. These learning rules are most likely to have evolved during the past hundreds of thousands of years of human evolution and, thus, to have conferred a biological advantage on those who conformed to them with greatest fidelity (Wilson, 1978, p. 119).

Morris, Lorenz and Wilson in their different ways all search for roots of our social behaviour in biology and particularly in our evolutionary history. Against this project it is widely argued that evidence does not support the contention that biological determinism provides the basic patterns of behaviour which are then worked out culturally in different ways. Alland (1972) argues that biological explanations of human behaviour are inadequate for the reason that human behaviour is based not simply on capacities for aggression, territoriality and creativity but on customs which develop in specific environments and social circumstances. Aggressive and non-aggressive behaviour are both possibilities within the behavioural capacities of people. What dominates and in what form is largely determined by cultural patterns. No one type reflects the real innate nature of human beings. Human nature is largely open, and it is this very openness that gives the human species its great advantage in the biological world (Alland, 1972, p. 24). Enmity in human beings, argues Alland, has its roots in the social process. It is frequently the result of a situation in which people exploit other people often without aggressive feelings or hatred, indeed with little awareness of feeling for the exploited (p. 61).

The response of Leakey & Lewin (1977) is similar. They place the emphasis on social environment. 'Evolution produced an animal capable of tackling whatever challenges the environment might offer' (p. 245). They belittle the role of innate drives and stress our voracious appetite for learning instead.

Sahlins (1977), in a wide ranging critique of sociobiology, particularly as presented by E. O. Wilson, also emphasises the role of culture in human behaviour. He says 'culture is biology plus the symbolic faculty' (p. 65). By this he means that although human behaviour operates within constraints which are biologically established, this behaviour is primarily characterised by its cultural component. This component is dependent upon symbols of communication such as language, gestures, artistic forms and so on. Sahlins also attacks Wilson's hypothesis of the evolution of altruistic behaviour in humans. The essence of Wilson's proposition is that altruistic behaviour which is of negative survival value for the individual who practises it could, nevertheless, be selected (assuming it had a genetic base) when those who are advantaged are the kin. Kin selection is a thoroughly reputable concept in biology. However, Sahlins argues that the 'kin' in the anthropological groups with which Wilson is concerned are not biologically related in the close way that Wilson supposes.

These criticisms of the claims for the dominant role of genetic determinism are indeed cogent, but hardly decisive. We just do not know enough about complex behavioural traits of humans to be able to say with any confidence the extent to which genetic and cultural components are involved. Studies with young children may shed some light on these matters. Those of Piaget, for example, seem to indicate non-cultural factors in the organisation of perception at various ages. Recent studies of boys and girls seem to indicate that males are in infancy more aggressive than females and that females have superior verbal skills. Chomsky's cross-cultural work on language seems to indicate that there are basic structures to which all language conforms. Cross-cultural commonality in the meaning of a smile may indicate genetic determination. There are surely genetic bases for sexual attraction. Perhaps eventually, out of cumulative data, the distinctive contribution of the genes can be determined.

For the present, however, the reliable results of the sociobiology debate are rather small. There is no definitive evidence that human beings behave as they do because of the genes bequeathed by animal ancestors. It seems clear that the genetic constitution sets constraints on behaviour but that within those constraints culture has a decisive influence on what people do and do not do. The definition of the constraints is a matter for the future. Our position is similar to that of Ruse (1979, p. 214) who concludes his careful evaluation as follows: 'I am far from convinced that the sociobiologists have yet made their case. What I do plead is that their sins are not as grave as their critics argue. Human sociobiology should be given the chance to prove its worth. If it cannot deliver on its promises, it will collapse soon enough.'

The fall upward

The debate about the genetic influence on social behaviour takes a somewhat different form when the focus is upon genetic selection in the early human community instead of on its prehuman animal ancestry. Although Robert Ardrey, like Morris, Lorenz and Wilson, sometimes emphasised what human beings share with other animals, his most distinctive thesis was that our ancestors differed from other apes specifically in their violence. In *African Genesis* he argued that the human species came from killer apes and inherited their qualities (Ardrey 1961). Although Ardrey modified his position somewhat in later writings (Ardrey, 1967, 1970, 1976), his general position is that only a thin veneer of civilisation inhibits our inherited tendencies to violence against those whom we regard as outsiders. Accordingly his picture of pre-civilised human society is a very negative one, and this is shared by Lorenz.

This picture of a natural human barbarism partly tamed by civilisation is radically challenged by other students of early human beings. Alland (1972) and Leakey & Lewin (1977) picture our ancestors of the long hunting and gathering period as remarkably non-violent. Indeed, they question whether hunting played a major role in these cultures, arguing that the non-violent gathering was the primary economic activity. According to them, human beings became warlike only with the agricultural revolution, some ten thousand years ago. The capture of slaves and the conquest of territory were meaningless or worthless during the millions of years of hunting and gathering. The period in which violence and war have been common is too short to have affected us genetically. Hence violence and warlike characteristics are a matter of culture. Shepard (1973, 1978) shares this conviction that the destructive features of human life began with civilisation. He envisions the hunting and gathering age as a time when people lived in harmony with each other and with their environment. Unlike Alland and Leakey & Lewin, he stresses that hunting was a very important component of life, but it was not characterised by ruthless slaughter. There was respect for the prey that took on a religious character both in the killing and in the eating ritual. It was the agricultural revolution that introduced opposition between human nature and human activity. People were forced into drudgery in order to live. In consequence they became acquisitive and violent. Contented hunters became unhappy farmers.

Alland, Shepard and Leakey & Lewin probably exaggerate the harmlessness of our hunter–gatherer ancestors. As Wilson (1975*a*, Chapter 27) points out, when true men of the genus *Homo* began employing stone axes, some of the species of large African mammals became extinct. It is reasonable to suppose that this impoverishment was due to excessive predation by the increasingly competent bands of men. Again when Stone Age hunters entered Europe and later North America, whole species disappeared in the so-called 'Pleistocene Overkill' (Martin 1967). This does not mean that Alland, Shepard and Leakey & Lewin are entirely wrong. Hunting and gathering societies have at times developed ritual relations with animals which have preserved the species. The Aranda tribe of Australian aborigines today has a complex mythology of animals and sacred places that results in only certain people and places being used for hunting kangaroos. The aborigines have a keen understanding of kangaroo ecology and their mythology and taboos are closely connected with practices that result in conservation of their main source of food, the kangaroo. It includes a taboo on hunting in one area which the aborigines since time long past have set aside as a sort of national park. This area happens to be the one most favoured part of the total environment for the kangaroos (Newsome, 1980). But though examples like this can be multiplied, the evidence indicates that our ancestors could be butchers too. The modern annihilation of species is not so out of character!

The debate between Lorenz, Ardrey and Wilson, on the one side, and Alland, Leakey and Shepard, on the other, is a familiar one in Western theology. It is the debate about the source of evil in human society or what constitutes the 'fallen' condition of human beings. Both sides recognise the old distinction between nature and spirit or nature and culture. And both identify the fundamental problem with this split. For the former group, the remnant of nature within cultural life is the reason for evil within us. For the latter group, nature is good and it is culture which has introduced our problem. Human beings have fallen from an original paradise.

Waddington (1960) saw this 'fall' as inevitably bound up with progress. He drew attention to the essential role of authority and authority acceptor if there is to be any cultural evolution. This is the role of the teacher and the taught. The authority necessary to ensure that a message is received can be derived from the world of fact within which the content of the message can be verified. But a great deal of social transmission takes place at a time when the recipient is much too young to apply any verification procedures. Human social life is built up primarily on socially-transmitted information which is accepted, rather than on information which has been tested and verified. If one compares social transmission with heredity, then verification is analogous to natural selection. But there are problems which the intuition of humankind has always realised. The essential points are enshrined in the myth of the fall. The plucking of the fruit of the tree of knowledge in the Garden of Eden led to the fall.

As Waddington (1960, p. 164) interprets this myth, humans gained access to the socially-transmitted store of knowledge only at a cost of a process which involved also the consciousness of good and evil. The cost is heavy. The authority which seems to be necessary if we are to be a receiver of socially-transmitted messages seems to be produced by a mechanism which usually leads to its over-development. Without an external system of authority which becomes internalised in the acceptor, *Homo sapiens* could not become a human person. The price he paid is excessive development of this authority, both as the external authority and when it becomes internalised in the acceptor. The consequence is feelings of guilt, anxiety and despair. Waddington quotes as an extreme expression of this human dilemma Kierkegaard's statement from *Sickness unto Death*, 'With every increase in the degree of consciousness, and in proportion to that increase, the intensity of despair increases: the more consciousness, the more intense the despair.'

Christian theology has long been characterised by this ambivalent view of the fall. Humanity is not, as Lorenz and Ardrey suppose, in the grip of an inexorable evil because of an inheritance against which the forces of civilisation must struggle. On the other hand, it is not simply fallen from a perfect state to which the ideal would be to return, although Christian theology agrees with this second view to the extent that the human species *is* fallen and that the 'fall' is itself a source of suffering and sin. In this respect it corresponds with Shepard's view of the neolithic settlement. But on the other hand, it was because of the fallen human condition that God undertook the redemption of humankind. The hoped-for New Jerusalem was not to be a return to the Garden of Eden but a movement to something of incomparably greater worth. The work of redemption would not restore primitive innocence but bring mature fulfilment. In some way the history and culture of the human race, with all its evil, is, paradoxically, a 'fall upward'.

Perhaps the most important theological work on the fall in recent times was Reinhold Niebuhr's (1941) *The Nature and Destiny of Man*. Although he wrote before the time of the debate summarised above, he described its essential outline. Westerners have long understood themselves in terms of the duality of nature and spirit or genetic endowment and culture. They have recognised a tension or incongruity between them, and many have viewed this as the fundamental human problem. Some, with Lorenz and Ardrey, have seen the power of nature or genetic heritage as the source of evil and have sought redemption in the strengthening of culture. Others, with Leakey and Shepard, have seen original human nature as good and identified the source of evil with culture. Niebuhr saw that although the Bible recognised the tension, it refused to identify either side with evil. Neither nature nor culture is bad, and that we participate in both is our glory. But this glory is at the same time the condition that leads us continually into sin.

One main consequence of locating evil either in nature or in culture is that it makes the human problem something outside of the human will. It is not we who sin. We are simply placed in a situation that produces evil. We are victims or spectators. Whereas Waddington's doctrine avoids the simplistic identification of evil with nature or culture and appreciates the ambiguity of the fall, it still does not address us in our personal responsibility. In his analysis reality is simply such that consciousness and guilt go together. Niebuhr agrees that they are closely interconnected, but he teaches, nevertheless, that we are responsible. We are not simply caught in a situation. It is not the situation but what we freely do in the situation that is the basic evil. In short, we must reckon with sin and not primarily with fate. We do not have to be victims of circumstances. Human beings are always free to take a stand.

The fall in Niebuhr's sense is a phenomenon of human existence as such, everywhere. We agree. But the symbol of the fall and specifically the fall upward can be properly used to describe particular occurrences in the evolutionary historical process. It identifies the occurrence of a new level of order and freedom bought at the price of suffering. In this sense the coming of animal life was a fall upward. Until animal life came there was no suffering, but also there was little value. Animal life introduced greatly increased instability into an otherwise regular world. But the instability was transcedence. The emergence of the distinctively human was another fall upward, well depicted by Niebuhr's doctrine. Waddington in another way is describing the price paid for that supremely important phenomenon – cultural evolution. 'There are always the dream of youth,' says Whitehead (1933, p. 381), 'and the harvest of tragedy'. It may well be that with every new excursion into knowledge there is a cost which comes from the overdevelopment of authority as the group is inducted into the new ways. Each new liberation in technology, politics, education and sex, produces new forms of enslavement.

The price paid for neolithic culture and, *a fortiori*, for urban civilisations was enormous. These built on the destruction of the environment, the enslavement of masses of people, the reduced status of women, the hierarchical and exploitative organisation of society and the institution of war. Nevertheless, their achievements are impressive, the great increase in the human population which they made possible should not be simply deplored. The industrial revolution and the recent revolution in communications were falls upward as well. In each case the new has brought with it suffering and evil that has not occurred before. But the new also brought with it new stimuli, new hopes, new intensities of experience and new possibilities. We deplore the tendency of the new totally to displace the old. It would be a richer world if the agricultural experiment could have been a more limited one. It would have been a richer world if the industrial revolution had not been repeated so faithfully in every developed country and were not chased now by developing countries as though it were a pattern for all peoples. The extermination of the last remnants of hunting and gathering societies which is taking place in our time, as for example in Brazil, expresses the fall without the redeeming upward element. We fear a vision of evolution that lacks all sense of the fall.

In a prescient way A. R. Wallace, the co-discoverer with Darwin of the principle of natural selection wrote of the ambiguity of human progress as he contemplated the possible fate of the King Bird of Paradise in the Malay Archipelago. On finding his first specimen Wallace wrote:

> I thought of the long ages of the past, during which successive generations of this little creature had run their course – year by year being born, and living and dying amid these dark and gloomy woods, with no intelligent eye to gaze upon their loveliness; to all appearance such a wanton waste of beauty. Such ideas excite a

> feeling of melancholy. It seems sad that, on the one hand, such exquisite creatures should live out their lives and exhibit their charms only in these wild, inhospitable regions, doomed for ages yet to come to hopeless barbarism; while on the other hand, should civilized man ever reach those distant lands, and bring moral, intellectual and physical light into the recesses of these virgin forests, we may be sure that he will so disturb the nicely-balanced relations of organic and inorganic nature as to cause the disappearance, and finally the extinction, of those very beings whose wonderful structure and beauty he alone is fitted to appreciate and enjoy (Wallace, 1869).

Do animals experience?

In the preceding sections we considered the implications for understanding ourselves as human beings which follow from full acknowledgement of the evolutionary continuity between ourselves and other animals. It is equally important to ask the opposite question. What does the evolutionary continuity between ourselves and other animals tell us about other animals?

This question has been sharpened by the fuller clarification in the first section of this chapter of what is entailed in the ecological model. The ecological model assumes that each living thing is actively taking account of its environment and that in human experience this taking account is clearly expressed. Indeed, taking account of the environment from the point of view of the one, human or otherwise, that is taking account, must always be something analogous to human experience. The question now is not whether the ecological model applies to human experience but whether, in its fully expressed form, it applies to other animals. Is it plausible to think of other animals as experiencing their worlds?

Many scientists have wanted to exclude experience from the animal's world. Even in the study of human beings there have been efforts to deny that there is any such thing as experience. When materialism and the mechanistic model are applied to human beings, human subjectivity is ignored or denied. We take this position seriously for its historical importance, but we reject it. We will not repeat the arguments as they bear upon human experience. The question now is whether experience is a function only of the human organism or whether it is a category of wider applicability.

A main difficulty with such a question is that there is no consensus on how to use the word. Some identify experience with fully self-conscious awareness. Others speak comfortably of unconscious or non-conscious experience. We belong to the latter group. For us to say that something experiences is to say that it is not merely an object in our world of experience but also a subject of relations in its own right. It is acted upon and it acts. The ecological model is a model of living things which are acted upon and which respond by acting in their turn. They are patients and agents. In short they are subjects. Instead of asking whether living organisms experience, we could ask whether they participate in subjectivity. For us an organism is either a subject or composed of subjects or else both itself a subject and composed of subordinate subjects. We cannot, by shifting language, evade the ambiguities of our terms, but we hope our readers will have sufficient sense of what we mean to be able to follow our discussion.

There are still some who assert that animals other than the human exist only for human beings and in human experience – that they are not subjects at all. Yet many of the forms of behaviour that lead us to assume that other human beings have subjective feelings of pain and joy are found also among other animals. The behaviour of these animals is sometimes similar to that of human beings who, we believe, are thinking and reasoning. The clever behaviour of a skilful Australian sheep-dog in rounding up the flock gives all the appearance of an animal that is weighing up the pros and cons of attacking from this flank or from that. There are moments when it stops and appears to work out its strategy for the next series of moves. It was Montaigne who said: 'When you are playing with the cat how do you know the cat is not playing with you?' Any one who has a cat or dog knows that they appear to suffer when injured and in other circumstances to be happy and contented. Biologists are becoming more willing to confess their suspicion that some birds sing because they enjoy it. This is not to deny the survival value of bird song any more than it is to deny the survival value of play of young mammals whilst at the same time admitting the possibility that they too enjoy play. Enjoyment of play, song or sex is the proximate cause of the particular behaviour. The ultimate consequence may be the survival of the species.

Armstrong (1969, pp. 353–4) writes of 'the jubilant birds' as described by poets in various cultures. Lewis Thomas (1974, p. 51) after describing, in his own poetic way, the song of the thrush in his back yard, remarks that he has the strange impression that the bird sings for its own pleasure. 'I cannot believe,' he says, 'that he is simply saying "thrush here"'. At a more rigorous level of argument, leading students of bird song have come to much the same conclusion. Thorpe (1961, Chapter 1) argues for the musical value of much bird song. Indeed, birds seem to do about as much musically as can possibly be done with very short units of design and on the whole without polyphony. He cites many instances of songs which seem to 'transcend biological requirements' and suggests that the bird is seeking new vocal and auditory experiences 'playing with sounds' (p. 6). Hartshorne's (1973) *Born to Sing* is a sustained and highly documented argument in favour of the view that many birds have some musical feeling.

The European blackbird, whose song is most musical to the ordinary listener, has been studied over many years by Joan Hall-Craggs. Her work, says Thorpe (1978, p. 51),

> provides very convincing evidence that an individual progresses during the period of its song production and 'improves' both the form of the song and the relationships of the individual notes in a manner which conforms to human aesthetic ideals of balance and movement. She finds that if a blackbird is singing 'well' (from our aesthetic point of view) and a neighbouring blackbird approaches its territory as a potential rival, the singer may sing more vigorously, but certainly not more musically, in order to intimidate the intruder. In fact, on the contrary, it becomes a little upset and the song temporarily loose and disjointed; phrases are left unfinished and pauses in between the phrases become even longer than normal. Thus it appears that the bird has to attend to the form of its song in order to be able to sing well by our standards.
>
> If one records the song of a particular blackbird daily, throughout the singing season, changes of apparently aesthetic significance are detected. First, in the early part of the reproductive period, the song may appear highly functional; but later in the season, when the functional needs have been fulfilled, the song becomes organized more closely, and in a manner so nearly resembling our own ideas of musical form that it is difficult to deny that it is musically improved. So we appear to be moving towards the type which we call 'art music', where our experience of musical scores enables us to guess what kind of change is about to happen next. This sense of form seems to fit a number of bird songs in a most remarkable way.

From these remarks it would seem that the bird exceeds the functional needs of its song. For what other reason, it might be asked, than its enjoyment of what it does? Another way of looking at this is that it may be necessary to have musical feelings in order to sing musically and to sing musically is necessary for the bird's survival and reproduction.

Our common-sense judgments based on behaviour are supported by the fact that animals are equipped physiologically with the necessary organs to have experiences much like the human. Their sometimes original responsiveness to their environment, their ability to solve problems, their apparent memory and anticipation, all make it easy for us to assume that basic features of human experience can be generalised to them.

For those willing to take the first step, the question is how far can they go. For some the limit is set by a highly developed brain. The brain is extremely important for experience and similarities between human experience and that of other organisms with developed brains is likely to be much greater than similarities with organisms not so equipped. But what is remarkable in the light of this enormous difference are the similarities still to be found. We confront a gradation of differences but no point at which a great gap appears.

Karl Popper expresses this view in a congenial way: 'if the evolutionary story applies to life and to consciousness, then there ought to be degrees of life and degrees of consciousness. When we look for evidence as to whether there are degrees of life and consciousness, I do think we find reasonably good evidence for both' (Popper & Eccles, 1977, p. 438). Even in the amoeba 'there is a centre of activity, of curiosity, of exploration, of planning; there is an explorer, the animal mind' (p. 30).

Popper is not alone in his willingness to attribute subjectivity to the amoeba. Early in this century one of the foremost students of behaviour in lower organisms, H.S. Jennings, was so impressed with the variety of behaviour in the amoeba that he summed up his impressions as follows:

> The writer is thoroughly convinced after long study of the behaviour of this organism, that if amoeba were a large animal, so as to come within the everyday experience of human beings, its behaviour would at once call forth the attribution to it of states of pleasure and pain, of hunger and desire, and the like, on precisely the same basis as we attribute these things to the dog (Jennings, 1906, p. 336).

Since Jennings' time countless workers have studied both rigid and plastic behaviour in invertebrate animals including the phenomenon of learning, an ability now recognised all the way down the animal kingdom to and including single-celled organisms. A synopsis of these studies up to 1973 takes up three volumes (Corning, Dyal & Willows, 1973). Perusing these studies one is impressed with the extent to which these lowly animals resemble the higher ones in what they do, so much so that Jennings' impressions seem amply reinforced by the years of experiments since his time. For example, Best & Rubenstein (1962) demonstrated with simple planarian worms something which had hitherto only been known for animals like the rat. When a rat is placed in a strange situation its feeding is suppressed until it becomes familiar with the novel environment. This is generally considered to be related to some sort of state of anxiety in the animal. Experiments with planarians showed the same sort of delay in feeding when worms which had learned to feed in one place were put in another that was only slightly different. It appears that something like human emotions are important in the lives of very primitive animals as well as among our close relatives. These indications extend to single cells.

J.Z. Young (1978, p. 19) finds it necessary to attribute choices even to bacteria. 'When we make "choices" they involve selection from a vast set of possible actions, for example, in speaking. The difference in scale is so great that the comparison of human choice with that of the bacterium may well seem ridiculous. But there is a complete set of intermediaries between them and us. All life involves selection between alternative possibilities.'

These quotations do not prove that the ecological model is applicable to all living organisms. But they do indicate that many who take evolution seriously and have considered the evidence find nothing implausible in such a move. Indeed, the burden of proof would appear to be upon those who would draw absolute lines between different sets of living things, holding that while some are agents and patients in their own right, others exist only in the experience of the observer or are merely particles of matter.

Many who might accept the extension of experience as a possible speculation would still deny its relevance to science. This is important. A theory that fails to guide scientific research or provide more adequate interpretation of scientific data is readily ignored. The previous chapter claimed that an ecological view of living things already guides a good deal of research and could illumine much more. The question now is whether the further clarification of the model by the insistence that it entails viewing all organisms as subjects has any additional relevance for science. We believe that it has.

Already much work has been done with animals that at least implicitly assumes that they are subjects. Jane Goodall's (1971) work with chimpanzees included relationships with the animals which are clearly mutual. The current effort to teach chimpanzees how to communicate with us linguistically assumes that they are subjects who experience their world in their own right. The view that some birds sing because they enjoy it has provided hypotheses for studies of bird song discussed above. Some such hypotheses were substantiated empirically by Charles Hartshorne (1973, Chapter 7).

Griffin argues in his book, *The Question of Animal Awareness* (Griffin, 1976, p. 14), that it is high time that students of behaviour relaxed their behaviourist stance and led the way to an experimental science 'dealing with the mental expression of other species' besides humans. He is led to this conviction from studies on orientation and communication between animals. Griffin discovered that bats orient themselves in relation to objects including flying prey by means of ultrasonic echo-location. He found that when flying through familiar surroundings many bats seem to rely heavily on spatial memory. Although their orientation sounds continue to be emitted in an apparently normal way, the bats collide with newly placed obstacles and turn back from the former location of objects that have suddenly been removed. The bats seem to pay more attention to their 'internal images' of spatial relations than to the new echoes from newly placed objects. This kind of behaviour demonstrates that some sort of 'internal map' must exist in the bat's brain. Not only bats but other animals, such as bees, appear to form these 'cognitive maps' of their surroundings. In some cases these maps are quite remarkable. The honey bee, for example, uses the sun for its orientation, but it is able to compensate for the movement of the sun through the skies as the 'map' changes with time of day. The evidence that animals employ some sort of internal imaging of their surroundings suggested to Griffin 'the need to reconsider the general question of subjective mental experience in animals' (p. 14).

Another example Griffin (p. 23) gives is the swarming of honey bees when in need of a new location in which the colony can continue to exist. The bees exchange information about the location and suitability of potential locations for the new hive. This they do by means of complex dances which symbolically trace out on a vertical surface of the honey comb (and in the dark) the direction and distance of the new locations. Individual bees are swayed by this information to the extent that after inspection of individual locations worker bees change their preference and dance for the superior place rather than the one they first discovered or was communicated to them by their mates. Only after many hours of such exchanges of information, involving dozens of bees, and only when the dances of virtually all the scouts indicate the same hive site, does the swarm as a whole fly off to it. 'This consensus results from communicative interactions between individual bees which alternatively "speak" and "listen". But this impressive analogy to human linguistic exchanges is not even mentioned by most behaviour scientists' (p. 23). The bees do not appear to be acting as programmed robots. It is not a totally stereotyped form of behaviour. The bee does not always respond to a dance. If the 'language' were in words rather than in dances and were the bees closer to the size of a dog, then we would be strongly inclined to attribute to them similar 'experiences' to those we humans have when we communicate about whether to go to this or that place.

Griffin (p. 58) has drawn up the following list of terms in order of their current acceptance to behavioural scientists:

Acceptable	Pattern recognition
	Neural template
	Sollwert
	Search image
	Affect
	Spontaneity
	Expectancy
	Covert verbal behaviour
	Internal image
	Concept
	Understanding
	Intention
	Feeling
	Awareness
	Mental experience
	Mind (mental)
	Thought
	Choice
	Free will
Taboo	Consciousness

Strict behaviourists might stop after affect. Others might venture further down the list. Their reservation stems from a viewpoint that Griffin believes may have served science well for some 50 years or more by focusing attention on phenomena amenable to experimental analysis. The message, Griffin comments, has been: 'As a working strategy of research, assume no mental states or subjective experiences, and see how much of animal behaviour can be accounted for on this parsimonious basis. This has now been done on a large scale, and some of the results suggest that it is time to review our perspectives and strategies in the light of the new discoveries' (p. 55).

The most reasonable view, which Griffin supports, is that the gradient represented by the list of terms is a true continuum without sharp discontinuities.

The prejudice against considering the subjective aspect of animals had had its effect also upon the dominant expressions of evolutionary theory. Jacques Monod is a recent example. As our quote from Monod in Chapter 2 (p. 58) indicated, he does not deny that choices by individual animals play a role of immense importance in evolutionary theory. Yet this does not shake him from his central conviction that the total evolutionary process is 'the product of a vast lottery, in which natural selection has blindly picked the rare winners from among numbers drawn at utter random' (Monod, 1974, p. 131). In short, having accurately described and even emphasised the role of purposive animal behaviour in directing the course of evolution, he continues to insist that only chance and necessity play any role. It is hard to avoid the conclusion that neglect of the subjectivity of animals prevented him from formulating the more adequate theory that his own statements call for.

It is striking that in contrast, while making no more claim than Monod for the importance of animal choice, Hardy (1975) develops the theory for which this understanding calls. In contrast to Monod, Hardy is very sensitive to the subjectivity of animals, and assembles extensive evidence of its importance. It can hardly be doubted that as the purposive behaviour of individual animals is recognised as a third factor alongside chance and necessity in the evolutionary process, new doors of research will open.

The claim that viewing birds and mammals as subjects can have some effect on the way they are studied does not seem to be particularly strange. It has already influenced the study of birds and mammals widely. But can the view that all organisms are subjects affect research with simpler organisms? Even here the answer must be that implicitly it already has done so and that if the model is fully accepted it can open up large new areas for research.

The objection is warranted that hypotheses about the subjective states of other organisms, even human ones, cannot be directly proved. But science does not require such direct proof. A major function of theory is to raise new questions and suggest new inquiries. The model of living things as experiencing subjects can certainly do this. It can also suggest specific falsifiable hypotheses. And finally, it can provide a more adequate conceptual framework for interrelating the results of many different inquiries, by which in turn it can generate still new questions.

Our major concern is with living organisms, and we hope to have shown that an ecological model can advance our understanding of them. There remains, however, a further dimension of our theory which receives less support. Even those who support the continuity among forms of life often deny similar continuities between the living and the non-living. The first three chapters showed how difficult, indeed arbitrary, it is to draw a definitive line between what is alive and what is not, and wherever the line is drawn the ecological model applies on both sides. It is as applicable to an event in an electromagnetic field as to the behaviour of a cat. This is important for biology; for if the cat were finally analysable into material substances, the relevance of the ecological model would be only to more superficial levels of the world. But if the internal relations, essential to the ecological model, imply that all true entities are subjects as well as objects, then more is entailed than was made explicit in Chapter 3.

Karl Popper was cited above as a supporter of the view that all organisms are subjects. But when it is suggested that a further continuity can be established between simple living things and such inanimate entities as atoms, Popper attacks this as panpsychism (Popper & Eccles, 1977, pp. 68–71). Our position is not identical with the panpsychism Popper attacks. We do not think all things are or have psyches or souls or minds, or that all events include an element of consciousness. But we do share with panpsychism the rejection of radical discontinuities. Hence it will be well to note Popper's objections and to reply briefly.

1. Popper notes that in nature there are abrupt emergents. For example, solidity in crystals does not presuppose some similar property in liquids. Hence the emergence of consciousness does not presuppose some similar state in antecedent entities.

Our reply is that indeed consciousness emerges from conditions where it is lacking. Indeed, we are more hesitant than Popper to attribute consciousness to all living organisms. The experience or subjectivity of simple organisms is probably non-conscious. We would not speak, as Popper does, of a mind in an amoeba. Those that have what are properly called mind, reasoning, psyche and consciousness all evolved from simpler organisms which lacked these. But not every characteristic of living things emerged from entities wholly lacking in these characteristics. Specifically, there is no evidence that entities which take account of their environment or are internally related to other entities, emerged from entities which are related only externally or mechanically. Internal relations do not require consciousness. Indeed, even in conscious human experience, consciousness illumines only limited aspects of the way the experience takes account of its environment. But the term experience can be used to include not only human and amoebic experience, conscious or non-conscious as the case may be, but also the non-conscious taking account of the environment which characterises molecular, atomic and quantum events as well. We fail to see any reason to attribute experience in this sense to the amoeba and to deny it totally to living cells in multi-cellular organisms, to organelles, to viruses or to DNA molecules.

2. Popper thinks that 'the main motive of post-Darwinian panpsychism was to avoid the need to admit the emergence of something totally novel' (p. 69). But he rightly states that it must admit 'a great step from pre-psychical to psychical processes'. Thus Popper insists that panpsychists are forced still to admit the emergence of something really new.

We share the motivation Popper deplores, but certainly not out of opposition to the view that novelty is introduced in the evolutionary process. A proper evolutionary theory should reckon with the continual influx of novelty, not only with the emergence of life and consciousness, but throughout. What we oppose is the idea that wholly different mechanisms are involved in these and other instances of emergence. That is why we do oppose the idea that what emerges is 'totally novel'. Conscious mental states are very different from non-conscious mental states, but they are not *totally* different, and we see no reason to say that non-conscious mental states are totally different from still more primitive modes of taking account internally of the world.

3. Popper says: 'We should not assign inside states, or mental states, or conscious states to atoms' (p. 71), because they have no memory, even unconscious memory. Popper's argument is that 'consciousness, and every kind of awareness, relates certain of its constituents to earlier constituents' (p. 70). Hence the lack of memory entails the absence of consciousness.

The argument has weight against the attribution of consciousness to atoms, and we agree that atoms are not conscious. But it is not clear how the argument counts against 'inside states'. It might well show only that inside states cannot be conscious if there is no memory. We do not care for the language of 'inside states', but we do believe that atomic events are internally related to other atomic events and have offered evidence for such relations in Chapter 3. We do not posit that internal relations require or entail memory. But we do believe that internal relations characterise all events and this provides the continuity in internality among all things. We do not find that Popper's arguments count against this view.

Thorpe (1977, p. 7) is more open than Popper to what he also calls 'panpsychism'. He admits that 'it is easy enough to assume some sort of psychic element in the ultimate physical particles', but he rejects it nonetheless. His trouble with panpsychism is that he can see 'no conceivable scientific possibility of investigating its significance' (p. 7). We, too, would have hesitated to assert that our view, that all individual things are subjects as well as objects, has scientific import were it not for the remarkable work of David Bohm (1969, 1973, 1977, 1978). He writes: 'Our physical theories are at present in a state of flux, that may lead to radical changes in them, such that current fundamental ideas, based on measure and metric, may also have to be replaced by new ideas based on order' (Bohm, 1969, p. 18). These new ideas, he suggests, will have to involve the notion of order in a way that is more fundamental than the notions that now exist in the theories of physics. The conclusions of a reductionist mechanistic biology are dependent upon the assumption that the ultimate particles are exclusively mechanical in their properties. 'Therefore,' says Bohm (p. 29), 'the question of whether the basic laws of physics are in fact mechanical or not is of the utmost importance in biology.' Here the physicist is arguing on his own grounds that if physics comes to a non-mechanistic view of its subject this will be of the utmost significance for biology. 'It does seem odd therefore,' says Bohm (p. 34), 'that just when physics is moving away from mechanism, biology and psychology are moving closer to it. If this trend continues, it may well be that scientists will be regarding living and intelligent beings as mechanical, while they suppose that inanimate matter is too complex and subtle to fit into the limited categories of mechanism.' But, as Bohm points out, such a position cannot stand up to critical analysis, for the molecules studied by biologists in living organisms are constituted of electrons, protons and other such particles, from which it must follow that they too are capable of behaving in ways that cannot be described in terms of mechanical concepts.

In Bohm's (1973, 1977) view, science as we have known it describes the objective aspect of things, the exterior aspect or what he calls the explicate order of the world. What this science fails to see is that this explicate order is dependent upon what he calls an implicate order, an inner aspect of things. An example from physics is the visual image on the television receiver: points that are near each other in the ordered visual image are not necessarily 'near' each other in the radio wave from which the image is translated. The radio wave carries the visual image in an implicate order. The function of the receiver is to explicate this order, i.e., to unfold it in the form of a visual image.

Generally speaking the laws of physics refer mainly to the explicate order, which can be described in precise detail by reference to Cartesian co-ordinates. Bohm proposes that primary emphasis in physics must now turn to the implicate order. The logic of this is that explicate order is not understood except with reference to implicate order. Since Descartes, people have lost sight of the implicate order and have come to think the explicate order is self-sufficient. The sort of understanding physics has given us of electrons and atoms is of their explicate order. It leaves hidden the implicate order, of which the explicate order is an expression. Furthermore, part of the implicate order of the electron and the atom is the subjective aspect of these entities: what they are in themselves to themselves.

Even if the scientific advantages of this view turn out to be less than we anticipate, the philosophical advantages remain. Hartshorne (1977, pp. 92–3) suggests seven such advantages. They are:

We get rid of the problem of how 'mere matter' can produce life and minds. Instead the problem becomes how did higher types of experience develop out of lower types?

We do justice to the fact that between the 'life-less' matter and primitive forms of living matter there is only a relative difference, not an absolute one.

We are enabled to construe causal connections between events by generalising the concepts of memory and perception. What is remembered and perceived is internally related to the remembering and perceiving experience, and a cause must be internally related to its effect if Hume's strictures are to be avoided. These clues to the connectedness of events are lacking in materialism and dualism.

The way is opened up to give some explanation for the special case of how mind and body are related in animals. Why is it that one's thoughts and feelings affect one's body and vice versa? The mind-body relationship can be seen as one of sympathy. We share in the feelings of our cells (enjoyment, discomfort and so on) and they in reverse direction respond to our emotional life.

We solve the problem of relating primary and secondary qualities in the scheme of things. Primary qualities are what physics normally describes – causal spatio-temporal relations. Secondary qualities characterise events internally.

We can give an account of the relationship between perception and behaviour and explain why animals act as they do. When we try to explain why an animal tends to eat sweet things and avoid bitter things we may reply that this is the appropriate response built in by evolution to ensure survival. That is only part of the story. The animal has an experience when it tastes sweet and bitter. One taste it likes. The other it dislikes. That subjective side is ignored in the physico-chemical account.

The doctrine of mere-matter, mere-mindless and feelingless stuff puts limits to things with which we can sympathise or empathise. But if in physical nature also there is experience, then there is a universal community for mutual participation in sympathy.

The evolution of experience

If all living organisms are subjects experiencing their world as well as objects of human experience, then the most important story of evolution has not yet been written. In concluding this chapter we hazard some speculations about this subjective side of evolution.

In the ecological model, as has already been discussed, cellular events are considered to take account of their environment subjectively, analogously to the way human beings feel theirs. In this sense it is quite appropriate to say that cells 'feel' their environment. Presumably the feelings involved are rudimentary in comparison with human feelings, and entirely lacking in consciousness.

In plants the assemblage and mutual co-ordination of cells may enrich these feelings in the individual cell, but there is no indication of any new type of experience emerging. The life, and hence also the experience, in the plant is the life and experience in the individual cells which make it up. Animals, on the other hand, have a more centralised co-ordination of the movements of their parts based on awareness of more or less distant objects. Sense organs appeared first without much central co-ordination, if any at all. The development of the central nervous system and co-ordination of the sense organs must have brought with it a new level of experience. The experience of the animal is this new unified experience to which the experiences of the cells in some region of the brain contribute directly. This animal experience can be partly conscious. The emergence of conscious experience was the crossing of a great new threshold, which we associate with the development of the central nervous system.

In the simplest case conscious experience of the environment arises as needed to give direction to the movement of the organism. There may not be much connection between successive experiences. But there are such connections and they increase in importance. Memory and anticipation arise. The unified experience that arose for purely functional purposes enjoys its own existence, as all living things do. It becomes interested in enhancing this enjoyment in the future. It begins to direct the body not only for the sake of the body but also for the sake of the self-enjoyment of the unified experience itself. This experience is enriched by its integration with past experience and anticipation for future ones, and it is further enriched by the stimuli that it can seek out and enjoy. To some extent all animal experience functions for the sake of purely bodily needs and to some extent it influences the body to actions that benefit only the central experience itself. As brains become more complex the self-enjoyment of the animal experience becomes increasingly important. Generally there is no sharp distinction between the two goals, since what is good for the animal body also contributes to the unified animal enjoyment. But the animal also takes risks for the sake of enjoyment. Still, apart from human beings, the primary function of most animal experience is its service to the body.

With human beings this is not the case. A great deal of their bodily activity is performed without regard to its benefit to the body. As we sit at our desks typing it is not for the sake of the comfort, strength, health or safety of our bodies. We are disciplining our bodies to put up with some weariness and some discomfort because of concerns that belong to our human experience, not to our cells as such. And this is not a recent human development. Paleolithic peoples disciplined their bodies extensively for the sake of their distinctively human purposes.

Too little is known of our ancestors to give a date for such a change. Like all changes associated with human evolution, it was gradual.

Presumably it could not have occurred apart from a large brain. But its consequence and expression was in the realm of culture. Wilson (1975*b*, p. 22) states that a very 'rapid phase of acceleration began about 100 000 years ago. It consisted primarily of cultural evolution as opposed to genetic evolution of brain capacity. The brain had reached a threshold, beyond which a wholly new, enormously more rapid form of mental evolution took over.' If Wilson is correct, then the evolutionary threshold was crossed then. In any case, when unified human experience began to make its own enjoyment its primary end, and to instruct the body accordingly, the human being had arrived. The unified human experience in its continuity through time and its dominance over the body is the human psyche.

The emergence of the human psyche is a case of the fall. In the sense discussed earlier in this chapter this may be the most important part of the fall upwards. As long as the animal experience functioned in accordance with the needs of the body, there was unbroken harmony. The body acted as it needed and wanted to act, so far as its environment would allow. But the human psyche directed the body to actions for which it was not programmed by its genetic endowment, actions which might be dangerous or uncomfortable. Whereas before, the body could only be restricted from its spontaneous expressions by external forces, now it could be inhibited from within. The human psyche introduced a fundamental 'dis-ease' or 'dis-harmony' into the world of living things. Yet by doing so it made possible far richer experience.

The cultural revolution of which Wilson writes involved changes in tools and weapons. But more important than the development of artifacts, it introduced questions about the meaning of what was done. Shepard stresses the respect shown for the prey and the ceremonies that grew up around the killing and eating. Dobzhansky (1967, p. 70) points to the ceremonial burying of the dead in Neanderthal man as the first evidence we have of death awareness and therefore of self-awareness (p. 76). In short, with the appearance of the psyche, religion also appeared. What was being done anyway required some interpretation. Ritual paved the way for myth and myth in turn gave rise to more elaborate ritual. The meanings were not like modern scientific explanations. They were more like the meanings that govern our dreams. Perhaps they expressed what structuralists call the deep structures of the mind. Experiences of the public world were incorporated into these meanings. There may have been a certain tension between what these meanings required and the immediate preferences of the body, but the psychic life itself was whole.

The human psyche crossed another threshold with the agricultural revolution and the rise of cities. As has already been pointed out, this was an advance with a great cost and a cost which was unevenly borne. The new threshold was the emergence of rationality as an important factor in psychic life. The flow of water in canals had to be controlled. Land had to be surveyed. Complex construction projects were undertaken. All this required social organisation and planning. This was not an advance in intelligence. The Stone Age hunter was as intelligent as the Egyptian architect. It was a new use of intelligence. It involved the birth of the professions and paved the way for science and philosophy.

The crossing of this threshold, too, involved pain and loss. The old wholeness of the psyche was gone. The new rational structures of thought did not fit well with the inherited mythical ones, which were rooted deeply in the non-conscious aspects of experience. These mythical structures continued to provide much of the meaning of life. Hence the psyche was torn between pre-rational sources of meaning and the rational ordering of segments of life. That division has never been healed. Religion and politics and much of ordinary life were based on the old structures of meaning. But a part of one's existence was alienated from that. The psychic unity of paleolithic existence was shattered.

During the first millenium BC yet another threshold was crossed – another fall upward. In the middle part of that millenium, apparently quite independently, spiritual leaders arose in China, India, Persia, Greece and Israel who expressed and called for a quite new psychic development – the crossing of another threshold. Full self-consciousness appeared.

From the perspective of rationality these leaders attacked the old religion and politics based on the archaic system of meanings and proposed new ways of ordering the whole of existence. Thus emerged 'rational religion'. The new ways differed among themselves. They have become for us now the great traditional Ways of Zoroastrianism, Jainism, Buddhism, Greek philosophy, Confucianism, philosophical Taoism, Hinduism and Judaism with its later offshoots in Christianity and Islam. Again much was lost. Prior to this time religion was a unifying communal force. Through the rise of these new ways, faith became something to be decided about and chosen. It became divisive. One's faith could be in tension with the demands and expectations of society. Social order could be criticised and rejected on the basis of faith. Much of unity and wholeness was destroyed. But in its place there arose possibilities of experience for the believers that were quite beyond anything that had existed before. In this fall, too, the movement was upward.

The teachings of the founders of these new ways were honoured more in the breach than in practice. Nevertheless, for more than two thousand years they provided for most of the civilised world the norms in relation to which people took their bearings. Many there were who did not participate in the new kind of existence to which they were called, but even most of these recognised the call. That in itself gave some meaning and direction to life.

In the past two centuries this situation has changed. More and more of the world's intellectual and cultural leaders have given up seeking guidance and direction from these Ways. They have turned to science, philosophy, art, psychology, even drugs. Or they have denied the need of any direction at all. Where interest in these Ways remains, Easterners have turned West and Westerners have turned East (see for example Cox, 1977). The ancient Ways are far from dead but they are in turmoil. This turmoil may be the death-throes of all that we have known as civilisation. There are some who would welcome this, hoping that it would lead to a renewal of a more whole, pre-civilised existence. The current religious turmoil may also be a forward move in which the achievements of all the Ways, enlightened by all that the sciences have taught us, can be united in a convincing synthesis. Perhaps such a new faith could provide the conviction and vision required to arouse us to the drastic changes we need in our intellectual life and our relations to the rest of nature.

We believe that some of the traditional Ways, and especially Christianity, show signs of the ability to be transformed once more into the new vision the world so urgently needs. Whether this will happen, and whether it will happen in time, no one can tell.

Conclusions

The idea of a gulf between human beings and the remainder of the universe has been encouraged by both humanists and scientists. Scientists have encouraged it by picturing the rest of the universe as mechanical and objective through and through. Humanists have encouraged it by stressing the radical uniqueness of the human mind and spirit. The result has been the widespread acceptance of a dualistic way of thinking.

If in fact nothing is really merely mechanical or merely objective, then a blow is struck against this dualistic edifice, an edifice which is already crumbling because of the many difficulties it poses to coherent thinking. It is our intention to strike this new blow. We propose that the ecological model applies to human beings as well as to all other things. Indeed, the full grasp of what is involved in the model is possible only when human beings take themselves in the concreteness of immediate experience as illustrating and illuminating the model. The argument is not that understanding the non-human world provides the categories by which human experience can be understood. But rather the non-human world can only be adequately understood in terms of what human beings know directly and immediately – human experience.

Still, to recognise that the human species is continuous with the rest of nature does have an effect on human self-understanding. Some wish to draw conclusions about our similarity with other animals which go far beyond the evidence. But the activity of transcending, which is present in all living things, is dominant in humanity to such an extent that human beings are products more of culture than of biology. Nevertheless, human beings *are* conditioned by their genes, and each argument about the role of the genetic heritage in shaping disposition and behaviour should be considered on its merits.

The heart of the contemporary argument is about human aggressiveness. Is it a result of cultural conditioning, or are people biologically determined to be aggressive? Views of the possibilities for human life in the future are deeply affected by conclusions at this point. An analogous argument is familiar to Christian theology in the debate between those who stress individual freedom and those who emphasise original sin. The former are often hopeful that cultural improvements can greatly reduce destructive conflict; the latter believe that society must organise itself so as to control and limit a conflict that will reappear in any social order. The ferocity of the attack on contemporary socio-biology is motivated by the fear that people will be persuaded by it that a truly just and free society is impossible.

The issues here are peculiarly difficult to resolve. On the one hand, socio-biology has claimed or implied far more than its evidence warrants. The possibilities of cultural transformation are greater than those who stress either genetic determination or original sin are likely to grant. On the other hand, only those who attend carefully to the power of the tendency to aggression will be able to propose realistic changes in society. Humans are transcending beings, but they are deeply conditioned by what they transcend. Also, the act of transcending always involves loss. We have pointed to some of this ambiguity of progress in our image of the 'fall upward' and have suggested briefly how that image may be used to illuminate the development which has led to the dangerous situation in which humanity now finds itself.

The continuity of humanity with the rest of nature has implications also for the understanding of animals. This chapter has proposed that generalisation of some features of human experience is plausible and appropriate. A biology which accepts and thematises this generalisation will be enriched and freed from arbitrary limitations which it has placed upon itself. This use of categories generalised from human experience may be useful even to physics.

But a changed view of non-human things has implications beyond the sphere of scientific study. The human treatment of other living things has heretofore been directed in large part by assumption of utter discontinuity. If there is continuity between human experience and that of other animals, then relations to them should reflect this truth. Accordingly an ethic is needed that takes into account the subjective as well as objective reality of other living things. This is the topic of the next chapter.

5

An ethic of life

> In the relationship between man and animals, the flowers and all created things there exists a grand pattern of morality, as yet hardly apprehended but which, in the end, will be manifest and which will be the complement of human morality ... This was doubtless the first duty – this must be where it all begins, and the various legislators of the human spirit were right to neglect all other considerations for this one – Man had first to be civilized in the direction of Man. This task is already advanced and progressed daily. But Man must also be civilized towards Nature. In this direction, there is everything to be done.
>
> Victor Hugo (quoted by White, 1969, p. 116)

Every view of reality has ethical implications, and since the one that has emerged in the preceding chapters is different from that on the basis of which our inherited systems of ethics developed, its implications call for revision of these systems. Before directly attempting to formulate an ethic of life, however, there is need first to justify the enterprise against those who believe that ethics as such, and not simply the specific teachings associated with its past forms, is obsolete.

The preceding chapters argued for the ecological model of life in which substantial entities are finally composed of events. In this model subjectivity or experience is attributed to all such events. Since experience is always valuable, events have intrinsic value. All things therefore have some intrinsic value either in themselves or in their constituent parts. This view is not reconcilable with the dominant anthropocentric tradition in ethics. The second section of this chapter develops this argument.

When intrinsic value is attributed to things other than the human there is a strong tendency to treat all things as having infinite value. From Albert Schweitzer's 'reverence for life', for example, it follows for some that any living thing should not be killed when this can be avoided. This view is criticised later in this chapter. Instead of holding that every life is of infinite or absolute value, we deny that any finite thing is of infinite worth. This entails abandonment of the absolutist ground for ethics and is sketched out in the third section of this chapter in relation to the rights of animals.

The abandonment of the absolutist ground for ethics requires a clarification of the shape that a relational ethic can take with respect to human rights. This is the subject of the fourth section. Having considered rights in terms of individual species, consideration is given to the more comprehensive claims of all life upon us in the final section.

Why an ethic?

Not infrequently we find when speaking to high school students on the subject of hunger in the poor world that, sooner or later, someone says that the right thing to do is to let the Indians starve because to do anything else is to go against the law of nature, which is survival of the fittest. In the preceding chapter we saw that it has not only been high school students who have drawn this conclusion from their view of evolution. Darwin's chief protagonist, T. H. Huxley, himself believed that this expressed the law of nature or the law of the jungle which got us where we are, but he was not willing to project this principle into the future. In his Romanes Lecture (in Huxley & Huxley, 1947) he analysed his dilemma and came to the conclusion that human society had to reverse the direction of biological evolution and promote human values. We are warned that we should not too easily derive an ethic from an account of evolution!

The debate reported in the preceding chapter about the role of our genetic heritage in shaping our present and our future was notably free of ethical reflection. Indeed, most of the key participants are sceptical of the role of ethics in human affairs. Ardrey is caustic.

> Conscience as a guiding force in the human drama is one of such small reliability that it assumes very nearly the role of villain. Conscience has evolved directly from the amity – enmity complex of our primate past. But unlike civilization it has not acted as a force to inhibit the predatory instinct. It has instead been the conqueror's chief ally. And if mankind survives the contemporary predicament, it will be in spite of, not because of, the parochial powers of our animal conscience (Ardrey, 1961, p. 355).

This objection to the proposed task of this chapter need not be a deterrent. The purpose is not to appeal to what Ardrey identifies as conscience but to clarify, in his terms, what civilisation has to tell. Ardrey himself seems to believe that the application of reason to human affairs is desirable. By ethics we mean the principles of such application.

> Shepard's critique strikes closer home (Shepard, 1978, p. 248). Ethically there have been discourses against the enslavement, torture, and killing of people since civilization began without ending war, tyranny, or cruelty. There is no evidence that crime, brutality, or murder have diminished at all. If human behaviour is not improved by the incorporation of such ethics into the dominant religions, what reason is there to suppose that such a new ethic can save animals. The extension of the humane idea to the wild can only produce mischief, for it will see in the behaviours and interrelationships among animals infinite cruelties and seek to prevent them. The sucking of the host's blood by the parasite, the competition among scavengers to eat a carcass, the exclusion of the weak and sick, predation itself, the enormous mortality which removes the majority of the newborn every year from nearly every species – all these and more, humane action will try to prevent, just as it prevents dogs from eating cats and men from eating dogs.

Shepard's first point with respect to the impotence of ethics is overdrawn. For good or ill societies do behave differently according to their ethical teaching. Still Shepard is correct that the influence of such teaching could easily be exaggerated. His second point is directly important for our project in this chapter. An ethic growing out of the preceding chapters will be one that extends 'the humane idea', that is, the principle of moral responsibility, beyond the scope of the human community so as to include the humane treatment of animals as well. Shepard is right in indicating how easily this could lead to absurdity. But surely ethical reflection can be more intelligent than he supposes. There is certainly danger that we may offer poor principles. They may sentimentally encourage the relief of suffering in a few individuals while ignoring the real needs. But it is precisely to counter such bad ethics that we need to reflect about ethics. It is not wise to abandon a discipline to the practitioners whose errors we see. It is better to contest the field.

The trouble is that when one contests the field one engages issues as they are defined in the field. The focus of an ethical chapter must accordingly be on rights and duties. How these are conceived is certainly altered by a view of animals as true subjects. And it is these alterations to which the heart of this chapter will in fact be devoted.

But this is only one aspect of the ecological model of living things. Another is the deep sense of the interconnectedness, even interpenetration, of all things. It is difficult to express that understanding in the inherited categories of ethics. This aspect of the ecological model comes more to the fore in later chapters.

Lorenz (1966, p. 257) is more balanced and judicious than either Ardrey or Shepard in his appraisal. Yet he, too, notes the weakness of action done from a sense of moral responsibility rather than from inclination. In times of stress the man who behaves socially from natural inclination 'has huge reserves of moral strength to draw upon. But the man who, even in everyday life, has constantly to exert all his moral strength in order to curb his natural inclination into a semblance of normal social behaviour, is very likely to break down completely in case of additional stress.'

Lorenz is correct; it is far better to move people to action through influencing their inclination than by appealing to moral duty. This is, in fact, Shepard's goal also. Both know that human beings are moved more by their belief about what is beneficial to themselves, by their sense of what is real and fitting, and by their sensibility, than by ethical appeals. Most of the language and mood of ethical reflection belongs to a sensibility that is bound up with the anthopocentrism and rationalism of the past from which our emerging sense of participation in nature can and should free us. In that case, the argument might run, instead of seeking to modify the ethical teachings, the laws, and the economic practices of the inherited system, we should be participating in the shaping of a quite new one. From this point of view, whereas some portions of this book may be judged as useful, a chapter on ethics would be judged as retrograde.

There is much truth in this criticism. The changes here proposed are not radical enough. As long as rights of animals are viewed as demands upon human beings which are costly to us, they are almost certain to give way in practice. Other animals will be respected only as they are genuinely experienced in a different way, and that change will involve a change in the way human beings experience themselves as well.

Even so, it is regrettable that the two tasks are too often seen as in opposition. Changes in sensibility cannot be legislated, but they can be encouraged by legislation. The actual experience of Blacks by Southern Whites in the USA would not have changed nearly so much had there not been federal laws mandating changed behaviour. The legal requirement of environmental impact statements accelerates the development of new habits of attention to the interconnectedness of things even though many of those who write them would prefer not to be bothered. If we delay political and legal action for the sake of the whales until a new sensibility has developed globally, there are unlikely to be any whales left. Although the way into the new consciousness is led by persons of deep intuition and sensitivity, their influence can be extended or limited according to whether their new ideas about behaviour are also rationally defensible.

Lorenz, alone among these critics, recognises that ethics has a place, even if a limited one. But even he misses the central point of ethics because he, like the others, takes the spectator point of view. They ask, what actually moves people to action and to what action does it move them? They ask, what role do ethical pronouncements play in society? They debate whether human behaviour is affected more by genetic determinants or by environmental conditioning, by nature or by nurture. They do not attend to the moral freedom of the individual to be bound by neither. Hence they do not ask the question, what, given all that I know about myself, should I, as an individual, do? Still, people do ask this question, and there is nothing in our vision of life that suggests we should do so less in the future. Indeed, when we see things in a new way we are likely to be confused. We want to change to more appropriate ways of acting, but we do not know what they are. What we lack is not the inclination but the guidance. At such times ethical principles are relevant not only to individuals but even to legislators. Such times are more frequent than the critics of ethics imply, and in our judgment we are now in such a time.

Waddington (1960), more than any other biologist in recent times, has again looked at the relationship of evolution to ethics by analysing the nature of cultural evolution which, after all, is the most potent evolutionary force in human societies. He considers that ethical beliefs have an essential role to play in cultural evolution in the teacher–learner relationship. 'It is not merely the case that human values have in fact emerged; what I am claiming is that the specifically human mode of evolution, based on socio–genetic transmission of information, essentially requires the existence, as a functional part of the mechanism, of something which must have many of the characteristics of ethical belief' (p. 202). He continues: 'The general anagenesis of evolution is towards what may crudely be called richness of experience' (p. 204). This chapter is devoted to the clarification of appropriate ethical belief guided by the conviction that our goal is the enhancement of 'richness of experience'. We believe that what this general anagenesis of evolution is towards may equally well be called the enhancement of life, for life *is* experience. To have richer experience is to be more alive.

Ethics beyond anthropocentrism

Well before the environmental crisis aroused world-wide concern for conservation of life on earth, Aldo Leopold anticipated the need for what he called a conservation ethic in the following story. When godlike Odysseus returned from the wars in Troy, he hanged all on one rope some dozen slave girls whom he suspected of misbehaviour during his absence.

> This hanging involved no question of propriety, much less justice. The girls were property. The disposal of property was then, as now, a matter of expediency, not of right and wrong.
>
> Criteria of right and wrong were not lacking from Odysseus' Greece. The ethical structure of that day covered wives, but had not been extended to human chattels. During the three thousand years which have sinced elapsed, ethical criteria have been extended to many fields of conduct, with corresponding shrinkages in those judged by expediency only.
>
> This extension of ethics, so far only studied by philosophers, is actually a process of ecological evolution. Its sequences may be described in biological as well as philosophical terms. An ethic, biologically, is a limitation on freedom of action in the struggle for existence. An ethic, philosophically, is a differentiation of social from anti-social conduct. These are two definitions of one thing. The thing has its origin in the tendency of independent individuals and societies to evolve modes of co-operation. The biologist calls these symbioses. Man elaborated certain advanced symbioses called politics and economics. Like their simple biological antecedents, they enable individuals or groups to exploit each other in an orderly way. Their first yardstick was expediency.
>
> At a certain stage of complexity, the human community found expediency yardsticks were no longer sufficient. One by one it has evolved and superimposed upon them a set of ethical yardsticks. The first ethics dealt with the relationship between individuals. The Mosaic Decalogue is an example: Christianity tries to integrate the

> individual and society, democracy to integrate social organisation and the individual.
>
> There is yet no ethic dealing with man's relationship to land and to the non-human animals and plants which grow upon it. Land, like Odysseus' slave girls, is still property. The land-relation is still strictly economic, entailing privileges, but not obligations.
>
> The extension of ethics to this third element in human environment is, if we read evolution correctly, an ecological possibility. It is the third step in a sequence. The first two have already been taken. Civilized man exhibits, in his own mind, evidence that the third is needed. For example, his sense of right and wrong may be aroused quite as strongly by desecration of a nearby woodlot as by a famine in China . . . Individual thinkers, since the days of Ezekiel and Isaiah have asserted that the despoilation of land is not only inexpedient, but wrong. Society, however, has not yet affirmed their belief. I regard the present conservation movement as the embryo of such an affirmation (Leopold, 1933, p. 643).

Earlier in Western history we had a sense that the value of things was not only their value to human beings but also their value to God. In the Genesis account of creation, God finds goodness in things even before, and quite apart from, the creation of Adam. Jesus stressed the divine concern for the sparrows and even the grasses of the field. If a man is worth many sparrows then a sparrow's worth is not zero. But the Christian tradition did not live up to the conviction of its founder. Augustine argued that since beasts lack reason we need not concern ourselves with their suffering. And that is why they have no rights. Passmore (1975), in an interesting history of our treatment of animals, considers that the characteristic teaching of the Roman Catholic Church on the treatment of animals derives from Aquinas. We owe no duties to animals, as God has given us complete dominion over them. However, animals can feel pain and men can feel pity for their pain. Indeed, says Aquinas, the one who feels pity for animal suffering is more likely to have compassion on his fellows. Cardinal Newman reiterated this view in the Catholic Encyclopedia. 'We may use them, we may destroy them at our pleasure, not our wanton pleasure, but still for our own ends, for our own benefit or satisfaction, provided we can give a rational account of what we do' (quoted by Passmore, 1975, p. 203). And in the late nineteenth century when asked to permit the setting up of a Society for the Prevention of Cruelty to Animals, Pope

Pius IX replied that 'a society for such a purpose could not be sanctioned in Rome' (quoted by Passmore, 1975, p. 203).

Schopenhauer complained of 'the unnatural distinction Christianity makes between man and the animal world to which he really belongs. It sets up man as all-important, and looks upon animals as merely things ... Christianity contains in fact a great and essential imperfection in limiting its precepts to man, and in refusing rights to the entire animal world' (Schopenhauer, 1890, p. 112).

Fortunately, this is not the whole story. Chrysostom in the fourth century urged Christians to extend their gentleness to animals. Saint Francis is the most famous example of a Christian lover of nature though, as Passmore (1975) points out, the case is not altogether clear since the image given of the saint varies from chronicler to chronicler. And in our own day numerous societies for the rights of animals have been created, some of them by the church, such as the Catholic Study Circle for Animal Welfare in Great Britain and in the US the National Catholic Society for Animal Welfare, which has significantly renamed itself the Society for Animal Rights.

The recognition that animals are subjects and not just objects was sometimes included as an unassimilated element in an anthropocentric system. Bentham and Mill, for example, believed that animals too experienced pleasure and pain, and since they took pleasure as the inherent value, they saw that it was inconsistent to exclude animals altogether from ethical consideration. They acknowledged this, but they did not deal with the theoretical questions this raises for the utilitarian system. As a cultural corollary the movement for the humane treatment of animals won widespread support, but it was viewed as a sentimental luxury rather than as having fundamental prophetic import for the structure of ideas by which society lived.

In the East the desirability of compassion for animals is asserted by many religions. For example, the *Institutes of Vishnu* contains many regulations relating to the humane treatment of cows and domestic animals such as dogs. Penances are laid down even for the killing of insects. For millenia in India the concern in both Hinduism and Buddhism has been for the salvation of all sentient beings. Zoroaster also is said to have had great compassion for animals (Passmore, 1975).

Albert Schweitzer remains the one great Western twentieth-century thinker who took seriously the value of all living things. For him, 'Ethics is the infinitely extended responsibility toward all life' (Schweitzer, 1949, p. 241). His principle of reverence for life deeply shaped his own existence and has had a spreading influence on others. But it did not gain expression in practical ethical guidelines.

Indeed, Schweitzer consciously opposed the development of a differentiated evaluation of forms of life which would make such guidelines possible. He wrote:

> To undertake to lay down universally valid distinctions of value between different kinds of life will end in judging them by the greater or lesser distance from which they seem to stand from us human beings – as we ourselves judge. But this is a purely subjective criterion. Who among us knows what significance any other kind of life has in itself, and as part of the universe? . . . The man who is truly ethical . . . makes distinctions only as each case comes before him, and under the pressure of necessity, as, for example, when it falls to him to decide which of two lives he must sacrifice in order to preserve the other (Schweitzer, 1933, p. 271).

Judgments of value among species will have a subjective element, and similarity to human beings is likely to play a distorting role at times. But it does not follow that no generalisations are possible or that human beings will show greater wisdom in this area if they make each decision *ad hoc*. Indeed, Schweitzer's own *ad hoc* decisions, far from being models of wisdom, reflect a prejudice against predators and an insensitivity toward the harm done to wild animals by captivity, which could support much of the worst of human action toward animals (Shepard, 1959, pp. 26–9). An adequate ethic of reverence for life requires the development which Schweitzer refused to give it.

In the absence of an ethic of reverence for life the two most influential ethical views in the modern world remain utilitarianism and the categorical ethics of Immanuel Kant. Utilitarianism teaches that each of us should seek to maximise human welfare. The welfare of each person is to count equally. Nothing else counts. In the classical forms of the theory, pleasure was taken as the measure of the good, but this is secondary to the basic form of the theory. More neutral terms such as satisfaction, or meeting needs and desires, will serve.

Kantians reject utilitarianism because it judges actions according to their consequences rather than according to the principles that guide them. By utilitarian principles, lying and stealing and killing may at times be justified as more likely to lead to the greater good of the greater number than refraining from these actions. From the Kantian point of view this is to abandon serious morality. Life is to be guided by principles. These principles are rules of behaviour which one can will to be universal. One cannot will that lying be universal, hence it is always wrong to lie. But in another way Kantians accept and strengthen the utilitarian focus on human beings, for every human being is always to be treated as an end, never as a means.

Both ethical theories in their consistent application imply that everything other than the human is appropriately considered only as means to human ends. This is to say that the welfare of a dog enters into ethics only in so far as the dog is important to some human being. In, and of itself, the dog's pain or pleasure is irrelevant to ethics.

Modern economic theory is based on the same assumption. It provides an objective basis for determining the value of any entity. An entity is worth what someone will pay for it. Human beings are assigned no value in this scale because they are the determiners of value. Value means value *to* human beings. The value of a dog is what it is worth to some human being.

Even so these ethical theories and the economics and politics they support are a great advance over much that has occurred in history. During much of our history many, if not most, human beings have also been treated as having value only as they were of use or importance to other human beings. Slavery is the clearest expression of this theory, but even beyond the institution of slavery, vast numbers of people have been treated as sources of wealth and power for the few, not at all as ends in themselves. The problem is still pervasive today, and the anthropocentric ethical theories continue to be important in the effort to gain recognition in practice as well as in theory of the worth of every individual in and for herself or himself.

Since this point of view still has an important positive role to play, it is with some hesitancy that we call attention to its fundamental limitation. Nevertheless, it is time to work seriously on a new ethics that is congruent with contemporary understanding of the human and the natural, and relevant to our dealing with a range of issues on which our received, anthropocentric ethics has too little to say. The anthropocentric bias of Western ethics is reflected in the sorts of justification which are provided as standard for the preservation of the natural environment. Godfrey-Smith (1979, p. 311) lists four such arguments, which he calls the silo argument (for maintaining a stock-pile of useful organisms), the laboratory argument (for scientific enquiry), the gymnasium argument (for leisure) and the cathedral argument (for aesthetic pleasure). All are arguments to look after nature because nature looks after us.

If subjectivity were confined to human beings, if all else existed only as objects of human knowledge and use, then anthropocentric ethics and economics would conform to reality. But such a view simply does not fit with what we know about life and about ourselves as one form of life. We are subjects in a wider community of subjects as well as objects in a wider community of objects. If we, as subjects, are of value, and we are, there is every reason to think that other subjects are of value too. If our value is not only our usefulness to others but also our immediate enjoyment of our existence, this is true for other creatures as well.

The recognition that every animal is an end in itself and not merely a means to human ends explodes the assumptions of our traditional ethics. If all animals are ends, Kant's doctrine of the Kingdom of Ends must be vastly expanded. This is what Schweitzer undertook to do with his principle of reverence for life. But when all animals are regarded as ends, it must be admitted, against Kant, that these ends are also means. What is needed is a new ethic which recognises in every animal, including humans, both end and means. Similarly, the utilitarian calculus breaks down. If the sharp distinction between human beings and everything else is abolished, then the pleasure or satisfaction of all animals must be considered. But if each is to count the same, if every sparrow is to be counted equally with every human being, the calculation can lead only to absurdity.

Similarly, the foundations of our economic thinking are exposed as inadequate. If each animal is an end as well as a means to human ends, then the value placed upon the animal by humans is not an adequate measure of its worth. Further, non-human animals have value to each other. Worms are of value to birds as well as to human beings. In so far as the laws of the market place ignore all this they are inappropriate to the nature of reality. To some extent we have recognised this in our laws for the preservation of endangered species and the protection of whales. But we have not yet begun to rethink the basic theories of the economic world so as to incorporate our intuitions. Some may protest that this is a trivial matter compared with the need for the rethinking of economic theory in the light of a proper evaluation of human beings, but we think not. Concern for human beings, especially for the oppressed, and concern for animals are closely related. Aquinas tells us that the person who feels pity for the suffering animals is more likely to have compassion on his fellows. The righteous man will treat his beasts well (cf. Passmore, 1975). Clark (1977, p. 197) points out that in the eighteenth century the same people who spoke against slavery 'were also active in the cause of animals'. We will not have a proper evaluation of people until we have a theory of value that can include other living creatures.

The general ethical principle that follows from what we have said is that we should respect every entity for its intrinsic value as well as for its instrumental value to others, including ourselves. Its intrinsic value is the richness of its experience or of the experiences of its constituent parts. We deal with entities appropriately when we rightly balance their intrinsic and their instrumental worth. No one can provide mathematical formulae for determining the relative worth of different entities. Ethical living is an art and not a science. But ethical judgments are informed by basic perspectives, and the view of life we have offered is part of such a perspective. Its implications for ethical judgments can be clarified further by developing the understanding expressed in earlier chapters.

In this understanding there is continuity between all levels of existence. Accordingly, if there is intrinsic value anywhere, there is intrinsic value everywhere. But the intrinsic value that can be attributed to the subjective experience of events at the sub-atomic, atomic and molecular levels, is so slight that for practical, and therefore ethical, purposes, it can safely be ignored. The same is true of mere aggregates of such events such as rocks. The intrinsic value in a rock is only the sum of the intrinsic values of the molecular, atomic and sub-atomic events that compose it. Furthermore, the effects of most human actions upon such events is trivial. Entities of all these types may reasonably be treated as means, or in terms of their instrumental values only.

A living cell is also composed of events at the sub-atomic, atomic and molecular levels, but in the living cell these are so structured that the resultant cellular events are by no means mere aggregates of these entities. A cell, unlike a stone, has an inherent unity and cellular events are constituted by a new level of internal relations with their environment. This means that a cell has experience of its world in some dim way analogous to our own, although we doubt that its experience is conscious. In short, its intrinsic value is far greater than that of its constitutive elements or of a stone. If a choice were to be made between a completely inorganic universe and a universe in which there was cellular life, there is no question but that the latter should be chosen. The value of such a universe would be incomparably the greater of the two. Hence, the intrinsic value of cells is not entirely negligible from an ethical point of view. Even so, it is a rare circumstance when the perspective of cells as such would loom large in ethical consideration. The primary value of cells is instrumental.

Plants are not mere aggregates of cells. They are marvellously complex societies of cells. They perform numerous functions which cells outside of these societies cannot perform. Nevertheless, we do not attribute to plants the kind of unity we think we discern in the cell. The life of the plant is the life of the cells which compose it. The plant has enormous instrumental value for human beings and other animals which cells in isolation do not have. But its intrinsic value is the sum of the values of the cells. We judge that plants, like the cells which compose them, can appropriately be treated primarily as means – extremely important means which we abuse at our peril!

In animal life, in distinction from plant life, a new level of experience arises. At some point in the evolution of the central nervous system, conscious feeling, as distinct from non-conscious feeling, entered into the world. This is a qualitatively different experience which presumably increases progressively with the development of the nervous system. With the increased complexity of the nervous system and the development of the brain we have every reason to suppose there is increased capacity for richness of experience. Hence animals, especially those with highly developed nervous systems, cannot rightly be treated as mere means. They are entities which we must respect as ends as well. Their existence and enjoyment is important, regardless of the consequences for us or for other entities. In proportion to their capacity for rich experience we should respect them and give consideration to making this experience possible. In short, they make a claim upon us, we have duties toward them.

Animal rights

What then can we say of animal rights? One widespread view is that talk of rights cannot be separated from mutuality of duties or contracts. From this it is argued that since, in general, animals cannot enter into contracts with us, it is inappropriate to attribute rights to them. There are obvious weaknesses in this theory since most of its advocates do attribute rights to members of the human species even when there can be no question of such mutuality. But without pursuing these matters, we would be prepared to regard the question of 'rights' as a terminological one if those who denied rights to animals were nevertheless willing to speak in some other way of their ethical claim upon us and our duties toward them. But this is very rare. On the whole the denial of rights to animals is taken as also entailing that we have no obligations toward them, that we are free to exploit them for our private pleasure or advantage without limit. Hence it seems better to insist that the idea of rights is not bound only to the context of mutual duties and contracts. It seems quite natural to use it wherever duties come into play. Whenever I have a duty toward something, that is, whenever I ought to treat it in a certain way for its own sake, I can make the same point by saying that the entity in question has a right to be treated in that way. This does not mean that my neighbour's wallet has a right not to be stolen; for my obligation not to steal the wallet derives entirely from my obligation to my neighbour. In this case, it is my neighbour's rights that are in question. But I ought not to torture my neighbour's dog even if my neighbour does not object. The dog has the right not to be tortured. This point is well summarised by Linzey (1976, p. 21): 'If A has a duty towards B, it should follow that B has a corresponding right against A.'

However, animals are also means to the satisfactions of the needs of others. Their instrumental value in most cases is more important than their intrinsic value. Most of them constitute food for other animals. They play a role in the replenishing of the soil for plants. Many of them serve human needs and enhance human enjoyment. It has not been wrong to view animals as means to our ends. What has been wrong has been to view them as *only* means. Ethics, law and economics should take account not only of the uses of animals, but also of their rights, which are correlative with their potential for richness of experience.

As Waddington (1960) recognised, richness of experience is a crude concept. Even if we worked hard to refine it, our success would inevitably be limited. But even in its crudest form it offers us considerable practical guidance. Although our evidence is indirect, we do have reason to think that porpoises have a high degree of communication with one another and are probably quite concerned for one another. They seem to have deep emotions. They are adept at learning complex tasks as they demonstrate in oceanariums, and they remember what they have learned for years. They are also capable of relatively abstract thinking. For example, rough-toothed dolphins were trained to perform a new trick for a reward of fish. After several days of training they exhibited ever-different types of leaps and contortions, apparently 'realising' that the forms of behaviour they had displayed previously would not be rewarded (Würsig, 1979). All this suggests a rich subjective life which we have little reason to attribute to a tuna or a shark. We can confidently affirm that the intrinsic value of a porpoise is greater than that of a member of these other species. Accordingly porpoises make claims upon us beyond those made by tuna and sharks.

Many of the topics important for the discussion of human rights are irrelevant to the discussion of animal rights. Individual human beings realise their capacities for richness of experience to very different degrees. Our ethical task is to help one another to realise individually and collectively as much of this richness as possible. In dealing with animals this is largely irrelevant. For the most part there is little humans can do positively to enrich the experience of animals. The focus of concern is much more on avoiding those actions that prevent rich experience. It is possible to reduce the amount of pain and suffering we inflict upon other creatures.

Since pain and suffering are felt by animals individually, the societies for the prevention of cruelty to animals and the humane movement have concentrated their concerns for generations on the plight of individual animals. With respect to these issues we have few reservations about the statement of principles of the Humane Society in the US (Morris & Fox, 1978, p. 236).

> It is wrong to kill animals needlessly or for entertainment or to cause animals pain or torment.
>
> It is wrong to fail to provide adequate food, shelter and care for animals for which man has accepted the responsibility.
>
> It is wrong to use animals for medical, educational, or commercial experimentation or research, unless absolute necessity can be demonstrated and unless such is done without causing the animals pain or torment.
>
> It is wrong to maintain animals that are to be used for food in a manner that causes them discomfort or denies them an opportunity to develop and live in conditions that are reasonably natural for them.
>
> It is wrong for those who eat animals to kill them in any manner that does not result in instantaneous unconsciousness. Methods employed should cause no more than minimum apprehension.
>
> It is wrong to confine animals for display, impoundment, or as pets in conditions that are not comfortable and appropriate.

It is wrong to permit domestic animals to propagate to an extent that leads to overpopulation and misery.

The serious application of these principles would enormously reduce the suffering now inflicted by us on our fellow creatures. Sensitive people who investigate our zoos, laboratories (see Ryder, 1975) and slaughterhouses will understand the passion with which others devote themselves to the reduction of this suffering. But probably the area of greatest suffering, and greatest current increase in suffering, is factory farming. It is there that the view that animals exist only as means to our ends, ends which are measured in purely economic terms, is most consistently and destructively expressed. Harrison (1964) and Singer (1976) describe the morally intolerable practices that are now commonplace. Animals are *not* machines. Meat should *not* be manufactured with complete indifference to animal suffering. The failure of the major traditions of Western ethics to address themselves seriously to the treatment of animals by human beings reflects a false view of reality. It also shows how far our natural sensitivities can be blunted by an educational process based on erroneous ideas.

One may reasonably argue against the apparently unqualified or absolute formulations of this humane view of animal rights, with Narveson (1977, p. 173), that the utilitarian principle should be employed more consistently. Inflicting suffering on animals is legitimate if there is more than compensatory advantage. Raising animals for food can be justified if 'the amount of pleasure which humans derive per pound of animal flesh exceeds the amount of discomfort and pain per pound which are inflicted on the animals in the process'. But Narveson recognises that such calculation would still support a great reduction in factory farming methods and that analogous formulations would eliminate much experimentation on higher animals. It is in this non-absolutist spirit that we support the principles of the Humane Society.

The principles espoused by the Humane Society would require major changes in our educational system and our agricultural practices as well as in our entertainment. But there are other believers in animal rights who are convinced the Humane Society does not go far enough. The Humane Society does not declare that animals have the right to live. Here it accepts a fundamental distinction between human beings and other animals. From the point of view of the Humane Society we are at liberty to kill other animals if we can render them unconscious instantaneously and produce a minimum of apprehension. Today the question is being vigorously raised as to how we can be justified in killing other animals at all (Godlovitch, 1971; Linzey, 1976; Regan, 1976). If it is wrong to kill other human beings, on what grounds is it morally acceptable to kill individual members of other species. Is this not speciesism, that is, an irrational prejudice against other species?

Much of the argument of those who affirm that animals have the right to life is directed against those who hold that there is some absolute distinction between human beings and other animals. We support this part of the argument. We agree that animals are not mere means to human ends. We agree that animals have rights which are being pervasively violated by human beings. But we are not persuaded by the argument that among these rights is the absolute right to life.

Those who argue for the right of individual animals to life begin their argument with an analysis of the basis on which the right to life can be attributed to all members of *Homo sapiens*. They claim that the usual arguments in terms of rationality, free will or self-consciousness are not the basis on which we justify the right to life of all humans. Since we attribute the right to life even to those human beings who lack these characteristics, the right must not be contingent on such variable characteristics. Perhaps it can be based on the view that all human beings, regardless of the variety of their capacities, have interests and that they have, accordingly, the right to pursue these interests which they cannot do if they are killed.

Regan regards this as the most plausible justification of the view that all human beings have the right to life. But he immediately points out that this 'argument for the view that humans have an equal natural right to life . . . seems to provide an equally plausible justification for the view that animals have this right also' (Regan, 1976, p. 202). Regan does not argue that there are no circumstances under which an animal may be killed with morality, but he does argue that the justification required for killing an animal would be similar to that required for killing a human being.

This argument has considerable force. It hinges upon the view that all human beings, however brain-damaged and permanently incapable of distinctively human functioning they may be, have an equal right to life. Once that absolute egalitarianism is accepted, its grounds must be such as to allow that animals have the same right. But this conclusion seems to us to demonstrate the dubiousness of the premise. It requires, it seems, that sharks and porpoises have the same right, and, unless carefully defined, that mosquitoes have the same right to life as human children. Fortunately for the argument, Singer has undertaken to draw some limits. For him the limits are the limits to which we can reasonably attribute capacity for suffering.

> The grounds we have for believing that other mammals and birds suffer are, then, closely analogous to the grounds we have for believing that other humans suffer. It remains to consider how far down the evolutionary scale this analogy holds. Obviously it becomes poorer when we get further away from man. To be more precise would require a detailed examination of all that we know about other forms of life. With fish, reptiles and other vertebrates the analogy still seems strong, with molluscs like oysters it is much weaker. Insects are more difficult, and it may be that in our present state of knowledge we must be agnostic about whether they are capable of suffering (Singer, 1973, p. 18).

The recognition of degrees of analogy of animal behaviour to human behaviour is commendable, but it is unfortunate that it seems to be associated in Singer with the view that everything depends upon whether there is any capacity to suffer at all. To us it seems more plausible that there are degrees of the capacity to suffer and that it is much worse to inflict suffering on creatures with highly developed capacity to suffer than on those where this capacity is rudimentary. We agree with Narveson that there are qualitative as well as quantitative distinctions in suffering (Narveson, 1977, p. 168). Furthermore, it is not clear that an argument can be drawn from any capacity to suffer whatever, or the possession of any interests whatsoever, to the right to life.

If this is not a basis for animal right to life, and we do not believe it is, then human capacity to suffer and possession of interests alone cannot be the basis for the right to life of individual human beings either. At the same time, this right cannot be based on any absolute difference between human beings and all other species. Hence a new formulation of its grounds is needed. That formulation can begin with the conviction, reiterated often above, that every human being, like every animal, is an end as well as a means, or that each one has intrinsic value. But this alone is still not sufficient grounds for affirming the right to life. How, then, does the recognition of the intrinsic value in other human beings lead to prohibition of killing? Three factors need to be considered.

First, the actuality of the intrinsic value of the human being lies in more than present and past experience which the killing does not destroy. What the killing does is to prevent the occurrence of further experience inherited from and building upon that person's past and present. This loss is an evil, the more so because each individual person is unique (see Narveson, 1977, p. 166).

Second, human beings know that they will die. There is no way death can be prevented. However, the effect of anticipating a natural death at the end of a full life is likely to be much less negative than the anticipation and fear that one's life will be cut off in mid-stream. For the sake of quality of life, that fear must be reduced as far as possible.

Third, in the great majority of cases the killing of one person, however painless to that person, is the infliction of suffering on others. The death of one human being has consequences for others. There is always sadness when anyone dies. When death comes at the end of a full life, the experience of relatives and friends can be a positive one. When it comes abruptly in earlier life, it is disruptive and destructive in many ways.

If these are reasons for opposing the killing of human beings, what about chickens? Is our failure to apply the prohibition of killing to chickens mere anthropocentric prejudice? It seems not. We can examine the case of the chicken under the three headings in which we discussed the human prohibition.

First, it is true that the killing of a chicken prevents the occurrence of the additional experiences that chicken would have had. But it is not clear that the distinction between those prevented experiences and the experiences of another chicken is of much consequence. If the death of one chicken makes room for the raising of another, the values lost are largely replaced by the values gained. The quality and amount of chicken experience remain largely unchanged. In the human case, the future experiences that are cut off are unique and irreplaceable. That is because they derive from a unique individual with a unique history whose particular capacity to generate new experience is forever destroyed. In the chicken's case the element of uniqueness is trivial.

Second, the quality of a chicken's experience is not pervasively affected by the anticipation of its death. It may be terrified by immediate danger and, as the Humane Society states, it should be our ethical responsibility to minimise this terror, but the suffering of a chicken attendant upon old age and death may be as great as that occasioned by violent death earlier in life.

Third, chickens do have some social life which is affected by the death of its members, for one thing pecking order status changes. But in the normal course of events no hardship is inflicted on the others by the death of one. There seems to be little or no grief.

To establish a clear contrast which justifies a totally different attitude to killing in the two cases, we have taken a creature some way down the scale of animal life. The case is far different with chimpanzees. Among them, there are indications of an individuality resembling our own and of social relations which lead to grieving for the dead. What kind of anticipation of death may be found among such creatures, and how this affects the quality of their lives, we do not know. The killing of such an adult animal in the wild often leads to the suffering of its offspring. Hence the considerations that lead to opposing the killing of human beings count also against the killing of some animals, especially some wild animals.

However, there are many cases in which judgment is more difficult. The rights of animals and the rights of human beings are often in conflict. Those who recognise the intrinsic value of other creatures as requiring consideration for them were shocked by the Japanese fishermen who were bent on slaughtering porpoises. World-wide criticism in turn shocked the fishermen, who took their anthropocentric ethical principles for granted. To them the porpoises were competitors for the fish. Every fish eaten by the porpoises reduced the number available for the fishermen to catch. The natural anthropocentric ethical response is that the slaughter of the porpoises is appropriate.

We are siding with the criticism of the Japanese fishermen, even though they claimed, in 1979, that porpoises caused a loss to them in that year of two million dollars worth of fish. The dolphins they caught were sold in Japan as food for other animals. That itself must constitute a considerable source of revenue for the fishermen. Even so these economic advantages do not justify the slaughter.

But is it legitimate and desirable to side with criticism of the Japanese fishermen only as long as the hardship consequent on human beings is relatively minor? What will be the ethical response as the resources of the sea continue to decline and the demands upon it increase? At that time the competition between humans and porpoises may become significant. If at that point our fundamental ethical thinking is still anthropocentric, and if we allow ourselves exceptions only in areas where we can afford the luxury, it is easy to predict that if we do not exterminate porpoises altogether we will reduce their numbers drastically.

The present pattern of global practice is hurtling us toward a position in which we shall be forced to view such creatures as porpoises as our competitors for sustenance. The baleen whales, such as the Blue

Whale, may come into this category when the krill, which is their main food, is extensively harvested for human food as has already started. If we feel sufficiently disturbed by this destiny, we may yet find motivation to change. That change would be for the sake of porpoises and whales as well as our own children. To feel and change in that way is not mere sentimentality. It is appropriate to the real nature of ourselves and of our fellow creatures.

Whereas it used to be argued, as by Cardinal Newman, that the least human good compensates for any possible cost to animals, we now claim that it requires a great human advantage to compensate for animal suffering and loss. How great an advantage to balance how great a loss will be a matter for each case in point. For example, it is possible that the production of a vaccine to combat hepatitis B virus may prove a fatal conflict between the value of people and of chimpanzees. Chimpanzees are the only species, other than humans, in which the safety of the vaccine can be tested. Each chimpanzee can be used for one test only. If the vaccine is produced in quantity – and some 150 million people are thought to be carriers of the virus – many of the 50 000 chimps who remain in the wild may be killed or captured for the vaccine makers' colonies. If chimpanzees are not used it may be impossible to safely test and hence manufacture the vaccine (Wade, 1978). The world has a growing population of four billion people and a dwindling population of some 50 000 chimps (already classified as threatened). Since the disease is rarely fatal we think the larger population should find some way of solving its problem that is not to the detriment of the threatened population.

In African countries today the question arises as to whether wilderness should be preserved in sufficient quantity to secure the survival of many species of wild animals. Land set aside for this purpose cannot be used for the feeding of Africa's growing human population. Ruanda, for example, has to choose between having elephants in national parks and land that can be farmed for people. Such choices are never easy. If Ruanda's population continues to grow, if the production of adequate food becomes a critical problem and if the responsibility for maintaining habitat for elephants is left entirely to the country in which that habitat is available, then in time we would reluctantly judge that Ruanda would be morally obligated to abandon its protection of elephants. But precisely for reasons of this sort, we judge that it is an urgent imperative that human beings restrain their growth, that imaginative efforts be made to deal with needs for food without bringing additional wilderness under the plough, and that international agencies be set up to compensate poorer nations that are willing to contribute to the total global need for the preservation of endangered species.

Human rights

Homo sapiens is a species of animal, and what has been said of other animals can be said of humans as well. Human beings too, are both ends and means. Nevertheless, despite continuity between human beings and other animal species, there are also characteristics of human beings that are lacking in most, if not all, other species. Whereas in animals in general the unified animal experience primarily functions in the service of the body, in human beings it aims at its own richness of experience and frequently subordinates the body to these aims. This is associated with the new predominance of cultural evolution, the vast explosion of individual differences and whole new dimensions of intellectual activity. *Homo sapiens* is a species of animal, but not only that. In human beings consciousness has become conscious of itself. Because of this, human beings are primarily ends and only secondarily means. Human beings have, of course, great instrumental value for one another and for other creatures, but to treat any individual chiefly in terms of this instrumental value is inappropriate. Ethics has not been wrong to emphasise that human beings should be treated as ends. The error has been to pretend to ignore that human beings are after all means to one another's ends and to try to make absolute the distinction of ends and means. It is proper for human beings to serve as means not only to the welfare of other human beings but also to the welfare of other creatures. This is not sheer sentimentality. It is a duty. It becomes sentimental only when one loses a sense of proportion, exaggerating the intrinsic worth of the creatures one serves. It becomes evil only if it leads to undervaluing human beings. Of necessity we use people and are used by them. What should properly be avoided is not this use, as such, but a use that is insensitive to the intrinsic worth of the person who is being employed as a means. When people are treated in this way they feel 'used' and speak properly and in indignation of exploitation.

In relation to other animals we do well if we can arrive at some rough measure of the richness of experience enjoyed by members of a particular species. It is enough to judge, for example, that porpoises as a whole are of much more intrinsic worth than sharks. There is little practical reason for raising further questions about the relative value of the experience of one member of the species in comparison with another. With human beings, however, questions of the relative value of different experiences and types of experience are inevitable. Education, religion and psychotherapy all assume that some modes of human experience are inherently preferable to others. Although most of us prefer experience that contains a considerable amount of suffering to no experience at all, this is not absolute. There are many people in our hospitals who would much prefer death to the continuation of the only kind of life that seems possible for them. There is no escape from the conclusion that some human experiences are of vastly more intrinsic value than are others.

But does this mean that some people are of more value than others? No one doubts that some people are of more instrumental value to society than are others, but the dominant view of Jews, Christians and Western humanists is that all are of equal intrinsic worth. This equality is based on a view of the human 'psyche' or 'soul' as something discontinuous with all other forms of life. Whereas the intrinsic worth of a porpoise, if it is recognised at all, is given a finite value in the sense that it is less than that of a human and more than that of a mosquito, the intrinsic worth of a human is held to be infinite. Human life is said to be of infinite value or to be sacred; that is the meaning of sacred in this context. In short the question of comparative intrinsic worth does not arise because human worth is given an absoluteness that bears little relation to experienced worth. Yet what gives intrinsic worth is richness of experience and capacity for richness of experience, neither of which is infinite for any human being.

It is clear that no society has consistently lived in terms of an absolute reverence for every human individual. Far from it. Nevertheless, this deep-seated belief has played a pervasive role in the shaping of Western attitudes and continues to do so long after its religious and philosophical basis has eroded. Furthermore, much of what is best in Western civilisation is informed by it. Much of the purest motivation for dealing with needs of the poor of the world still stems from this view as does much of the pressure toward more egalitarian forms of society. There is almost a taboo against questioning it, and when it has been rejected, as by the Nazis, the consequences have been so shocking as to confirm the importance of the taboo.

Still, the topic does need discussion. If there are good reasons for holding the egalitarian view, it should be possible to express these in an open forum. Since the view of human life developed in the preceding chapters is incompatible with the traditional grounds for the egalitarian position, it is necessary to broach the subject frankly to determine where the chips fall.

If the locus of the intrinsic value of the human is in human experiences, and if these vary from person to person, then it seems clear that in fact there are differences of individual worth. It can be stated clearly and frankly. There is no substantial reason to believe that all persons have equal intrinsic value. What does this imply? Does it open the gates to all sorts of inhumanity? Surely not.

The first principle of ethics that follows from this view of the human situation is that one should promote richness of experience wherever possible. There is a gap between what is potential and what is actual in each person. Each falls short of achieving the quality of experience that his or her genetic inheritance allows. In part this is because cultures function to restrict people as well as to encourage them. In part it is because people do not realise the values that their culture makes available to them. We are called both to develop the potentialities allowed us by our several cultural environments and to develop those cultures so that they will encourage fuller personal development.

Two groups of people need special attention. First, what happens in early childhood is of great importance throughout life and is peculiarly influenced by the social environment. The child needs to develop habits and disciplines that will support continuing growth in adulthood. Second, there are some who have been peculiarly deprived by adverse cultural situations who are yet eager to grow. For example, by material circumstances, some are prevented from completing what is now regarded as a minimal education. Society owes to them whatever help it can provide to overcome such handicaps.

Inequality is the human predicament. Each person comes into the world with unequal genetic endowments unless they have an identical twin. Some have many talents. Others have few. Some are burdened with genetic defects that incapacitate them for life. Others are burdened with a deprived upbringing that leaves them stunted and hurt. What then is to be our individual attitude to this unequal distribution of the goods and bads in our endowment?

If I am more or less normal, it is through no merit of my own. That others have enormous burdens to bear may be no fault of theirs. The child with Down's syndrome could have been me. The fact that others are lacking, through no fault of theirs, changes for me the character of my having. If there is any meaning to the unity of humankind then all are called to share the cost.

A community becomes responsible for the deprived through every responsible member of that community. This is the strong meaning of justice. Justice does not require equality. It does require that we share one another's fate.

One major danger which history might teach us to avoid is the assumption that the form of rich experience with which we are acquainted is the only one or the best. Many people have come recently, perhaps for the first time in history, to appreciate a genuine pluralism. They know now that it is not necessary for Easterners to adopt Western styles in order to achieve the highest values. They have come to appreciate the unique values of pre-civilised forms of experience as well. The different levels of achieved value are now seen not so much between cultures as within cultures. This is a great gain and greatly reduces the dangers entailed in recognising differences of intrinsic worth among people. In the past such differences have been culturally, racially and sexually defined. The doctrine of ultimate equality of worth mitigated against such judgments of superiority, but it did not prevent their taking of a terrible toll. Today one may hope for better results from the reliable evidence that these judgments are not warranted.

It would be arbitrary to assert that all cultures are equally successful in encouraging the realisation of rich experience. What can be asserted is that each realises some values better than do others and that none realise all the possible values. This should not deter us from learning from each other and seeking in this way to transform our several cultures.

Among the critical judgments to be made of almost all cultures is that they have denied to many of their members the opportunity to grow. In many cases they have adjudged some, if not most, members inferior in their capacity to grow and have thus justified denying them the means to do so. The result has been an actual inferiority of experiences on the part of slaves, peasants, women, ethnic minorities and other classes. This actual inferiority has been appealed to as justifying the practices which created it. One of our major ethical imperatives in the modern world is to reverse this practice by providing equal, and in some cases compensatory, opportunities for those sub-cultures to which these opportunities have for so long been denied.

Equality before the law is an elementary requirement for moving toward equality of opportunity. As a formal principle it is widely employed, but even where it is employed in good faith it cannot achieve its goals apart from the attainment of equality in other areas. The poor and ignorant cannot employ the courts as successfully as the rich and favoured. Even if they do, they will find that inevitably the structure of laws favours those who write them. Legal equality cannot be attained without political equality, but this is even more difficult to achieve. Equal political rights are only a beginning. It is inevitable where there is any freedom at all that some will be more skilfull than others in employing the political processes for their ends. As long as inherited inequalities among classes exist, it is inevitable that those with money and leisure, experienced in political activity and expectant of success will largely control the government. The ethical principle will call on us to exert energies to counteract these inevitable tendencies.

These conclusions drawn from our understanding of value do not differ substantially from conclusions often reached on the basis of the absolutist view so far as human relations with humans are concerned. At the fringes, however, there are some differences. The traditional absolutist view has led to judgments about abortion and medical practices with the dying which need to be challenged.

When human life is viewed as sacred or as of infinite value, there follows a prohibition of its destruction. Of course, qualifications are almost always introduced, but the principle is still powerful, and it has worked in praiseworthy ways against the casual destruction of people perceived as socially useless. On the other hand, it has forced discussions of abortion into artificial distinctions from which these discussions need to be freed.

If the human soul is of infinite worth and nothing else has any intrinsic value, then the question of the legitimacy of abortion is the question of when and whether the foetus receives a 'soul'. Before it receives a 'soul' it is of no value at all; afterwards, it is sacred. One problem is that what is known of the development of the foetus does not suggest that there is any one point at which so total a transformation occurs. One can, it is true, identify a stage of development at which no central nervous system exists and a later one in which it is well established, and so it is possible to posit that unified foetal experience emerges somewhere toward the latter end of this development. But viewed from one perspective, the fertilised egg already has some intrinsic value, and the much greater value of the fully developed foetus is still not absolute or infinite. The foetus should be recognised as having rights, increasing rights as time passes, but these rights are less than those of the newly born infant, and they must always be considered in connection with the rights of others. The notions that the mother should have the absolute right to decide about the contents of her womb, or that the foetus is no more than an unneeded appendage of her body, are not justified by this account. On the other hand, the foetus does not have the same rights as a fully developed human being. To apply to the killing of a foetus the same language that is used for the killing of a human person is an obstacle to reasonable reflection on a difficult topic.

Reflection about abortion is ethical when it considers the rights of the foetus, of the mother, of the father and of other relatives. Such reflection can never lead to the indifferent destruction of the living foetus. It certainly cannot support a casual attitude toward pregnancy. If children are not wanted, pregnancy is to be avoided rather than initiated and terminated. But where it is too late for avoidance, and where the rights of the parents and siblings are in conflict with the rights of the foetus, abortion is often the lesser evil. Since much abortion will take place under any circumstances, its legalisation and financial support by society will lessen the special suffering of the poor and safeguard the health of the mother.

The question of rights is not exhausted by the rights of the foetus and the persons most intimately involved. There is also the right of the larger society to safeguard its interests. During much of history, in many places, these interests counted in favour of bringing the foetus to birth even against the mother's wishes. Today the context has changed. The world can afford only a limited number of births. It is better that most of these be children whose birth is desired by their parents. Such global considerations do not overrule the rights of the foetus and the relatives. They should, however, become a factor in shaping public policy.

Because of modern medicine many people are alive today who would have been dead without it. Many of these people are restored to health and to contributing to the happiness of others. These are clear gains for which we are thankful. But the same type of medical advances often make it possible to delay death without offering the hope of renewed health and usefulness. Where the patient desires to be kept alive in this way and where the resources required are not excessive, we certainly do not object. But where patients are kept alive against their will solely because of the view that human life is sacred or of infinite worth, we protest. Here, too, matters should be discussed without absolutes. The issues are complex. The right of patients to die when they are ready must be balanced against the interests of family and friends, as well as those of the doctor and the hospital and of society as a whole. The question of dangerous precedents cannot be ignored. But when all this is considered there will surely be many times when the wishes of the patient should be respected. We show more true respect for human life when we acknowledge the right to die than when we force a fellow human being who wishes to die to endure useless suffering instead.

Further, there is no rational reason to keep alive human bodies when there can no longer be significant human experience or any human experience at all. There is an indignity in the conception of the meaning of human life revealed by vigorous attempts to maintain its outward, visible and entirely trivial signs. If one seeks primarily to save life, one is likely to forget the purpose of life. It is not breathing and urinating and similar functions that make a human being important. There is no intrinsic value in these mechanistically maintained appearances. There is nothing sacred about every breath and every heartbeat if they are merely maintaining a shell without the possibility of any purpose being fulfilled. The intrinsic value of human life lies in the capacity for feeling and in the experience itself. When that capacity has gone, so has intrinsic value. When there is no memory, no anticipation, there is no life in any human sense of the word. The wise man, said Seneca, lives as long as he should, not as long as he can. Of course there is some value in the living body cells, but this is relatively trivial. The possibility of unique human experience is not being preserved by them. Only suffering can be added to present experience by the expectation of being treated in this way in the end. The grieving of family and friends is only worsened. A rational ethic would argue not only that it is admissible to allow such bodies to die but that it is a moral obligation to do so. Death is a natural and necessary part of life. The right to die without unwanted interference must be an important element of any adequate ethic of life.

Biosphere ethics

In ethics dealing with human relations, principles governing individual behaviour and rights are important. But even here individualistic principles by themselves are inadequate. It can even be charged that individualistic ethics has worsened the condition of human beings by blinding us to the fact that our condition is more dependent upon social, economic, and political factors than upon how we treat one another as individuals. People tend to concern themselves with victims and symptoms after the event instead of with the circumstances that produce victims and symptoms.

Just as an individualistic ethic is inadequate by itself in the human sphere, so too is it inadequate in our relations with other animals. This is in part the point of Shepard's sharp attack on the extension of the humane ethic to the wilderness. The limitations of such an ethic can be illustrated in many ways.

For example, it would be quite possible to fulfil all the prescriptions of the Humane Society while genetically changing many species so that they become less and less animal and more and more like milk- or meat-producing machines. A display of the effect of modern methods of chicken-raising on chickens was given to the citizens of Bern during a protest by farmers about the price they received for chickens. The demonstration was held outside the government administrative head-quarters. Thousands of chickens were released into the streets. To the horror even of the people involved, and certainly to passers-by, the chickens, unable to walk, scrambled helplessly under the wheels of cars or fell helpless on the pavement. This no doubt resulted in part from violation of the fourth of the rules of the Humane Society opposing the denial to animals raised for food of the 'opportunity to develop and live in conditions that are reasonably natural for them'. But it resulted also from the genetic selection of chickens for egg production and meat rather than for normal animal functions.

It would also be possible to follow the principles of the Humane Society while 'developing' the remaining wilderness areas, thereby destroying the habitats of numerous animals. We could continue to pollute our lakes, rivers and oceans, greatly reducing aquatic life. Important as the humane treatment of individual animals is, there are greater threats to animal life on the planet. Indeed, chiefly as a result of human destruction of natural habitats, about 450 species of mammals have become extinct in the last 300 years. Similar figures can be quoted for birds (Eckholm, 1978). About 1000 birds and mammals now hover on the brink of extinction. And multitudes of unsung populations and obscure species of invertebrates are extinct or endangered (Ehrlich & Ehrlich, 1981). These threats are better recognised in 'The Universal Declaration of the Rights of Animals' adopted in 1977 than by the

Humane Society in the USA, but the accent even here falls one-sidedly on the suffering of individual animals.

One response to these problems is to consider species rights. In the case of many species, the individual counts for very little. Individual insects seem often to have their significance almost entirely as servants of the species. Even in the case of chickens we have argued that the death of one is not a great loss if it makes way for the life of another. The right of the species to survival is much more fundamental than the right of the individual member to life.

Even where individuals do have great importance, as in the case of the chimpanzee, the right of the species to survival is also an important factor. In expressing our views above about the ethical apppropriateness of using chimpanzees to test a new vaccine, we introduced the argument that there are only 50 000 chimps still in the wild. The threat to the species adds to our concern that its individual members not be killed in large numbers to test this vaccine even though the human advantages would be considerable.

This discussion of species rights could move on to a discussion of the rights of ecosystems. But this multiplication of rights is not the best way to deal with the need for ethical guidance in the treatment of the living environment. All things have a *right* to be treated the way they *ought* to be treated for their own sake. But it is not possible to decide how to deal with an individual animal, species or ecosystem apart from a general view of the sort of world envisaged as being worth bringing into being. At this point a sense of the web of life, the value of diversity, and the way people are constituted by internal relations to others, will all guide the imagination. In addition it is necessary to take into account the relation of intrinsic and instrumental value. All this can be subsumed under the question of how to enhance the total richness of experience, or maximise life itself, which is the anagenesis of evolution. Four different views on this can be identified.

A return to precivilised life. One view of the maximisation of richness of experience would be the return to the condition of the planet before human beings became successfully destructive. No one knows when that was. Tens of thousands of years ago human ancestors were already destroying intrinsically interesting and valuable species in Europe and North America. But human destructiveness of the biosphere as a whole did not become important until animals were domesticated and crops were planted. With that began the desertification that in our time has reached such alarming proportions. In this view the planet was most alive before this began, when a larger area was covered by a more vigorous vegetation which provided habitats for a much larger population of wild animals. At that point the human species was part of this rich biosphere. And, in this view, it is to such a relationship that humanity needs to return to attain an enduringly viable life-system.

Adherents of this view do not suppose that the world can in fact revert to just the condition it had ten or fifty thousand years ago. Some of the changes that have occurred are irreversible and changes in weather are not controlled. They also know that the shift from our present human population to one that could live in balance with the environment in the old way would involve enormous suffering and loss of life. But some believe that this loss of life is simply inevitable, that what we have called civilisation is not sustainable. They think the human species is exhibiting the lemming syndrome of overpopulation and collapse. They hope that when the collapse occurs the race will have learned its lesson and will be prepared to live as a part of the biosphere rather than playing lord and master.

Maximising the quantity of human life. A second view of the maximisation of richness of experience is that we should move in just the opposite direction. For it, the evolutionary process is one in which new species rise and become dominant. When they do so they become the bearers of the evolutionary advance in which, from time to time, there are fundamental breakthroughs. Such a breakthrough occurred in the emergence of the human being. Once this had occurred the other species became means to human ends. This need not deny that they have some intrinsic worth, but since the worth of the human is of a fundamentally superior order, the lesser should be subordinated to the greater. Furthermore, the evolutionary process operates only through the human bearer of the advance. The task of humanity is the conquest and populating of the entire planet. The maximisation of human population is the way to richness of experience. This involves the replacement of much of the biosphere with human artifacts. Something is lost, but what is gained far more than compensates. Some animals will be preserved, but only those that serve human needs efficiently. These will be limited, since the most efficient use of the planet for the production of food for people is directly through plants. In this view, where the choice is between animals and an additional human being, the rational choice is always the human being.

Each of these two extreme views has a certain consistency. But each points to the mistakes of the other. The first fails to recognise that the most direct way of increasing the value of the biosphere is by increasing the human population. It minimises the distinctive richness of human experience and its special place in any scale of values. It treats the extent of transcendence over the rest of nature in human beings as a temporary aberration rather than as a distinctive and essential human achievement. In terms of the account in the preceding chapter it sees the long development of human culture only as a fall and fails to appreciate that it was also a fall upward.

The second view fully recognises that the movement is upward but fails to note that it is also a fall. It projects this movement indefinitely forward, failing to recognise that, even if this were to prove to be possible, there are human needs which would be increasingly starved. It exaggerates the human transcendence over the rest of nature to mean human independence from nature. Life on these terms may be possible, but it would be terribly impoverished. This is precisely the direction from which so many of the most sensitive people are today drawing back in horror. If each human life were of identical value regardless of how few of its potentials were realised, this might be justified anyway. But if human experience can vary greatly in value, then the multiplication of human population at the price of impoverishment of the quality of experience is a poor bargain.

Maximising the quality of human life. A third view follows from these criticisms of the second view. In this third view, to maximise life is to maximise the quality of human life. This is not attained by maximising the quantity of human life. Once we can determine what is required for a life of highest quality, we will then posit that the largest population that the planet can support at this level is the optimum condition for the earth. Some believe that the present population or an even larger one could attain to a high quality of life if the world's resources were properly managed. Others believe that we must aim at a gradual reduction of world population toward a lower optimum figure.

In this view, too, other living things are valued chiefly for their contribution to human life. But unlike the second view, adherents of this one are likely to emphasise the contribution of a varied and living environment to human enjoyment. Hence the preservation of endangered species and the expansion of protected wilderness can be justified. The maintenance of diverse human cultures also can be seen as contributing to the best world.

Whilst there are features of this ideal that are commendable it is unsatisfactory from the point of view of the ecological model of life. First, it tends to choose quality of life against quantity in too easy a way. The practical result is usually the preservation of islands of comfort in the midst of a suffering world. The tacit assumption is that those in the overcrowded and poverty stricken world will be the ones whose deaths can bring the whole planet into a happier balance. There is usually little emphasis on the importance of frugality and of finding ways in which the resources of the earth can be made to serve the needs of more people. The quality of life can too easily be seen hedonically. The value of other living things, even if verbally acknowledged, is not seen as a major claim upon us capable of demanding sacrifices on our part. A fourth view is therefore proposed.

Maximising richness of experience. To maximise the richness of experience is to maximise the quality of human life with minimum impact on non-human life. Emphasis is put on the quality of human life but not without serious consideration of the cost to other life. This is not an absurd idea. Many of the great explorations of the human spirit have involved austerity with respect to the world. Frugality is no impediment to richness of experience. The goal then is one in which a large human population learns to live comfortably yet frugally and thereby relaxing pressure on the biosphere as a whole. Not only will this move toward the abolition of degrading poverty, it will also release land for the use of other species. Less exploitative farming together with ingenious technology would be used in the service of a just and sustainable society, not an affluent and profligate one. Every step that could lead either to making room for more people or for more animals, especially wild animals, without lowering the quality of life would be a victory. The balance between human beings and other animals would always be a matter of judgment.

As mentioned above, one factor favouring space for the other animals is that the existence of such wilderness contributes to the quality of human life. Hence making space for wild animals even at some sacrifice to our own comfort and pleasure is compensated in part by the human enjoyment it makes possible. Also some use of some of these lands by humans is not incompatible with their function as animal habitat, so that a considerable amount of food and lumber can be produced without destroying habitat. Indeed, many areas that have been turned into desert by excessive human use could recover if allowed to return to wilderness and could actually produce more food for human beings than is now the case. Hence it is not always a matter of making painful choices between human welfare and that of other animals. The congruence is illustrated especially clearly in the case of the tropical rain forests. These are being destroyed at a rate which is contributing carbon dioxide to the atmosphere equal to that produced by the burning of fossil fuels (Train, 1978; Woodwell, 1978). It is in the interest of all nations to preserve as much of these forests as possible. It is in the interest of the animals that inhabit these forests as well.

Whenever possible human beings should try to find ways to meet their own needs which at the same time support other forms of animal life. Often with a sufficiently long view there is congruence between the two goals. But there are times when the conflict is not subject to resolution in favour of both parties. At such times the ethical requirement is a compromise of human interests for the sake of the interests of other species. Probably this means that lawyers should be designated to represent the interests of these species.

The two factors that have controlled this projection are quantity of rich experience and variety of types of experience. The two considerations can generally be jointly met despite the tension between them. For example, we generally favour allowing forests to recover much of the land they have lost to the brush, and the brush to recover much of the land it has lost to the desert. The assumption here is that the quantity of life would thereby be increased and its quality improved. But elimination of a unique desert ecosystem is not to be favoured even if this were replaced by one that had a larger quantity of high grades of life.

There will be times when conflicts will arise between our concern to maintain the plurality of human cultures and our concern for endangered species. A recent instance occurred when the United States government agreed to much larger quotas of whale kills for other countries in order to allow an Eskimo tribe to maintain its distinctive traditions. We suspect that too much was sacrificed here for too small a gain. If the Eskimos could have been allowed their hunt without raising the quotas of other nations we would have supported the decision. There are no absolutes here. There is the general principle – so to act as to maximise value in general which includes the value of the non-human world.

Conclusions

Despite the restricted role that ethical principles and arguments play in shaping behaviour, they remain important. It is unwise for those who are critical of inherited ethical and legal systems simply to abandon the fields of ethics and law to anthropocentrism. The need is to argue explicitly against such anthropocentrism and to take up the task of formulating principles which do not depend upon it.

The rejection of anthropocentrism entails the rejection of absolutist arguments for human rights. Human beings, like all other living things, are both ends and means. But this does not mean that we should be less concerned than in the past with human rights. It means only that these should be worked out without the appeal to absolutes which has led to distortion and inconsistencies in the past.

The rejection of anthropocentrism also entails that animals should be taken seriously as ends as well as means. This does not mean that all animals should be accorded the same rights as human beings. It does mean that the tasks of identifying appropriate rights for different types of animals lies before us. This requires that we consider carefully what characteristics of living things give them particular claims upon us and that we make some rough estimate as to which species have the characteristics in question. Through such analyses it is logical to reject the extension of the right to life to all individual animals while holding that some animals do have the characteristics which warrant this right.

But discussion of individual rights can obscure the fact that the more important issues about the human treatment of other animals cannot be dealt with at that level. Even if, unlikely as it is, such principles were followed, tens of thousands of species could be wiped out within the next few decades by continuing the ruthless destruction of their habitat in tropical rain forests. There is need on a global basis to come to a new vision of the importance of preserving habitat for wild animals. A start has been made by the International Union for the Conservation of Nature in its call for a world conservation strategy (IUCN, 1980). In general the need is to display how the human future and the future of other species can mutually support one another; for the ethical call upon human beings to sacrifice themselves for other species is unlikely by itself to stop present destructive practices. But it should also be understood that a responsible ethics does call for compromise between human interests and those of other animals.

6

Faith in life

> Living without zest is sad, but to do so forever . . .
> Charles Hartshorne (1962, p. 261)

> The idea that worship is love with the whole of one's being is correlated, in many high religions, with the idea that what we thus should love is itself also love, the divine love for all creatures, and for God himself as including all. And this in my opinion is not simply a pretty sentiment but is, in cold logic, the most rational way to view the matter.
> Charles Hartshorne (1967, p. 12)

> He who is grasped by the one thing that is needed has the many things under his feet. They concern him but not ultimately, and when he loses them he does not lose the one thing he needs that cannot be taken from him.
> Paul Tillich (1955, p. 160)

Some of the rational ethical principles which follow from an understanding of how human beings are related to other living things were presented in Chapter 5. But thought at this rational ethical level does not capture the full meaning of life. Human beings are more deeply moved by the way they experience their world than by the claims ethics makes on them. Historically it is true that in most societies ethics has been grounded in religion, be it Judaism, Christianity, Islam, Taoism, Buddhism or some other faith. Religion in its turn has expressed for its believers the way they believe things really are. The reflections on life in preceding chapters lead to a new view of how things really are, that is, to a new cosmology, and as Whitehead said (1926*b*, p. 141), 'whatever suggests a cosmology suggests a religion'. It is well to attend to the religion which flows from the ecological model of living things.

The neurophysiologist R.W. Sperry (1977, p. 692) said about the new cosmology he saw emerging from modern science, 'Man's creator becomes the vast interdependent and interwoven matrix of all evolving nature.' Another thoughtful neurophysiologist, J. Z. Young (1978, p. 259), calling for 'a religion of life', wrote:

> We can each make the effort to see ourselves more or less clearly as individuals and as part of human life and of all life. We all have our awareness of what the theologian Tillich called 'the depth of your being, of your ultimate concern, of what you take seriously without

any reservation' (Tillich, 1949). This is what I mean by the experience of the central part of life, which is perhaps as close as we can get to an understanding of God. The knowledge we have allows us to understand the aims of our lives and yet recognize our weaknesses and imperfections. It suggests that we can continue to try to improve on them and gives us the faith to do so.

Most of our concerns, even those we take to be ultimate concerns for ourselves, are but secondary concerns. They do not ground and fulfil life. Only that which grounds and fulfils life is of ultimate concern in the sense in which Tillich (1955, p. 152) used this phrase. We go along with Tillich when he says that the one appropriate response to our 'ultimate concern' is 'infinite passion'. We know what is of 'ultimate concern' by the passionate response it arouses in us.

We believe with Young that the new cosmology calls for a 'religion of life'. But we are impressed with the difficulty of stating rightly what that means. As Medawar pointed out, 'life' is an abstract noun; yet an abstraction cannot be our ultimate concern. If we use 'life' to refer simply to the sum total of living things, that can hardly be the creator to which Sperry refers. Young comes closer when he speaks of 'the central part of life', but of course that remains vague. In some ways a religion of life must distinguish the central part of life, the principle of life, the ground of life, or life itself from every particular form and manifestation of life and also from the sum total of individual living things. In doing so it must be guided by what we know of the whole course of life on our planet. To distinguish 'life itself' in this sense from life as an abstract noun or a collective reference to all living things and to indicate that the word is the central religious symbol, we will use a capital L.

The vision of Life in this book is very much dependent on Alfred North Whitehead, although he does not use quite the same language. Those who would like a more rigorous philosophical treatment of what is here called Life are referred to his writing, especially to *Process and Reality* (Whitehead, 1978).

This chapter begins with an explanation of the distinction between Life and living things in terms of Henry Nelson Wieman's (1946) analogous distinction between creative good and created goods. This shows the limitations of the ethical life of rational decisions in terms of presently apprehended principles and the importance of the life of trust.

The second section gives some concreteness to what has been said by presenting the story of a student who does trust Life and is transformed by it.

The third section works out in more discursive ways what it means to trust Life in relation to such matters as novelty, habit, the life of the body, nature and politics.

But this still leaves open the question, what is Life? In the fourth section Life is described as a cosmic power working everywhere that conditions will permit for the enrichment of experience.

But this same Life brings with it evil. The fifth section considers this evil and how human beings in East and West have responded to it. To many evil seems to have the last word.

The sixth section deals explicitly with the question of redemption, which is also the question of God. Whitehead's reflection about the full dimensions of God gives us hope that evil does not, after all, have the last word. This understanding of Life as God, while not a repetition of Biblical conceptuality, is a faithful witness to the God who in the Bible also is known as Life.

Created goods and creative good

All the worthwhile activities of civilisation are worthwhile because they contribute to richness of experience – nations, religious institutions, educational institutions, scientific institutions and other products of cultural evolution. Like our best ideals for the future, they are all created goods, as indeed are the people who are moulded by the cultural environment. Most of us give our loyalties in large part to created goods. But to give oneself to created goods, be they religious institutions or nation states or whatever, is to place ourselves in competition with those who give themselves to other such created goods. Too often efforts cancel each other out, or even lead to the terrible devastation of war.

There is a still more fundamental reason why it is insufficient to give one's loyalty to created goods. Wieman (1946) pointed this out in his distinction between created goods and the creative good. Created goods have been brought into being by the creative good. The creative process that brought these goods into being in the past is capable of bringing new goods into being in the future. But when we devote ourselves to preserving and enhancing the goods that already exist, we often block the emergence of new goods. What these goods will be cannot be known in advance. For example, any existent knowledge is a created good. We can bring the knowledge that exists for us into interaction with that of others. If there is genuine interaction, the knowledge that results will transcend the knowledge with which we began. We may trust this process to produce expanded or deepened knowledge. But we cannot know in advance what this knowledge will be.

So Wieman calls us to serve the creative good itself instead of the particular goods it has produced. But we cannot intelligently serve the creative good unless we can identify it. Wieman engages in a serious inquiry so as to describe this good for us. He limits his inquiry to the source of *human* good. Human good, he finds, grows in and through the interaction among people. To serve the source of human good, which is the creative good, is to promote and facilitate authentic human interaction. We can trust such interaction to bring new goods into being. This does not as yet identify the creative good, but it does identify how creative good can do its work in the world.

To trust the process of creative interchange is different from the ordinary ethical stance. Those who live in an ethical way are governed by their judgment of what is of greatest value or most worthwhile. They then seek ways of realising these goods. In the process they share in creating new goods. But they fail to realise that their own judgments of value are themselves created goods that need to be transformed continuously by the creative good. So these judgments become fixed and competitive with the judgments of others. Those who realise that their own 'righteousness is but as filthy rags' (Isaiah 64:6) are never satisfied with the good as they experience it but hope for its continual transformation by the creative good.

Wieman's restriction of his attention to the way creative process works at the human level was unfortunate. It tends to encourage the separation of the human sphere from the remainder of the world that is so characteristic of Western ethics and religion and that still makes many blind to the seriousness of the ecological crisis. More needs to be said about other entities besides the human so that the creative good may be identified wherever it does its work. But Wieman's basic proposition holds true. The style of life most appropriate to an understanding of life in terms of the ecological model is one of trust and service. The formulation of principles and ethical guide-lines has its place, as affirmed in Chapter 5, but this should be a subordinate place. Principles and guidelines too must be subject to transformation in the continuing process.

Created goods are living things with all their feelings, thoughts and valuings, and all the artifacts, communities and institutions they have created. Creative good is identified with Life. It is Life that is to be trusted and served. This is the subject of the remainder of this chapter.

A man of faith

What can it mean to trust Life as the creative good? Where is this Life to be found? How does trusting Life differ from other ways of organising human experience? To approach the answers to these questions consider the following example. A brilliant young graduate student had almost completed his doctoral thesis after four years of intensive study and research. During this time he had devoted his efforts very largely to exploring the nature of the nervous system and particularly the brain in higher mammals. He was working with one of the most noted neurophysiologists in the country. His study led him into the physics and chemistry of nerve impulses. He was fascinated. A whole new world of biology was opening up before his eyes. Furthermore, his studies had a significance that extended far beyond the nervous system. He was prying into the nature of life itself. He had as yet no clear picture of how all this new knowledge could contribute to the prevention or cure of diseases of the nervous system. That worried him slightly. Nor could he see how it would help to solve the problem that really set him on the path of studying the nervous system in the first place, that of the nature of mind and consciousness. That interest sprang out of an earlier deep religious conviction that profound human experience was as real as the brain in which it seemed to be registered. He wanted to know the connection between the two. But that problem too tended to fade into the background as he became more and more fascinated with the intricacies of the molecular reactions in the nervous system. Moreover, his supervisor was little interested in either curing diseases or understanding what mind and consciousness were. He persuaded the student to get on with the job in hand. After all, he was told, science produces knowledge which is of value for its own sake, whether or not it leads to the solution of practical problems.

While he was still writing his thesis an opportunity came for him to present his work before an international conference of neurophysiologists in the capital city of a large developing country. There for the first time in his life he was confronted with abject poverty, starvation, political oppression and torture – the whole works, as he described it. And all this alongside Western style affluence. He was shattered. What was he to do? The following possibilities occurred to him. He could return to his laboratory in the United States, complete his thesis and continue doing much the same experimental work. That had its attractions; the possibility of fame and the congenial environment of a university life continued into the indefinite future. Moreover, was it not just a bit irrational to become so disturbed by his new experience that he should change the orientation of his life? Most of his colleagues would support this first approach. A second possibility was that he could make his new found concern for the poor and oppressed a major concern that would change the whole direction of his life. To give up neurophysiology and join the Peace Corps would be a head-on attack. That would be to reject totally the career for which he had already spent so much of his life preparing.

A third possibility was to accept the new insight and take it seriously without rejecting all that he had come to value in a scientific career. He thought a great deal about this third possibility. In so doing he came to realise that the causes of poverty and oppression lay not just in the exercise of the greed and power of a few people over the many but also in the social, economic and political structures that make that possible. He came to see that the science and technology in which he had immersed himself were not themselves immune. The university in which he studied was largely financed by huge corporations which had first access to any discoveries. Some of these were trans-national drug houses that operated in the developing world. They produced tranquillisers and sedatives mostly for the rich. The motivation seemed not to be the relief of human suffering but the making of a profit. Sometimes when a drug was banned on the US market it continued to be marketed in developing countries. Our student's involvement in science no longer seemed to be as pure as he once thought.

His growing concern for a world outside the laboratory led him to rapidly complete his thesis. He got a post-doctoral grant to continue his studies on the nervous system. His university authorities agreed to his request that he spend half his time only on research in neurophysiology. The other half of his time he intended to spend on learning about the wider world in which science and technology operated. This meant studying in a neighbouring university where he could learn more about economics, politics and also philosophy and theology. He was asking more deeply than ever before what life is really for. He was not at all sure where his career now lay. He did know that he had need of more understanding than neurophysiology alone could give him. All this of course reduced his chances of an early job in his chosen field. Others would gain a march on him while he broadened his understanding of life. That was a real sacrifice but he was willing to pay it. He trusted his intuitions that life was more than professional expertise and advancement. That is where he is now.

In this example the contrast of the new insight with the old understanding was the occasion for the widening and deepening of our student's thought. A way of thinking emerged which assimilated the new through the transformation, and not the rejection, of the old. It is this third possibility that appropriately embodies the working of Life. To trust Life is to allow the challenging and threatening elements in our world to share in constituting our experience. It is to believe that they can enter into a creative interchange with what our past experience brings into the situation. It is to trust that the outcome of allowing the tension of the old and the new to be felt can be a creative synthesis which cannot be predetermined or planned.

Such trust is risky. The student of our example may not be able to continue his experiments with a clear conscience. If he shares his new understanding with colleagues, he may lose status in their eyes. If his new values are markedly different from his old ones, and if he acts upon them, he may suffer professionally. Trust in Life cannot mean that one trusts Life to support one's present projects and guarantee one's success! If that is one's aim, one is well-advised to try to control the process of life by screening out all threatening data. This is a widespread human strategy. But it leads toward stagnation and death. It tends to make successive experiences as much as possible repetitions of earlier ones. Such repetitions lose zest and intensity. By minimising the new they weaken the power of Life to work against entropy.

Suppose now that our student has allowed Life to work its creative transformation within, and that he has adjusted his professional career to his new understanding. Let us suppose that in this process he has been very much alive and that he looks back on his previous life as by contrast dull and misdirected. Now a new threat to Life emerges. His new understanding is for him of supreme importance. He is inclined to defend it against all criticism and to try to teach it to others. In short he gives himself to a particular created good. He tends again to screen out ideas or feelings that would threaten it. And though his sharing of his new understanding with others often functions as an occasion for new creative synthesis in them, his aim is rather to convert them to his view than to stimulate their independent growth. In short, the created good which is his new understanding becomes the object of his loyalty. If so, the new working of the creative good is blocked.

But this is not the only possibility. He may instead recognise that his initial mistake was not so much that he had a limited vision but that he was not open to that which might transform this vision. He will then realise that his new views, while clearly an advance over the old, cannot be assumed to be the final truth. He will share them with others who may be stimulated by them to re-think their own, but he will do so in the service of Life rather than as a form of indoctrination. And he will actively seek to be open to new challenges that can lead him to further creative transformation.

Trusting Life

The story of the student illustrates what it means to trust Life and to be transformed by it. It is important to emphasise that any story in which genuine growth occurs, any story in which people are adventurously alive, would serve the purpose. An alcoholic who joins Alcoholics Anonymous and through the interaction with other members sobers up is an example. Such a person is also in danger of treating the principles and achievements of AA as a final good and thus being closed to the future work of Life, but there is no necessity of such an outcome. Again, consider the case of a woman scientist who turns down a lucrative job in a munitions works because of a realisation of its social meaning. Such a person, too, having been creatively transformed in the process of this decision is in danger of becoming self-righteous or complacent. But she may instead continue to grow in her sensitivities and understanding through openness to new challenges and opportunities.

Indeed, what is highlighted in these dramatic examples is a pervasive feature of human experience. We are always bringing into the present the continuation of thought, feelings and purposes from our past experience. We are always receiving from and through our bodies manifold stimuli of new feelings and thoughts. The present experience is constituted by selective unification of compatible elements from all these sources. The issue is how much will prove compatible. If the unification is controlled by pre-established purposes, then most of the potential contribution of the body and the larger environment will be screened out. Only what fits our self-image and our personal goals will be admitted into experience. Of course, this is an exaggeration. We can never succeed in such perfect insulation of ourselves from challenges arising in the environment, though many of us do manage to order our lives in such a way that really new thoughts, major breakthroughs in understanding and sensitivity, or new ranges of perception hardly ever occur. But in every moment there is also a new possibility arising out of the totality of what is given for that moment. That possibility cannot be the complete inclusion of all that is given, for that is simply impossible. But it can introduce something new which allows for the inclusion of elements in the given which, apart from this novelty, are incompatible. For example, although joy and pain tend to exclude each other, they need not. In a more self-conscious experience they may coexist in such ways as to enrich the whole. This novel, unforeseeable and uncontrollable possibility is the working of Life that overcomes the force of entropy. If we trust Life we become more alive.

In each moment we are ordering our experience with a view not only to its immediate realisation of some value but also with a view to its effect upon the future. Sometimes we order the present with a view to exercising maximum control over the future. We try to determine now what our own future and that of others is to be. We want the future to conform to the values that we now appreciate and approve. We cannot, in fact, control the future in this way. When what we now anticipate as future takes place, it will have something to say about how it deals with its past, including our present. But we can lay a heavy weight upon it. A father cannot by his strong expectations, absolutely determine what profession his son will adopt. But the father may well be able to make a contrary choice painful and difficult. On the other hand, we may in any moment order ourselves toward the future so as to enrich the possibilities for that future and enhance the freedom of choice in later experiences. That is what we do when we trust Life. That is, if we believe that the possibilities that will arise in the future will transcend what we can now envision for it, if we believe that these possibilities are best realised in freedom when the given in that situation is as rich as possible, then we shall so order the present as to enable Life to work its fresh transformations in the future.

It is very important not to confuse the novelty which is the gift of Life with change in general. Novelty is the possibility for a specific kind of change, a type which runs counter to the vast movements of change in the universe. Change in general in the physical world is described as entropy. Life is that by virtue of which, when conditions allow, there are local counter-entropic changes. Furthermore, even within human experience much change is apart from Life. Such change is not neutral to Life; on the contrary, it is counter to Life.

The changes that occur when one resists novelty express in human terms the law of entropy. The range and quality of the content of experience and the level of intensity and richness of experience decline. But the change that occurs when one abandons oneself to varied experiences for the sake of stimulus and excitement also expresses entropy. Instead of cumulative integrations of contrasting elements subjecting themselves to the creative transformation of Life, one has successions of varied, poorly integrated experiences. This way too the range and quality of the content of experience decline along with the level of intensity and richness. Trust in Life is certainly not to trust that everything different is better or that the accumulation of diverse experiences in itself will enliven us.

The emphasis on creative novelty may nevertheless appear to do little justice to habit and discipline. Habit and discipline may either serve Life or oppose it, whereas because of our stress on the difference between trusting Life and living in terms of past patterns, habit and discipline have been cast largely in the negative role. It is time to consider their indispensable service of Life. Consider walking. As an infant learns to walk, Life is powerfully manifest. There are novel emotions, novel control, novel thoughts. The infant is absorbed in the creative transformation that is taking place. This is healthy and good. If one never learns to walk, one's possibilities for life are truncated. But if the processes of walking absorbed one's energies throughout one's life, life would be thwarted. What begins as creative transformation passes over into habits deeply programmed into the body. This frees us for new processes of creative transformation by Life.

An artist learns the rules through an arduous discipline for the sake of achieving artistic freedom. 'Whose service is perfect freedom' says the second collect for peace in the Book of Common Prayer. Freedom is the fruit of disciplined service to a cause. But the discipline which binds us becomes habitual through practice and makes us free. Human existence is impoverished and in chains until most physical actions become habitual by practice. So too with psychological habits. Patterns of courtesy, attentiveness to the feelings of others and sensitivity to the natural environment may all begin as conscious responses to Life and pass over into 'second nature'. As they do so they provide Life with fresh material for its synthesis. Only if we become complacent with our past achievements, turning them into ends instead of means of further ends, is it necessary to renew the polemic against even such laudable habits and disciplines as these. As Whitehead (1911, pp. 41–2) wrote: 'It is a profoundly erroneous truism, repeated by all copybooks and by eminent people when they are making speeches that we should cultivate the habit of thinking of what we are doing. The precise opposite is the case. Civilisation advances by extending the number of operations which we can perform without thinking about them.'

But Whitehead would agree with Medawar that there is another side that is equally to be emphasised. 'Civilization also advances by bringing instinctive activities within the domain of rational thought, by making them reasonable, proper and co-operative. Learning, therefore, is a twofold process: we learn to make the processes of deliberate thought "instinctive" and automatic, and we learn to make automatic and instinctive processes the subject of discriminating thought' (Medawar, 1957, p. 138). It is in this latter process that Life is most fully present.

Trusting Life is not a passive stance. It is not simply letting Life do its own thing within us. Yet it is sharply opposed to our ordinary images of activity as well. It is the renunciation of control. But that renunciation is itself an acting. To open ourselves to others and to allow challenges to enter into our experience is a kind of action. To let our defences go and put ourselves in the position to be transformed is a kind of action. To resist the temptation to make of the new synthesis achieved by Life, an end which limits the further working of Life, is a kind of action. The person who is most alive is not passively waiting for something to happen but alertly participating in that happening.

Life is present in us not only in our human experience but in the cells of our bodies as well. It is Life that has caused the growth from fertilised ovum into our present form. It is Life that has healed our wounds and restored our health after sickness. Of course Life does not prevent injuries and sickness, and even natural deformity. It functions always in specific conditions, and these may greatly restrict what it can do. Controlled human intervention can sometimes improve these conditions and enable Life to attain its normal goals. To trust Life in the body is not passively to allow events to run their course.

An attitude of trust in Life is likely to make a difference in our relation to our bodies. At very deep levels of our Western culture we have developed an estrangement from our bodies and the Life that works in them. We have viewed our bodies as objects to be used and controlled rather than as fully continuous with ourselves. The Cartesian view that human bodies are machines inhabited by souls both expressed and accentuated this tendency. To see the body as a society of living cells with which our human experience has a profound kinship can break us out of this objectifying attitude. We are parts of one society in which each has a role to play for the sake of the others. Life is inhibited in its working in the cells when we are tense and anxious or controlled. When Life cannot fully enliven the cells, its work of enriching our personal experience is also restricted. If we gladly serve the body, allowing it to enjoy the food and exercise and rest it needs, the whole living community of the body will in turn serve us. Joy and vitality can arise out of the body to inform our experience.

If we shift our understanding of the body from a machine we use to a living community with ourselves, our attitudes toward drugs will change also. We will try less to manipulate the events in the body and more to serve them. We will concern ourselves more with preventive medicine and public health than with the cure of diseases. This will involve reduced use of stimulants, depressants and hallucinogens and a heightened concern for the removal of poisons from our air, water and food. We will learn more about ways of enabling the body to heal itself and reduce our use of surgery. In our choice of medicine we will prefer when possible that which strengthens the body's ability to fight harmful bacteria to that which kills both harmful and helpful bacteria (see also Chapter 7).

Trust in Life leads to this shift from control to support in our attitude toward other living things. It is a reason for favouring wilderness over wildlife management wherever the choice is possible. It is a reason for favouring wildlife over the humanly manipulated genetics of domesticated animals and fanciful breeds of pets. It is a reason for preferring that the energies of biologists be directed to an improved understanding of ecology rather than the manipulation of living organisms in laboratories except where this is absolutely necessary.

In the political sphere trust in Life is the advocacy of human freedom. No one can force another human being to be fully open to Life, but on the whole we are most likely to be truly alive when we can participate in the shaping of our own destiny and experience responsibility for one another. Since life is the heightening of the importance of internal relations, the ecological model of life suggests that we truly participate in one another. A political system or an economic system based on sheer individualism is inappropriate to a trust in Life. Freedom is not freedom *from* one another but freedom *for* one another.

The responsible use of freedom may involve making projections of future states of society. Many such projections are made in this book. But to trust Life is to see the dangers involved in every proposal for ordering the future world. Any attempt to achieve a settled, perfected social structure is doomed. Ideals become oppressive if they do not allow for and encourage fresh, imaginative envisioning of a still better future by those for whom we are planning. Trust in Life calls for building a society in which each generation will have the freedom to conduct its own adventure to experiment with the new, to go beyond what we could now hope for it. This is frightening, because a world of change is an unstable world, a world which can destroy itself. But it would be no safer to try to suppress future life by planning the 'Utopia' in which people should live forever. The 'perfect' world would turn out to be hell precisely because it forbade change of its own basic structures. What is truly best for society is not an ideal state to be attained once for all but a process in which Life is freed to work its creative transformation.

To trust Life is not to be assured that nothing can go wrong or even that the entire adventure of Life on this planet will not be wiped out. It *is* to believe that if this happens it will be because we have not been willing to trust Life enough, because we have listened to other voices and marched to other drummers. That, too, is possible. Life does not force itself upon us. It does not prevent us from choosing rigidity, closure and death. It does not prevent us from immersing ourselves in an objective world and even ignoring the subjective inwardness within ourselves. But it does give us an alternative. Because of Life we *can* break from these deep-seated habits that are driving us to the brink of global catastrophe; we *can* choose the risk of new directions in preference to the near certainty of destruction that lies ahead on our present course. Whether we *will* we do not know.

Life as cosmic power

But what is Life? Trust in Life is trust in the emergent possibilities for creative transformation in each new situation. It is trust that such possibilities will arise and that their actualisation will lead to a richer future than will the sheer continuation of present intentions. Is Life then simply a name for the sum total of these possibilities or for the abstract features they have in common?

That does not seem adequate. Yet we noted before that for the biologist life is indeed an abstract noun denoting whatever it is that living things have in common. Biologists rightly insist that life is not another thing alongside the entities that make up living things. Instead it is the effectiveness of unrealised possibilities in shaping the actualisation of living things. It is a transcending of what has been by virtue of the relevance of what is not yet actual. Where the present is simply the product of the past and does not create something new, there entropy reigns undisturbed. But where, in the present, the past is turned into ingredients of a new order, there Life is present.

Life is known only on our own planet, but it can be assumed that where conditions allow in other planets circling other stars in this or other galaxies, Life has worked its magic. Having observed Life here one can assume that it is a cosmic principle. There is that in nature which actualises creative novelty whenever it can. This is not the fundamental energy of the universe as such. It is instead an aim at the realisation of novel forms and richness of experience. It is that by virtue of which Waddington (1960, p. 126) could call richness of experience 'the anagenesis of evolution'. The process of the world does not aim at some remote Omega point. Its teleology is simply the creation of values moment by moment.

Life as the cosmic principle that works for higher order in the midst of entropy is enormously powerful. When conditions allowed, it brought forth living forms upon this planet and multiplied them until it transformed the whole surface of the earth. It brought forth human intelligence and through that transformed the earth once more. Probably it has worked similar miracles elsewhere in the universe, perhaps at many places. Wherever it works it can be trusted to transform the inanimate into living matter and to quicken the spirit.

The power of Life is not limited to clearly living things. Indeed, there is no definite boundary between living and non-living things, and we may think of Life as exerting its gentle pressure everywhere, encouraging each thing to become something more than it is. But in vast areas of the universe such urging has little effect. Only where very special conditions exist, it seems, can life accomplish such spectacular transformations as have come about on the earth.

It is now widely accepted by physicists that the universe originated in a 'big bang' that occurred simultaneously 'everywhere' between ten and twenty thousand million years ago. The essential nature and development of our universe was determined during the first few minutes of that cataclysmic event. According to Weinberg (1978, p. 5) for about one-thousandth of a second the temperature was so hot (one hundred thousand million degrees Centigrade) that none of the components of matter, atoms or even nuclei of atoms could have held together. The matter rushing apart in the explosion consisted of various types of elementary particles. At the end of the first three minutes it was cool enough for the elementary particles to form nuclei, starting off with nuclei of helium and heavy hydrogen. Only thousands of years later was it cool enough for electrons to join these nuclei to form atoms of helium and hydrogen. Under the influence of gravitation the resulting gas formed clumps which condensed to form the galaxies and stars. However, the ingredients with which the stars began their existence were those formed within the first three minutes. The particular forces between particles in those three minutes, nay probably even in the first fraction of a second, were critical for what happened subsequently.

A slightly different history in those micro-seconds could have resulted in the universe being all helium with no hydrogen. Without hydrogen there would have subsequently been no heavy elements which were formed by the fusion of hydrogen nuclei. These heavy elements, such as carbon and iron, are essential for life. It is only under the intense gravitational attraction which led to supernovae explosions that these heavy elements formed and scattered throughout the universe. So a chain of events commencing (for this part of the story) with a very special sort of explosion of the size of the universe led to a universe of hydrogen. The explosion of supernovae led to heavy elements being formed that were essential if life were to evolve in this universe. In this sense we were conceived in the white hot temperature of the stars (Weinberg, 1978)!

Some people want to see in this fantastic sequence of events the hand of an all powerful supernatural designer just as the pre-Darwinists wanted to see in the 'design' of nature the clockwork God of a clockwork universe. Our view of Life leads us to different conclusions. Life had to await the appropriate circumstances set by a very particular history of matter before living organisms became possible. There may have been countless universes before this one, or even contemporaneous with this one, which ended up as 'dead' universes, just as there were countless experiments in living organisms of different forms which have become extinct.

There is a marvellous order in the universe. But it is not the order brought about by the power that a dictator wields. Life exerts its influence to bring forth that which is possible according to the circumstance. In the fullness of time the possible became concretely realised.

The evolution of the cosmos is the evolution of order at successive levels of the creation. It has the appearance of a fighting struggle for integration and order against influences that tend to disorder and chaos. Wieman (1929, p. 213 *et seq.*) had an image of the universe far ahead of his time when he conceived the evolution of the cosmos as order and stability achieved at successive levels. There was a time, perhaps 20 billion years ago, when the association of elementary particles into atoms had achieved no stability. That epoch is now past. The association of electrons and protons and the like into atoms has achieved a marvellous stability and adequacy of organisation such as to sustain their integrity through the shocks and strains of cosmic change. That frontier of the organisation of elementary particles into atoms has now passed.

The more complex and subtle association of atoms and molecules into cells in living organisms is not so firmly established. Here misfits occur. Nevertheless, there is an order and stability in living cells that has enabled them to endure the many shocks of chance and circumstance since their first appearance in some tropical sea some three to four billion years ago. The frontier of life at the level of the cell is now past. The organisation of cells into complex living organisms may have taken millions of years to achieve. There is a stability there, yet not as secure as the association of atoms in molecules and elementary particles that constitute atoms, but still a sufficient basis for further advance.

Now Life works for a new organisation of conscious individuals associated in human societies. Here at the human level we find mutual support and co-operation most perilously inadequate to the interdependence of people upon one another and upon the rest of nature. Here is where further integration is most urgently needed. Here is where the achieved integration is most incomplete and inadequate. Why, asked Wieman, should chaos and distintegration be so widespread and perilous in human life? And he answers that this is because human life is now the fighting frontier of the progressive integration of the universe, so far as we know. There may be other frontiers far in advance of human society in other parts of the cosmos, but we know nothing of them. So far as our knowledge reaches, human society is the utmost cosmic venture toward creation of richer integrations. Here is the great upreach toward values higher than any which have ever visited the realm of existence. Here the existing universe is groping out into that vast realm of possibility of as yet unrealised value. Here the cosmic venture is under way. Here is where heaven and hell shimmer in a mirage of possibility.

There was a time when the integration of electrons into atoms was the fighting frontier of progressive integration in the universe. There was a time, later, when the integration of atoms into cells and still later of cells into complex organisms was the outpost of organisation and increasing value. Those frontiers are long past. The storm now rages about that kind of association called human society. Religion of the noblest kind, says Wieman, is the human recognition of this creative cosmic struggle and our personal allegiance to the process of progressive integration. Therefore the religious person must be disciplined and equipped in body and mind for the task, he or she must have more calmness and mastery in the midst of peril and turmoil, more sensitivity and deeper insight into the bonds of interdependence that hold people together in rich community, a more passionate and richly integrating life purpose which can transmute the common things of daily experience. All this we must have if we are to be the shock troops of the integrating process of the universe. All this we can have, for Life is the source of these gifts. To fail to respond to the call of Life is not just a personal failure. It is cosmic tragedy.

Somewhere in the universe there may be wholly different conditions that also make entirely different forms of life possible, perhaps even in empty space, or in dimensions different from our own. Of such possibilities nothing is known. But if omnipotence means ability to bring about new things without relevance to the conditions, then certainly Life is not omnipotent. Some day this planet will no longer provide the conditions that make living things possible. Life will not prevent that eventuality.

Still, Life is the ideal power. It is the only power that creates value and freedom, and it works everywhere and always to this end. Indeed every power other than Life erodes order. Only Life creates it. If by the world we mean the encompassing meaningful order, then Life is the creator of the world. Life creates by bringing order out of chaos. It does not create the chaos, for the chaos is uncreated. Life creates by bringing new order out of old, which has become repressive or restrictive. Life is creator.

Life and evil

The affirmation that Life is creator brings with it the recognition that Life is the reason for most of what we call evil. If this were a lifeless universe, there would be no pain, no sickness, no suffering, no sin, no injustice, no oppression, no death. It is Life that has introduced all of this into the universe. There are some who have believed the price for life was too high. For them Life itself is the Devil. Without Life there would have been a kind of peace.

Most of us, however, even those who speak most eloquently of the horror of these evils, do not go so far. We share Life's prizing of living. Our complaint is that we cannot have the joy of life without the mutual slaughter in which living things engage. We are appalled that we cannot have human freedom without slavery and genocide. We rightly refuse to believe that in perpetuating such evils we are fulfilling the purposes of Life. For the sake of Life we reject much that is being done by human beings. But much of what we view as evil is inseparable from life. Living things compete for the resources of the earth and most lose. We may wonder whether Life could have adopted another strategy for evolution. Whatever abstract possibilities there may be, Life accepted the price of enormous waste. Further, suffering and death are the lot of all, and this includes human beings. If we affirm life, it must be a life that includes a continual process of dying and eventuates in death. Should biologists learn how to stop the ageing or dying process, and should that knowledge be implemented, it would only serve to hasten the death of the entire planet. In this sense evil is an inherent part of good, at least in this world and in this life. Even if we could imagine some future life on this planet free from all evil, there would remain the evil that the conditions for the existence of that perfect world would eventually disappear. The whole adventure of Life on this planet must come to an end.

The awareness of the inextricable mixture of evil with good in all of life has given rise to the great religious traditions of East and West. In India the major religious genius expended itself in finding ways of gaining release from this ambiguous existence. Some Hindu sages identified the ultimate dynamic power in the universe as Brahman.

They sought release from the suffering of existence in the recognition that one's own deepest self or Atman is nothing other than this same power. Gautama Buddha, on the other hand, believed that such metaphysical speculations were fruitless. He undertook a psychological analysis of the cause of suffering, identifying this ultimately with clinging or desiring. He showed that if we extinguish all desire and cease to cling to anything, we will find release from suffering. Later Buddhists encouraged this cessation of all desire by arguing, against Brahmanism, that there is no Brahman or Atman, only Emptiness. The realisation of this Emptiness was for them the avenue to freedom from suffering.

In Israel the response was quite different. Much of the Old Testament witnesses to an affirmation of life which counted long life the chief blessing but accepted death as its end. Dissatisfaction with particular evils was responded to by the promise of a better future. The expectation might be of a Promised Land or a return from exile. But as time passed, the sense of particular evils which might be overcome in history gave way to a recognition of how deeply evil is built into our historical existence. Hope then took new forms: an apocalyptic Kingdom, the resurrection of the dead, personal immortality. In diverse forms, the expectation that somehow, somewhere there can be an unambiguous life, in which evil will be no more, has coloured the whole history of the West.

In modern times many Western humanists have supposed that they did not need either the Eastern release from suffering or the Western hope of an unambiguous good. They have believed that people can live meaningfully with hope for the attainment of immediate values and relative justice without the illusions of religion. Indeed, they have held that only when energy is withdrawn from otherworldly goals and expectations will humanity take its destiny into its own hands and bring about a better world. Some Christians and Jews have joined with non-traditional humanists in this view.

The results have been ambiguous. Some have devoted themselves through a lifetime to public service without the apparent need to believe that any ultimate purpose is thereby served. Bertrand Russell is a striking example. But others have grown cynical. The history of culture becomes a fatal sequence of doors opening in Bluebeard's castle (Steiner, 1971). The withdrawal of confidence in an ultimate meaning has not led people to throw their energies into historical transformation. Instead it has led them to view such efforts as pointless. There are several astute observers who believe that without some grounding of meaning beyond the flux of events, with their inevitable ending in death, human beings cannot find the zest to motivate needed actions. The conclusion would be that trust in Life cannot be sustained unless one can trust Life to achieve some end other than extinction.

Life as God

Life as the central religious symbol is God. Hence this chapter is about faith in God. Yet the avoidance thus far of the word God has not been simply a game. That word carries so heavy a freight of associations that, however one explains one's meaning, the reader is almost certain to read other meanings into it. Perhaps some who do not believe in what they understand by God may be willing to consider the appropriateness of faith in Life. And perhaps others, who are quick to say that what is not absolutely controlling cannot be God, may be willing to recognise that it may yet be Life.

Nevertheless, the time has come to speak directly of God. We are asking the question of whether we can trust Life to achieve some end other than extinction. That is, we are asking the question of redemption. To speak of redemption is to speak of God. The question of God has always been bound up with the experience of evil and the question whether evil has the last word. If there is One who overcomes evil, then there is One whom we can trust and worship unreservedly. That One is God.

In very different language the exposition of Life in this chapter follows Whitehead's account of God. Most of it is based on Whitehead's doctrine of the Primordial Nature of God whose envisagement of all possibilities in their relevance for each situation in the world is the source of ordered novelty and novel order. In the last part of *Process and Reality*, Whitehead (1978) turned to the question of redemption in the light of the fact that in the world all things decay and perish. He wrote there of the Consequent Nature of God, the aspect of God in which what perishes in the world remains in God.

The Whiteheadian idea of God is appropriately called Life not only because the immanence of God in the world is the life-giving principle but also because the life-giving principle is itself alive. A lifeless principle could not ground or explain the urge to aliveness that permeates the universe. Indeed, God is the supreme and perfect exemplification of the ecological model of life. That model has as its distinctive feature the insistence that living things are internally related to their environment, that is, that their being is in part constituted by the environment. In different environments a living thing is different. There is no living thing in itself which, only incidentally, happens to have one environment or another.

Traditional Western theism has thought of God in quite the opposite way. Aristotle paved the way with his doctrine of the Unmoved Mover. For Aristotle, to be affected by other things and in that way to be dependent upon them was a weakness. Such weakness could not be attributed to that final good toward which all things move. Hence God, who moves all things by the divine perfection, is moved by nothing. Aristotle's God is pure activity, and that activity is contemplation, but the only object of that contemplation is God. This completely self-contained and self-sufficient deity seems quite different from Yahweh or the Father of Jesus. The biblical God seems to be bound up with the course of events, punishing and rewarding, promising and fulfilling promises, creating and redeeming and even becoming human. Yet the Aristotelian God fascinated the thinkers of the church, and Christian orthodoxy adopted the ideas of the divine immutability, impassibility and aseity – that is, possessing being in and for itself. Although they did not deny that God knew and loved the world, they did teach that this knowledge and love were timeless and had no real effect upon God. There is no real interaction between God and the world. According to their view, everything in the world is caused by God, but there is no reciprocal effect of the world on God. Some of the reasons for the decline of belief in God in modern times are bound up with features of this orthodox synthesis that are clearly not found in the Bible.

We agree with Whitehead that this synthesis is profoundly perverse. It pictures the living God of the Bible as an object having only external relations. True perfection consists not in excluding everything but in including everything. The Primordial Nature of God is the inclusion within God of the entire sphere of possibility, realised and unrealised in the universe. God so orders possibilities that they may be realised in the world, each in its due season. It is this ordering that establishes the effective relevance of novel possibilities and thus makes life possible in the world. But like all living things, God not only acts on others, but also takes account of others in the divine self-constitution. The Consequent Nature is God's perfect responsiveness to the joys and suffering of the world. God is not the world, and the world is not God.

But God includes the world, and the world includes God. God perfects the world and the world perfects God. There is no world apart from God, and there is no God apart from some world. Of course there are differences. Whereas no world can exist without God, God can exist without *this* world. Not only our planet but our whole universe may disappear and be superseded by something else, and God will continue. But since God, like all living things, only perfectly, embodies the principle of internal relations, God's life depends on there being some world to include.

But some readers may object that, even if all this can be believed, it is improper to call Life 'God'. They may insist that God must be conceived as a purposive, loving being, and that we have said nothing about Life that warrants such attributions. This is an important challenge which warrants a response with regard to Life as purposeful and Life as loving.

Life is purposeful. Indeed, it is defined by its purpose. It is not the sheer blind 'ongoingness' of things, but the cosmic aim for value. Its purpose controls all its activity and through it the whole course of living things on our planet. Yet this could easily be misunderstood. Life does not aim specifically at the creation of human beings. It has no one goal for the course of evolution on our planet. There is no plan for the future written into Life which it is our task to discern. Life aims at the realisation of value, that is rich experience or aliveness. In some measure it realises value in every living thing. But it aims beyond the realisation of trivial value for the realisation of richer experience. To this end it produces creatures in profligate abundance so that through the processes of selection some will emerge with greater intelligence and capacity for feeling. Life has achieved rich value in dolphins as well as in human beings. We cannot guess the forms it may have achieved on other worlds.

Life is not only purposeful in itself, but it is the source of all the derivative purposes in living things. Purpose involves a distinction between what is and what might be and an appetition for some form of what might be. It is Life that introduces into the midst of the sheer givenness of the physical world the attractive vision of unrealised possibility. Animal purposes, especially in the human species, have become the main agency of evolutionary development. But in the human case there is such freedom to diverge from the purposes of Life, that this supreme achievement of Life on this planet threatens to become the great destroyer.

Whereas there is no set goal for evolution, Life provides a specific purpose for each entity in each moment. It is the aim to achieve some optimum value in that instance. This is not an abstract aim, but quite specific in each case. In this sense Life is very personal. The gift of Life to each living thing is tailored to its particular needs and possibilities. Life is the supreme instance of love. Yet that, too, can be misunderstood. Precisely because every living thing is dealt with lovingly in its concrete particularity, some features of what we humans mean by love cannot apply. We ask for a love that is concerned for us not merely in our concrete particularity but as if we were somehow special, a love that favours us and our ends against others. Life favours all living things, and precisely for that reason does not take sides in our inevitably competitive existence. Life favours both the fox and the hare, supporting the success of both hunter and hunted. Life favours all the runners in the race, and hence has no favourites. This love that is the most intimate and particular of all loves, is at the same time awesomely disinterested and, in that sense, impersonal. For Jesus, too, the perfect love of the personal God is manifest in this: that the sun shines on just and unjust alike.

In many animals, Life seems to subordinate the individual to the species. The aims of the individual blindly serve the preservation and improvement of the species. Present attainment is sacrificed to the future. In human beings, however, the individual comes much more to the fore. Life's aims for us envisage the good of human societies, the species, and even the whole community of living things, but they include immediacy of enjoyment and personal greatness in the individual person now as well as in the future. In this sense Life's love for us is personal.

It remains true that the symbol 'Life' names God in a way that does not highlight the fullness of God's personal being or redemptive action. For that we have turned to Whitehead's concept of the Consequent Nature of God. It is only as the Primordial Nature is completed in the Consequent Nature that we can speak of God as conscious. It is only in the Consequent Nature that God knows us, loves us and redeems us in the fullness of our personal existence.

As he meditates upon the apparent final victory of evil, Whitehead writes as follows:

> The ultimate evil in the temporal world is deeper than any specific evil. It lies in the fact that the past fades, that time is 'perpetually perishing'. Objectification involves elimination. The present fact has

> not the past fact with it in any full immediacy. The process of time veils the past below distinctive feeling . . . In the temporal world, it is the empirical fact that process entails loss: the past is present under an abstraction. But there is no reason, of any ultimate metaphysical generality, why this should be the whole story (Whitehead,* 1978, pp. 340/517).

Believing that God, like all actual things, not only acts but is also acted upon, he continues: 'The consequent nature of God is his judgment upon the world. He saves the world as it passes into the immediacy of his own life. It is the judgment of a tenderness which loses nothing that can be saved' (pp. 346/525). And in the Galilean origin of Christianity he finds the same element: 'It dwells upon the tender elements in the world, which slowly and in quietness operate by love; and it finds purpose in the present immediacy of a kingdom not of this world' (pp. 343/520).

The Bible witnesses to a changing understanding of God among the Hebrew and Jewish people. Christian theology developed again quite different doctrines of God. No one today can simply believe in the biblical God. Certainly we do not pretend to do so. That would be to affirm a number of different concepts of God. Merely to repeat biblical ideals in the context of a quite different world view would not be faithful to the spiritual dynamic that is manifest in the Bible. We are more concerned that faith in our time give us the same freedom which, according to Paul, faith brought to believers in his day. Nevertheless, we do care about our continuity with biblical faith. We would like our witness to Life in this book to be an appropriate continuation of that witness. We believe it is.

Although the Bible uses many images of the divine, and we too favour doing so, no image is more central than Life. It is closely bound up with both the 'Spirit' and the 'Word'. The 'Spirit' or 'Breath' of God, in Hebrew understanding, was the Life of God. It is that 'Breath' that is breathed into us so that we became living beings. We put it in another way, saying that it is the immanence of the divine Life within us that makes us alive. The Bible sees the Spirit as the giver of life both in the sense of biological enlivening and in the sense of quickening our human experience. We, too, have insisted on the identity of the Life that does both of these things. It is the divine Life that makes us both

* The two sets of page numbers for Whitehead 1978 refer to revised edition 1978/original edition 1929.

physically and spiritually alive; for no more than Scripture do we think of distinct principles of biological and psychological life. Although the language of Scripture sometimes suggests that the Spirit is something other than God that is sent or given by God, Jews and Christians alike are clear that what is sent or given is in fact God. The Spirit of God is God. God is the Spirit. If it is the Spirit that enlivens us, we may equally well say that it is God who makes us alive. If the Spirit is the true Life within us, then God is that Life.

The second great image of God in the Bible is the Word. God 'speaks' and it is done. We are created by God's Word. That Word calls the prophets and addresses them. It gives shape to the life of the believer. It seems that the Word, like the Spirit, is pictured at times as something separate from God, but this is not so. God comes in the Word. To hear the Word is to experience God's presence. This is made unequivocally clear in the prologue to John's Gospel. There the Word which was together with God from the beginning, and through which everything was created that was created, is God. And it affirms that 'in him was life, and the life was the light of men' (John 1:4, RSV). It is this Word which, in the Johanine account, becomes flesh in the Jesus who affirms that he is the Life, the Life which was in the universe from the beginning.

Conclusions

There is much of the speculative throughout this book. Indeed the effort to free science from speculation simply binds it to the models produced by earlier and outmoded speculation. We saw in the sixties how little power the old paradigms had to shape the thinking of a new generation. We should not be misled by the more quiescent temper of the present youth into supposing that inherited forms of thought have recovered their convincing power.If the scientific and scholarly communities will not engage in responsible speculation, we will continue to find millions attracted to astrology and to both Eastern and Western cults that are cut off from the deeper wisdom of their own traditions. This book is an effort at responsible speculation that could help to shape a new home for both science and religion.

In this chapter that speculation has been carried further than elsewhere. Cosmology and religion require such extensions beyond scientific models. But responsible speculation is not a mere flight of fancy. It requires firm grounding in personal experience and scientific knowledge and theory.

A distinction has been made between living things and Life as that by virtue of which they are alive. That distinction has been explained in terms of another, between created goods and creative good. To be truly alive is to refuse to be bound to any created good, when we allow ourselves to be transformed by new experience and knowledge. This is the old distinction between idolatry and authentic faith. It has not lost its importance.

We have gone on to speak of Life as that which transforms and liberates us. We have judged that this Life is present throughout the universe and works its miracles wherever conditions permit. We have described the nature of its power and its efficacy, reading them off our perceptions of what has actually taken place. We have seen that the same Life which brings created goods such as ourselves into being also inevitably brings conflict, suffering and death in its train. For all our love of Life death seems to have the last word both for us and for all things.

At this point we have arrived where so much of the world's deepest religious thought has begun. What are we to say in the face of suffering and death? That they are inevitable concomitants of life? That is true. But what are we to say to the fact that it seems that it is death and not life that wins in the end not just for individuals but universes also. Shall we give up our faith in Life? Shall we shake our fist into the face of the universe? Shall we resign ourselves to the inevitable? Shall we eat, drink and be merry? Shall we seek release through meditation or mystical disciplines? Shall we cynically detach ourselves as observers of human pretensions? Or shall we, despite everything, trust Life? These are questions which our scientific knowledge helps us but little to answer. They are religious questions. One may refuse to ask them consciously. But at some deep level of the psyche those who see reality lucidly must live by answers, even if they are unconscious ones.

We find ourselves choosing to trust Life. That means not only that we try to open ourselves to what Life can bring us and do with and through us. It means also that we live in the expectation that death's word is not the last. What form Life's victory will take we do not know. The thought of Whitehead is deeply meaningful in this respect. It is the consummate speculation of this chapter and of this book. In the context of his own thought it is a reasonable and warranted speculation, although the warrants cannot be fully reproduced here. But for us, faith in Life means more the confidence that Life is not finally defeated than the belief that we can state accurately the form of its victory.

Faith that Life wins out somehow over death does not entail any complacency about the course of events on our planet. They may well lead to disaster soon. But faith in Life does mean caring deeply about what happens both in our own lifetime and thereafter. The future of life on this planet depends very much on how people in this present generation respond to its crises and challenges. Somehow, somewhere, we trust Life will triumph even if life disappears from this planet. But much, of greatest value, that could have been will then never be. The human calling is to respond to Life here and now so that life on this planet may be liberated from the forces of death that now threaten it. The remaining chapters of this book are reflections that express a response to this call.

7

The biological manipulation of human life

If science and ethics are separated, ethics always appears too late on the scene. Jürgen Moltmann (1979, p. 131)

Biology as a source of understanding nature has been the concern of preceding chapters. In particular, biology has been drawn upon to develop an ecological model of living things. Understanding of life gives a perspective not only on living organisms including the human organism but also on the non-human world. From this understanding we have drawn some ethical implications and have then gone on to speak of a whole life-orientation of faith in life.

This attitude is one of respect for living things and openness to being shaped by the power of Life itself. It is in many ways opposed to the attitude expressed in manipulation. Yet this is no call for passivity. To be open to life is to be attentive to the workings of Life everywhere, but it is also to be active in expressing what Life calls us to be and to do. This certainly includes ordering the environment and organising society. It includes the use of medicines to combat disease and surgery to repair malfunctioning parts of the body. But these manipulative acts are set in a context of respectful attention to the strategies of life and ecological interdependence.

Today advances in molecular biology, genetics, physiology, behavioural biology and bio-medical technology have opened up much more extensive and fundamental ways of manipulating human life. These new possibilities sharpen for us the basic question of our attitude toward manipulation and raise a host of ethical issues which have hitherto hardly existed. This chapter will deal with some of these.

In preparation for thinking about these issues the first section draws together the relevant ethical considerations derived from the preceding chapters of this book. Our attitude is not absolutist. That is, we assert neither that all manipulation is evil nor that there are no problems with manipulation if it is directed to worthy ends. An attempt is made to provide some guidelines which can be applied in the concrete cases considered in subsequent sections of the chapter.

The second section deals with the ethical problems raised by the scarcity of the expensive technology developed by medical research and of the organs the transplanting of which can keep people alive. Where some lives can be saved and others must be sacrificed, how can decisions be made responsibly?

Not only can people be kept alive with advanced medical technology, their experience can also be manipulated through chemical and electrical means. There are nightmare possibilities of control as well as beneficent possibilities of pleasurable experiences and of relief from destructive drives. The third section deals with the ethical issues raised by these new possibilities.

As more is learned about genetic diseases, the question arises persistently about the extent to which we should intervene to prevent the birth of diseased infants. If the human condition can be improved by reducing the incidence of genetic diseases, should a conscious attempt be made to produce a higher quality of human beings by genetic control? The fourth section treats these questions of negative and positive eugenics.

The fifth and final section deals with an area that now looms disturbingly, as well as hopefully, on the horizon – the area of genetic engineering. It has recently become possible to alter the genetic composition of living cells and thereby produce new forms of life. This has applications for good and evil in agriculture and in the treatment of genetic diseases. It poses dangers of vast proportions. Ethical reflection on these problems is still in its infancy.

Throughout these sections the ethical issue is twofold: how to deal with the problems and opportunities that scientific and technological successes have produced and secondly whether the direction which science and technology have been encouraged to go is the right one. Much of it does not reflect an ecological understanding of human existence or of living things in general, and in these cases we call for redirection of the public resources which support and channel the scientific enterprise.

The life ethic

The ecological model of life elaborated in previous chapters provides principles which could be a guide through the ambiguities of the ethical issues involved in the biological manipulation of life. The following aspects of the ecological model of life are relevant to the questions raised by bio-ethics.

Values and rights. All life has value but not all life is of equal value. Value is measured by richness of feeling and capacity for richness of feeling. There is a hierarchy of value from the simplest forms of life through to human beings corresponding to the richness of experience of each form of life. Of course, we can know only our own feelings, but in Chapter 4 we explained our reasons for attributing different levels of feeling all down the line of life. The ethical requirement is that we provide circumstances that promote richness of feeling. However, the ethical problem is not so simple. In addition to the intrinsic value of their experience, all living organisms including humans have instrumental value. This is their value, not to themselves, but to others either of the same species or of different species. These two sorts of value of living things are the basis of respect for life. From them flow certain responsibilities.

The intrinsic value of experience confers rights. Individuals who have the capacity for experience have the right to enjoy such richness of experience as they can. They have the right to have their lives respected and, where possible, preserved. Such rights are not absolute. The only absolute is respect for life itself. In the sphere of rights, there is a hierarchy corresponding to the hierarchy of intrinsic value. Ethical action will take into account this hierarchy and will be at pains to have it recognised and respected.

If we are called, ethically, to promote richness of experience, we are at the same time called to promote the ability to respond freely to situations rather than being passively shaped by them. Based on the ecological model, ethics requires that we respect and encourage the free decisions of those whose richness of experience we would promote.

The riddle of inequality. Within the human species we recognise differences in the capacity for richness of experience and therefore differences in intrinsic value. There is no one absolute value for all human life. The gap between what human beings actually experience and could experience is so great in each individual life that differences in capacity for intrinsic value are of little practical concern. Where these differences do become important for ethical decision is at the beginning and at the end of human life, in the foetal stage and, for some people, during terminal illness. Whilst there is equality in neither our genetic nor our cultural endowment (we are born unequal), nevertheless ethical action is committed to an equality in opportunity for all human beings. Each person who comes into the world should have the maximum opportunity to develop to the full his or her talents and to promote the richest possible experience for all.

From this last principle there follows another. Because we are unequal, justice requires that we share each other's fate. The cost of life, which is so unequally shared, is to be shared by all. We share that cost when we lift from another, to any extent, the oppressive weight of suffering, the intolerable weight of anxiety and the state of being overburdened which is the unfortunate lot of so many.

A distinction can be made between the weight which is overburdening, which we share, and the load which each of us can be expected to carry as being proper to one's capacity and vocation. Whatever the burdens from inheritance or circumstance that oppress us, we carry a responsibility for our decisions and actions. We are never solely victims of circumstance because we are always responsible beings. There is always something we can freely do. We can always take a stand. We cannot simply blame our genes or our parents for what we are. We are what they bequeathed us, and we are more, because the human person can both accept and transcend inheritance. This aspect of human nature, the ability to transcend, becomes an important component in understanding and encouragement in counselling. We wish it were more fully acknowledged. Some of us might learn to grieve less and endeavour more.

Our responsibility for our own decisions is a responsibility to act appropriately for our own interests and to enhance the richness of the life of others. These others are chiefly those near to us who are the ones most affected by our actions. But we cannot draw the line there. Our actions can affect the lives of persons we do not know. Indeed, their effects extend far into the future. We are responsible to generations yet unborn.

This ethical pattern of values, rights, sharing and responsibility is deepened by the ecological understanding of life. We are not merely separate individuals with external effects on one another. We are constituted by our relations with one another. The welfare of one is the welfare of all, and the suffering of one is the suffering of all. Of course, this interconnectedness of our lives is most apparent in our intimate circle of family and friends, and this is important. These intimate communities should be respected as such. We grow and prosper and enjoy rich experience as these communities grow and prosper. But there are no boundaries to these communities. In the words of John Donne: 'Ask not for whom the bell tolls, it tolls for thee.'

The ethics of risk taking. Life is an adventure, and adventure involves risk. The future cannot be guaranteed. Accidents happen and unforeseen circumstances develop. Lack of change may provide a sort of security that some of us find appealing, but such security is at the cost of the loss of further enrichment of our lives. The risk of destructive accidents must be faced. It is incumbent upon those who engineer change that they do what they can to assess the risks involved. Ethics requires that anticipated benefits outweigh the costs and dangers that risk incurs. People are willing to undergo some present suffering and danger because of the promise of greater subsequent enjoyment, as when they undergo painful surgery. But they need to assess both before they proceed. It is ethical to take risks. It is not ethical to ignore risks.

Addressing symptoms or causes? There are often two ways of dealing with such evils as disease. One is to treat the victims and symptoms of evil. The other is to work to eliminate the circumstances which produce the evil. The former way expresses the view of the world as composed of individuals who are only incidentally related to each other. Accordingly it attends to those individuals who are sick or injured. It intervenes to kill bacteria or set broken bones. Even in the treatment of the individual it tends to deal with sub-systems and to neglect their interrelations in the psycho-physical organism. The result is a pattern of manipulation in which great skill has been achieved and through which marvellous results are attained. It would be unreasonable to simply decry this way of dealing with evil or deprecate its successes just as it would be unreasonable to reject the theoretical science that is based on mechanistic models. But it is completely reasonable to point to its limitations and to the misdirection of energy to which it gives rise.

The second way of dealing with such evils as disease can subsume elements of the first way in a more adequate and inclusive approach. This ecological way sees the organism in its wholeness as fundamentally interconnected with its environment. The health of the organism depends on a healthful environment and a healthy life-style that relates the person to the environment. This suggests that primary attention should be directed toward a healthful environment and healthy behaviour for all rather than to the manipulation of limited sub-systems of the few. An ecological approach to human health will be far less manipulative than has been much of past medical practice and it will be more oriented to supporting the strategies of Life. It will also address the ethical issues that have come to the fore through the successful manipulative approach. But whatever guide-lines are developed can only be 'created goods'. They may be tested and applied, but they can only express the limited grasp of truth that is as yet accessible. So these guide-lines should be kept open for modification and transformation in the light of fresh insights, other perspectives and critical analyses.

Justice in the use of scarce medical resources

Today there is a special problem with the development of expensive, elaborate resources for saving lives. By their nature these resources are not available for all people who could benefit. Who then should benefit?

More than 20 000 people are alive in the world today who at any other time would have been dead. They have lost their kidneys and are kept alive with an artificial kidney, the renal dialysis machine. This machine extracts from the blood harmful substances that are normally excreted by the healthy kidney. Until 1960 artificial kidneys were used only in cases of reversible uraemia, that is, to replace defective kidneys for a few days or weeks until the patient's own kidneys could function normally again. There were technical difficulties in using the machine for longer periods. These were overcome, particularly by the invention of the Scribner vascular shunt which establishes a permanent short-circuit between an artery and a vein in the forearm. This device can be safely left in place and connected whenever necessary with the artificial kidney. From then on all that was necessary was an 8–10 hour dialysis session two to three times a week either in hospital or in the patient's home. This new treatment raised a series of ethical, social and economic problems for patient, family and doctor.

In 1960 when the artificial kidney was first used there were only a few such machines in France, the United States and Britain. But there were thousands of potential patients. Who were to be the fortunate ones? Who would be the unfortunate ones who would be allowed to die? Hospitals set up committees to sit in judgment on who should have priority for a scarce resource. They had to make agonising choices. Should the brilliant academic without a family be spared or the labourer breadwinner of a large family? That particular problem fortunately no longer exists. In industrialised countries, at least, there are now enough kidney machines for everyone needing treatment. But there is a second problem. In France the cost of an artificial kidney treatment is about $20 000 per year per patient. Every year 2000 new patients swell the number of those undergoing treatment. Given that the average life expectancy of a patient is ten years, this means that 20 000 people will receive treatment in the ten year period. This involves an expenditure of $4000 million. As a country's health budget cannot be increased indefinitely the $4000 million must be taken from some other part of the health expenditure (Crosnier, 1974). Priorities have to be established and decisions must not be left entirely to the bureaucrats and economists. Difficult ethical decisions have to be made when we have to choose whether to spend money helping this patient rather than that.

In 1959 the first successful kidney transplant was done. Removal of a kidney from a healthy donor involves some risk, about one in 2000 in young healthy people. So it is small but not negligible. The only other source of kidneys is from deceased patients. Obviously there are limits to both sources and now in most Western countries there is a shortage of kidneys for transplants. So the problem of who is to benefit from a short resource arises in another form. Patients who are kept alive with an artificial kidney machine tend to look on a transplant as their great hope. But if the patient is managing to lead a more or less normal life with the aid of a machine there may be little claim on the short resource. So medical criteria will exclude some applicants. Are there any other criteria on which to base a selection? Helmut Thielicke (1970, p. 172) contends that any search for 'objective criteria' for selection of patients for 'scarce life-saving medical resources ... is already a capitulation to the utilitarian point of view which violates man's dignity'. Furthermore, he says that a decision based on any criterion is little different from that arrived at by casting lots. Others too argue that random selection is the only way to determine who shall live and who shall die, when not all can live, on the grounds that it alone provides for equality of opportunity. To do otherwise is to threaten the trust which is inextricably bound up with the respect for human dignity and is an attitude of expectation about another, in this case that of the patient toward the doctor. A threat to this sort of trust was demonstrated in the billboard that was erected after the first heart transplants: 'Drive carefully Christian Barnard is watching you.' Not very different from the lottery system is the 'first come first served' approach which tends to be the way the lottery system would work anyway because applicants make their claims over a period of time and not all at once.

A kidney transplant is considerably less expensive than the use of a kidney machine. Even so, the question arises as to what medical resources of a country should or should not be devoted to transplants. The operation is performed in some hospitals in India, provided that the patient pays. To be able to pay you either have to be well-off or alternatively borrow funds. A person borrows and is helped by friends and charitable appeals, has the operation and pays for it. The patient then finds there are heavy expenses each year for the necessary drugs, if the new kidney is to keep functioning. So again there is an appeal to friends or charitable groups to defray these heavy costs. The person who is well-off, or has friends who are well-off, or who has the ability to collect from various resources, manages to get a new kidney. The poor person without connections goes without. How should the scarce resources of a developing country be used in its medical services? Who is to be helped and who is to be left without major help? The ethical problems are no different in principle from those in a rich country. They are more severe.

There is no rational way of solving these problems except by measuring alternative proposals in terms of the contribution each makes to enriching human experience. We believe that the consultative group in the large American hospital made the correct choice when they gave the one kidney available to a labourer with a family of five children and denied it to a professional man without a family to support. If the professional person's life was critical for the lives of many others, then our choice might have to be different. In making such a choice we have to deliberate in assessing both the intrinsic value and the instrumental value of individual persons. As we have said in Chapter 5, these values differ as between individuals, no matter how disturbing it is to have to acknowledge this fact. We will never be sure whether our judgments in these matters of life and death are correct, but since we must decide, we should bring to bear such wisdom as we can.

The impossibility of giving a satisfying answer to the ethical dilemmas with which the scarcity of expensive medical facilities confronts us forces us to consider the question more fundamentally. Is it possible that the resources that have been invested in developing kidney machines and the ability to transplant kidneys might have been better spent in reducing kidney diseases and the accidents which damage kidneys? This is difficult to say in face of the 20 000 people who are now kept alive by these medical gains. But we must remember the others who have died. In many cases their deaths could have been avoided if the availability of analgesics were controlled, as is the case now in Australia.

Our society is more oriented to heroic efforts to heal the sick, to deal with victims and to treat symptoms rather than to changing the social and other circumstances which result in people becoming victims. Preventive medicine is far less expensive than curative medicine and has a greater effect on the total well-being of the community. For example, the cost of a motor car accident with a casualty in 1970 in Australia has been estimated to be $7000. With 73 000 such accidents each year the total cost to the community becomes astronomical. Compulsory seat-belt legislation in Australia has reduced road deaths by 15 per cent and serious injuries by the same amount. The cost of such an accident preventive is quite tiny (B. S. Hetzel, personal communication).

A high proportion of cancers, some say 70 per cent to 90 per cent, are due to environmental factors related to a person's life-style such as smoking, alcohol consumption, overnutrition and exposure to industrial hazards (Wynder & Gori, 1977; Higginson, 1969, 1979). Such estimates are based on life-style differences in high-cancer incidence countries and low-cancer incidence countries. Japanese in Japan have a low incidence of both bowel cancer and breast cancer. But Japanese who live in the United States succumb to these cancers at the high incidence USA rates rather than the low Japanese rates. Japanese women seldom develop breast cancer. Their chances of doing so rise when they move to the United States. These differences attest to some environmental or life-style difference between Japan and the United States as the cause of the differences. Just which components of the Western life-style are the causes of the higher incidence of these cancers is not known but dietary differences are suspected to be important (Eckholm, 1977, p. 81). Hundreds of millions of dollars are spent each year on cancer treatment and the search for a 'cure'. Very little is spent on cancer prevention (Epstein, 1978). Prevention involves the identification of hazardous life-styles and the education of individuals to change their life-styles. The individual can do much in this way to improve his or her own health. But there are forces in society, such as the advertising of tobacco companies, which tragically work in the opposing direction.

A study of the lives of seven thousand men and women in California has demonstrated the importance of personal habits for good health. Physical well-being and longevity were correlated with adherence to the following seven practices:

sleeping seven to eight hours each night
eating three meals a day at regular times with few in-between snacks
eating breakfast every day
maintaining a desirable body weight
avoiding excessive alcohol consumption
getting regular exercise
not smoking

The health of those who followed all seven practices was consistently about the same as those thirty years younger who followed few or none of these practices (Eckholm, 1977, p. 215). Yet much of our social conditioning and advertising leads people to follow the unhealthy life-styles. Anyone who has been involved in educational programmes to promote a healthy life-style knows how formidable are the social and traditional obstacles to changing ingrained habits, even of the young.

We cannot avoid facing the specific ethical decisions with which the existence of expensive and scarce medical technology confronts us. But we can avoid being drawn into a mentality that is oriented to this mode of dealing with problems. We can insist that our primary task is to prevent disease, not to cure it, and that in the allocation of funds and resources this primacy be kept in mind.

Because so much energy has been directed to cures rather than to prevention, a situation has been produced in which distributive injustice in health services throughout the world is now a vexed problem. A rich country may spend a lot on organ transplants and sophisticated medical techniques, while a poor country still lacks the most basic health care. If confronted directly by the choice of giving a heart transplant to one man or saving 50 children from serious malnutrition, all would choose the latter. But we conceal from ourselves that we are in fact making the opposite choice. A redirection of medical research and international expenditure could eradicate many diseases that now curse hundreds of millions of people, especially in the tropical world. As Callahan says (see Levine, 1977, p. 5), 'It is fundamentally wrong that so much avoidable illness exists in the world.' To combat these diseases what is needed is not so much technological manipulation as attention to the patterns of human and environmental relationship that make for health.

Even within rich countries there continue to be preventable diseases affecting many while the few benefit from the advanced medical technology. Within the United States, for example, diseases that hit the poor the hardest include vitamin deficiency diseases, iron deficiency anaemia, protein deficiency diseases, metabolic toxaemia of late pregnancy, tuberculosis, pneumonia and rheumatic fever (Brewer, 1971). Preventable diseases, especially those that occur in women and children, are much more common among people who are poorly educated, poorly nourished and poorly housed than among the comfortably placed middle and upper classes. The prevalence of chronic bronchitis and coronary heart disease in Australia is about twice as high in low socio-economic groups as compared with high socio-economic groups (Hetzel, 1974, p. 59).

Only a total programme which combines health care with education and the elimination of poverty can overcome the problem of distributive injustice highlighted by the kidney machine and the kidney transplant. Within the healthier society that would result from such a policy the problems of healing the sick would continue, and there would still be instances in which painful choices would have to be made. But these would be set in a context of fundamental justice rather than representing in extreme form the injustice of the entire system.

The manipulation of human experience

Some years ago neurophysiologist José Delgado was led into a bull-ring to face a bull that was there waiting for him. He approached in the traditional fashion of a matador. The bull pawed the dust preparing for attack. But the moment it got close to Delgado it settled down calmly at his feet. Unknown to the assembled crowd, Delgado had previously implanted an electrode in the bull's brain, into a special region which inhibits aggressive activities. Delgado himself carried a radio transmitter which sent messages to the implanted electrode in the bull. All he had to do to calm the bull was to press a button. What he did to the bull has now been done to some humans who are otherwise uncontrollably aggressive. The person who has such an implantation can do his own button pushing as he feels the urge coming over him or he can be linked up with an outside control.

When electrodes are placed in other parts of the brain and then activated, pleasurable sensations can be produced. Delgado (1969, p. 185) commented on a human patient who responded to an implanted electrode by giggling and making funny comments saying the while that she enjoyed the sensation 'very much'. Repetition of the stimuli made the patient more communicative and flirtatious. Delgado says that many patients who have undergone this experience have not felt discomfort due to the presence of the electrodes in their heads. In a few cases where the electrodes were located in pleasure centres of the brain the patients had the opportunity to stimulate their own brains by pressing a button. The procedure was said to have therapeutic benefits. The possibilities of this manipulation obviously extend beyond the therapeutic. One can imagine a whole technology of pleasure, just as there is today a panoply of drugs which some people use for special experiences or just for 'kicks'. There was a report several years ago of experimenters who achieved with electrode stimulation of the brain a brain wave response in human subjects which they claimed to be identical with the response of Zen disciples in the state of Satori. This is a state which normally requires long periods, often many years, of intense meditational discipline.

Beside electrode implantation there are a number of other procedures which control human experience; psychotropic drugs, psychosurgery and psychological conditioning such as aversion therapy. While there is much controversy as to the effectiveness of any of these procedures, there is little doubt that they put into human hands power to alter the behaviour and experience of others. Medical interest in these procedures centres on the reduction of distressing symptoms. They provide as well Orwellian possibilities of control of dissidents and others by a powerful state or other group (e.g. Podrabinek, 1980).

A searing account of the use of psycho-chemical and psychological conditioning for this purpose is given by Zhores & Roy Medvedev (1974) in their book, *A Question of Madness*. After a career as a biologist and active dissident in the USSR, Medvedev was compelled to undergo psychiatric examination. He was 'accused' of suffering from 'schizophrenia without symptoms'. Once in a hospital he was threatened with the use of drugs, especially tofranil, first to be taken by mouth and then by injection if he refused. He was eventually released. Others were not so fortunate. Allegations are made that some have been treated with large doses of chlorpromazine. There have also been allegations of people being arrested in Western countries for various causes who have been offered treatment by drugs or conditioning as an alternative to prison (see for example Rose, 1976, Chapter 13). In any case, drugs are used extensively with hospital patients, not so much to heal them as to make them quiescent. A searing indictment of this practice was dramatically portrayed in Brian Clark's play *Whose Life is it Anyway?*

No one knows the extent to which the alteration of experience by drugs and psychological conditioning is dependent upon the continual application of the drug or the conditioning process. There is doubt that the various forms of 'brain washing' do more than alter overt behaviour, leaving the basic convictions or beliefs that motivate behaviour unchanged (Storr, 1966, pp. 160–4). Be that as it may, there is reason to fear control of behaviour in any form in so far as people are manipulated like machines. The fear is legitimate while this methodology is aimed at altering some particular brain mechanism and function. On the other hand, science fiction scenarios of whole communities being controlled by these means are probably groundless. It is extremely unlikely that they would be applied *en masse* in the foreseeable future partly because traditional methods of mass persuasion and mood change are likely to remain more practicable and more effective. Hitler changed the attitudes and actions of millions by old fashioned means without recourse at all to modern techniques of behaviour control. And today we know that the media, particularly television, have a profound effect on the behaviour of people.

The role of television in promoting violence in the USA has been the subject of numerous studies including two major government commissions. Murder is the fastest growing cause of death in the USA. The incidence doubled from 1960 to 1974. The age group most involved, both as perpetrators of the crime and as its victims, are in the 20 to 24 year age-bracket with teenagers 15 to 19 years of age coming next. Violence on television is part of the growing person's normal experience. One estimate is that between the ages of 5 and 15 years the average American child will view the killing of more than 13 000 persons on television. The two major commissions on this subject concluded that children learn aggressive behaviour from what they see on film and television. This has led to the promotion of the 'Family Hour' from 7 pm to 9 pm during which violence on television is largely banished. But as with the unhealthy habit of smoking there are strong forces that oppose this move. The industry is opposed to such regulations ostensibly on the grounds of censorship, though the real reasons are financial. For example, Paramount Pictures released to television networks in the USA the rights to 42 recent-release movies, including *The Godfather* and *The Godfather Part II*, for 76 million dollars. And NBC-TV paid 7 million dollars for a single showing of *The Godfather* in 1974 (Somers, 1976). Television is big business with a substantial proportion of this business dependent upon a young audience.

One might wish that the abilities to manipulate human experience had never been invented since they may be used more often in destructive than in positive ways. But this is an idle wish. Human beings have manipulated their experience with drugs such as alcohol for thousands of years. They have manipulated one another's experience through indoctrination and brain-washing for at least as long as recorded history. Means of manipulation have always been with us. The same questions are asked from one generation to the next. However, the new forms of manipulation raise these old questions with a new acuteness. These questions are even more fundamental than those raised by our increasing capacity to keep alive individuals who are suffering from debilitating diseases.

When a particular form of manipulation of human behaviour has negative effects that very largely outweigh any possible benefits, there is little difficulty in deciding that such manipulation does not enhance human experience. But what of the situation in which goals for enhanced human experience that have formerly been difficult to reach can be readily attained through drugs or electrical stimulation? Should one not encourage social policies which would make these means of attainment readily accessible? There is reason to doubt that the kinds of experiences which can be attained in these ways are in fact the true goals of life. For example, the experiences of pleasure gained by drugs or electrical stimulation are not, in so far as we can judge, particularly high-peak experiences. They may seem so to a person whose experiences of life happen to be in the valleys. But this is a delusion. The satisfactions associated with them are such that the quest for richness of experience is likely to be aborted as one can quickly discover by visiting any drug referral clinic. This applies to the old stand-bys of alcohol and tobacco as well as to modern innovations. Richness of experience is a matter of heightened freedom and novel relationships. Drugs and electrical stimulation of the pleasure centre of the brain ordinarily contribute to neither.

The case of the manipulative production of Satori is more difficult. Although our goal of rich experience is not directed to Satori as such, we would not deny that Satori is an unusually valuable experience. Its value is partly intrinsic but more in the state of existence which follows from it in one's life afterwards. If this intrinsic value and its consequences for life can be attained by manipulation, should not this type of manipulation be encouraged? Our answer to this very hypothetical question is a hesitant 'Yes'. But there is reason to be very sceptical that in fact anything of this sort is possible.

Even if it is supposed that Satori itself involves the activation of a particular part of the brain and that this same part of the brain can be activated in other ways, the actual experience of Satori is not simply the by-product of such activation. Every experience contains within itself the past out of which it grows. In the case of Satori this is ordinarily a past which includes great discipline and much rich experience. The relation of Satori to that past is one of contrast and release, with value located overwhelmingly in the immediacy of the new moment. But the contrast and release is part of that value. The new experience is the fulfilment of a quest, and in part its value is as fulfilment. In principle the direct stimulation of the brain cannot produce such a past with its necessary contribution to the richness of the new experience.

Further, it seems unlikely to us that the effects of Satori in subsequent experience can be separated from the disciplines through which Satori is realised. These are all issues worthy of further investigation. Hence the door must be left open to the possible discovery that Satori and its effects are nothing but the by-product of activation of a particular centre in the brain. If so, then to substitute instant manipulation for the long and difficult procedures now followed would make sense. But our 'yes' is qualified. If, indeed, Satori is of this character, then it is to be valued less highly. Responsible freedom is not the sort of thing that can be generated by brain stimulation, and if Satori turned out to be subject to production in this way, it must be lacking in this crucial element of richness of experience.

There are those who believe that a person's life is a continuous flight from pain and a persistent search for pleasure. Paul Tillich (1955, p. 144) remarked that he had never met a human being of whom that is true. Neither have we. As Tillich went on to say, it is true only of beings who have lost their humanity, either through complete disintegration or through mental illness. The ordinary human being is able to sacrifice pleasures and to take pain upon himself for a cause, for someone or something he or she loves. 'He can disregard both pain and pleasure because he is directed *not* towards his pleasure but towards the thing he loves and with which he wants to unite' (p. 145). Whilst such a person disregards pain and pleasure, he or she discovers joy. That experience is possible only when we are driven towards persons and things because of what they are and not because of what we can get from them. Such experiences bring joy because they fulfil what we are. The opposite of joy is not pain but emptiness and the boredom that follows from it. Emptiness is the lack of relatedness to things and persons and meanings. It is even lack of relatedness to oneself. Therefore we try to escape from ourselves and the loneliness of ourselves but without discovering a genuine relation to ourselves and the world. It is into this ecological vacuum that commercially manipulated fun steps. It is not the creative fun often connected with play but the shallow, distracted, greedy way of having fun that makes joy impossible.

To deny that the richest experience can be produced by manipulation is not to deny that obstacles to rich experience can be removed in this way. When this is possible, a qualified affirmation of its desirability can be made. The case of a person with apparently uncontrollable aggressive impulses is a good example for consideration. Such impulses are an obstacle to responsible freedom and rich patterns of relationship to others. They lead to destructive behaviour which, for the sake of others, must be controlled. To enable such a person to turn off the impulses by pushing a button is certainly superior to putting him in chains or drugging him into a vegetative existence. It enhances rather than restricts his freedom.

But this affirmation remains qualified. There is no certainty that aggressive impulses of this sort are truly controllable only by electronic means. Is there not some self-deceit in the claim of total uncontrollability? Might there not be patterns of social support and individual discipline which could enhance the freedom of the individual to gain control over what seems uncontrollable? Would that not be the way to greater richness of experience? This is yet another example of society favouring manipulative responses to crises rather than seeking to develop a healthier pattern of social life. If so, it expresses a misdirection of the energies of science and technology.

The end result of such misdirection is depicted in Aldous Huxley's prophetic novel, *Brave New World*. Of this Kass (1971, p. 785) writes:

> There we encounter, society dedicated to homogeneity and stability, administered by means of instant gratifications and peopled by creatures of human shape but of stunted humanity. They consume, fornicate, take 'soma' and operate the machinery that makes it all possible. They do not read, write, think, love or govern themselves. Creativity and curiosity, reason and passion, exist only in rudimen-

tary and mutilated form. In short, they are not men at all. True, our techniques, like theirs, may in fact enable us to treat schizophrenia, to alleviate anxiety, to curb aggressiveness. We, like they, may indeed be able to save mankind from itself, but probably only at the cost of its humanness. In the end, the price of relieving man's estate might well be the abolition of man.

Since it is not desirable to exclude all manipulation of experience, there is need to indicate some guide-lines. The most important proviso is that the subject must be free to choose or reject the procedure offered just as he or she is free to choose or reject other medical advice. If the subject is beyond the stage of making rational judgments, then the responsibility is taken over by a guardian who might be a parent or other person close to the patient.

All this is more easily said than done. Quite often doctors become so absorbed in their procedures that they give the patient little opportunity to learn about the risks of failure. As with so much of the manipulative approach to problem-solving, there is a concentration on means to achieving a particular end, neglecting that the action constituting that means has many other effects as well. The patient is often not sufficiently warned of these 'side-effects'. For example, in some cases of aversion therapy for sexual orientation the patient concludes the treatment with feelings toward neither sex. He or she might well ask whether that last state were better than the first. And society might well re-examine the sexual norms that drive its members to such lengths.

Negative eugenics, positive eugenics and cloning

There are two ways in which genetic knowledge can now be used to control the genetical constitution of people. Negative eugenics is the name given to procedures that reduce the incidence of deleterious genes in the community and therefore the incidence of genetic diseases. Negative eugenics is now practised in many countries, more particularly in developed countries, where genetic disease has assumed greater importance with the decline of infectious diseases. Positive eugenics is the name given to procedures that could increase the incidence of 'desirable' genes in the community and so increase the incidence of 'desirable' traits in people or particular persons. It is analogous to selective breeding in domesticated animals.

Negative eugenics can be regarded as an extension by humans of what happens already without human intervention. One out of every 130 conceptions end before the mother realises she is pregnant because the fertilised egg never attaches itself to the uterus. Some 25 per cent of all conceptions fail to survive to birth and of these a third have identifiable chromosomal abnormalities. Of those that are born alive one out of every 100 have some genetic defect, and this constitutes a social problem. More than 1600 human diseases caused by genetic defects have been identified. Some are very rare. Others, such as cystic fibrosis and sickle-cell anaemia, are relatively common diseases. About half of all cases of congenital blindness and about half of all cases of congenital deafness are due to defective genes (Birch, 1975*b*).

What can be done about reducing the incidence of genetic diseases? It is now possible to identify some of these diseases by making cultures of human cells. By studying the number and structure of the chromosomes and the enzyme make-up of the cells it is possible to detect certain genetic diseases. Some genetic diseases can also be detected by examining urine for biochemical compounds that are indicative of particular genetic diseases. The applications of such screening techniques fall into three classes.

1. Adults, married or unmarried, can be screened by these techniques. In some cases prospective parents can thus learn what are the chances that their child will have a genetic disease.
2. Amniocentesis is a technique that allows for identification of those embryos which can be predicted to have serious genetic diseases. Fluid is withdrawn from the amniotic sac during mid-pregnancy. The cells in this fluid are then analysed for evidence of particular genetic disease. If the foetus is diagnosed as having severe genetic disease parents can then decide if they wish to abort the foetus. Selective abortion of defective foetuses prevents the birth of severely defective children and so reduces the burden of disease on the family and on society.
3. Post-natal screening of newborn infants allows detection of certain genetic diseases which can be successfully treated from birth onwards. The defect called phenylketonurea (PKU) is one of the best known examples of this. Affected individuals lack an enzyme and as a result become mentally retarded. However, if the condition is diagnosed at birth, the child can be put on a special diet which overcomes the effects of this deficiency. In nearly every state in the United States laws require the screening of the newborn for PKU.

Tay-Sachs disease is the infantile form of amaurotic idiocy. It is due to a double-dose of a recessive gene which controls an enzyme necessary to fat degradation. Affected persons are said to be homozygous for the recessive gene. Individuals who have only a single dose of the gene are normal, but they are carriers of the gene and are said to be heterozygous for the gene. Biochemical techniques are available which make it possible to identify both a normal individual who is a carrier of the gene and the foetus *in utero* which has the double dose of the gene and is therefore itself afflicted with the disease.

The disease has tragic effects. It results in degeneration of the nervous system, mental degeneration, blindness and death, normally before the age of three or four years. The incidence of the disease is high in certain ethnic groups and could be completely eliminated by screening and abortion of defective foetuses. It occurs in one in 900 births among marriages between Ashkenazic Jews in the United States. It is estimated that the three million Jewish Americans under thirty years of age will produce thirty-three cases of Tay-Sachs disease annually. The total cost of carrier detection, prenatal diagnosis and termination of pregnancy of risk cases for all Jewish individuals in the USA who are under thirty has been estimated to be nearly six million dollars. The total hospital costs for the 990 cases of Tay-Sachs disease these individuals would produce over a thirty year period is about 35 million dollars (Harris, 1972). The disease has only one-tenth the incidence in the general population and there the cost and effort of identifying a single affected foetus could be enormous – quite out of proportion to the benefits obtained.

There are, unfortunately, some negative aspects of this and other screening programmes. Despite the fact that the carrier state is harmless, some people become greatly disturbed when they are told they are carriers. As a result some Tay-Sachs screening programmes have been abandoned. In Dayton, Ohio, 'The local advisory committee decided that the psychic burden on those 72 heterozygotes was too high a price to pay for the prevention of a single case' (Kuhr, 1975, p. 371). A responsible approach to screening must take into account the psychological effects of information becoming available to the people screened and the privacy of the information obtained. Failure to provide adequate means of dealing with psychological side-effects and the need for privacy is a violation of professional responsibility. The responsible approach involves counselling, education and discussion programmes (Shore, 1975, p. 170).

In the case of Tay-Sachs disease should a screening policy be recommended? A great deal of suffering on the part of infants and their families could be avoided along with great expenditure of resources on hopeless medical care. It is our judgment that this would be worthwhile provided the screening is done responsibly. We do not like abortions, but we believe it is pointless to speak of an absolute right of a foetus to live as long as possible when it has no possibility of developing fully human capacities and has no prospect but suffering and early death. The Ashkenazic Jews would have to be subjected to special restrictions and testing which would be inconvenient and unpleasant, but these problems may be no greater than many others which we now tolerate for the sake of public health. Richness of experience will be served by doing all that we can to eliminate this particular disease except where the disease is so rare that the cost and effort involved would be better directed to other health programmes.

However, we do not believe that decisions of this sort are properly made by persons like ourselves. We, like all other members of society, are involved and have the right to express our opinions. But the Ashkenazic Jews themselves are the people who are most immediately concerned. Whether they choose to undertake a wholesale programme to eradicate Tay-Sachs disease, leave it freely up to each couple to make its own decisions or actively oppose this intervention into the 'natural' course of things is primarily for them to decide.

That does not mean that we would not exercise persuasion where we can. If individuals are considering whether knowingly to bring a deformed child into the world, we would urge them not to do so. Professor C.H. Waddington put the argument strongly when he said:

> If I deliberately crippled a child, I am a monster and the community will lock me up. If a government lined up 1000 babies against a wall and shot them, a political bloodbath would follow. But what if I knowingly take a one in four risk of conceiving a crippled child? . . . Am I a victim of bad luck, to be sympathized with – or a gambler whose rashness is too cruel to be tolerated?

Those who argue against abortion of seriously defective foetuses might also be asked their attitude to preventing (say by a drug) the spontaneous abortions that now occur. If this were done it could involve an enormous increase in the number of children born mentally defective, since the great majority of chromosomally abnormal foetuses are spontaneously aborted (Clarke, 1969).

We have deliberately begun with a disease which makes normal life impossible and for which there is at present no cure. Consider now diabetes, a disease which allows for normal life and whose symptoms can be controlled by insulin. It is true that the larger the percentage of diabetics in a community the greater will be the health bill, but this is an expense the community can afford. Diabetic couples may want to consider adoption, but should they prefer their own child, this is surely a morally acceptable decision.

Between these extremes are many diseases less consistently disastrous than the former and usually more serious than the latter. Our general judgment is that in a world in which there are genetically healthy children who need parents, couples should be quite hesitant about bringing into the world infants who are likely to have serious genetic problems. In most instances couples will contribute more to richness of experience by adoption than by risking malformed children. But we are more confident that the decision should be made primarily by those most intimately involved. The mother in particular should be given maximum freedom to decide about her body and whether to give birth to a child.

One factor which parents should consider and which adds to the involvement of the whole of society is responsibility to posterity. The decision to transmit faulty genes is a decision which affects not only our children but the genetic future of the race. The ecological model of living things shows that although we are constituted most importantly by our relations with those closest to us, we cannot draw lines around this intimate group. We are all bound up with one another. Even if the decision is made by those most immediately involved, society has the right to encourage all of us to reflect about the wider ramifications of our decisions. In extreme cases, society may be forced to overrule the preferences of those most involved.

What is known as positive eugenics is the improvement of a species through breeding programmes. It is widely practised by plant and animal breeders to produce agricultural crops and domesticated animals to human specifications. Clearly the interest is not in improving the quality of experience on the part of the animals. High yielding grain and grain plants that are resistant to disease and drought are products of positive eugenics, as too are high milk-yielding cattle and high egg-laying chickens. Positive eugenics is not as yet practised on people, though in a bizarre sort of way the Nazis took some crude and horrifying steps to produce what they thought was to be a super race. There have been some serious advocates of positive eugenics elsewhere, but no programmes have followed their advocacy.

The most serious recent proponent of positive eugenics was the American geneticist, H. J. Muller. He advocated that sperm of men adjudged to be superior be used for artificial insemination, thereby increasing the incidence of favourable genes in society. Couples would be encouraged to use this special sperm in preference to that provided by the husband. In a similar way distinguished women could donate ova for storage in ova banks. The fertilised ova could then be raised in surrogate mothers.

Another proposal that amounts to a variety of positive eugenics is the cloning of superior individuals by transferring the nuclei of body cells to denucleated ova cells that would then be incubated in surrogate mothers. This has already been achieved in frogs where the procedure, though difficult, is much easier than it would be in the human because the incubating egg needs no surrogate mother. The supposed merit of this procedure is that any number of replicas of the future counterparts of Shakespeare and Leonardo da Vinci could be produced. This procedure would provide a much more direct and faster route to a chosen goal than selective breeding. There is no substantial evidence that a human being has yet been cloned. But the technical problems may not be insurmountable.

There are biological objections to positive eugenics. A substantial programme of positive eugenics could reduce the variety of genes in the genetic constitution of the human species. What affect this would have on the future of the species is not known. In the past genetic diversity has played a key role in evolution and this may also be true of the future. Hence many geneticists are reluctant to recommend programmes that reduce genetic diversity substantially. Secondly, little is known of the genetics of the sorts of characters Muller was interested in, 'intelligence, moral fibre, physical fitness and outstanding gifts'. Related to this is the fact that there is no such thing as the perfect genotype, that is to say an individual with a perfect genetic constitution. Every person, genius or otherwise, carries genes for traits that are undesirable by any standard, although some of them in particular combinations may play a role in survival. Thirdly, there is considerable doubt that selective breeding for several traits is at all efficient in terms of attaining a chosen goal. It is one thing for an animal breeder to arrange breeding programmes to increase the milk yield of cows, or the egg yield of chickens. It is another and much more difficult programme to breed for several diverse traits at one time. The process is slow and very inefficient.

The ethical objections to positive eugenics are numerous. If we did know how to breed for intelligence or artistic ability or moral fibre should we do it? Are these qualities so important that we should make an effort in those directions? The relevant question is whether such breeding would enhance richness of experience. In fact the answer is not known. Some evidence even suggests that there could be a negative correlation between such traits and richness of experience, at least in some circumstances. The talented Tolstoy at the height of his fame sat bewildered in his mansion experiencing what he called 'an arrest of life'. He had affluence, fame, knowledge and influence. Yet he asked, were not the villagers going about their daily round enjoying a richer experience than he could find? One cannot assert that the experiences of the villagers were richer than Tolstoy's. It is fairly certain that Tolstoy has contributed more to the richness of experience of others than have most of the villagers. But it is likely that richness of experience is very little correlated with specifiable genetic characteristics.

The concept of achieving superior humans by positive eugenics or by cloning is based on a non-ecological model of human existence. It sees human beings as largely what they are by virtue of their genetic inheritance. The ecological model indicates that we are what we are largely by virtue of our relationships. Of course, our genetic constitutions affect our relationships, but the community in which we grow and live is of prime importance. If an infant with Tolstoy's genetic endowment had been born to one of the villagers, he would have grown up to be an entirely different person from Tolstoy. Any serious programme for enriching human experience should focus on creating an environment in which we are each encouraged, with the talents we now possess, to grow toward what is possible for us instead of leading stunted lives.

Genetic engineering

'Genetic engineering' is a third way, additional to negative and positive eugenics, in which the genes of living organisms can be manipulated. As yet such manipulation has largely involved bacteria and other microorganisms though some success has also been achieved in manipulating genes in mammalian cells. This has led to the belief that one day genetic engineering may be applied to humans. That day may not be far distant.

In Chapter 1 the role of deoxyribose nucleic acid (DNA) as the chemical carrier of 'information' in the cell was discussed. It is this information that specifies 'cat' rather than 'dog' or 'oak tree'. It specifies the details of cat or dog such as coat colour, shape and so on. DNA does this by providing all the 'information' that specifies all the chemical reactions that go on inside the cell.

The term genetic engineering refers to the direct manipulation of DNA with a view to changing the genetic information in the cell. The research is called recombinant DNA research. In this work the long thread-like DNA molecule is cut into pieces with the aid of certain enzymes. The resulting segments of DNA are recombined with the DNA of a suitable 'vector' or carrier to reinsert the recombinant into an appropriate host cell. In this way the genetic instructions in a cell can be manipulated, new instructions can be added or instructions can be subtracted. What is added or subtracted are sequences of nucleotides of the DNA molecule (see Chapter 1).

The first successful transplants of DNA were made in microorganisms. Since then, transplants of DNA segments have been successfully accomplished in mammalian cells. For example, Paul Berg and his colleagues at Stanford University successfully transplanted a rabbit gene into monkey cells. The DNA they transplanted coded for a chain of amino acids that go to make up the haemoglobin molecule which is the red pigment of the blood. As a result of the gene transfer the monkey cells began manufacturing rabbit haemoglobin instead of monkey haemoglobin. The vector of the gene in this experiment was a virus called SV-40. Other investigators have transplanted mammalian genes into bacterial cells. For example, genes for the mammalian hormone insulin have been transplanted into bacteria. The bacteria then manufacture insulin.

The possible practical applications of this research are considerable. Bacteria with insulin genes could be used to manufacture supplies of insulin needed for the treatment of diabetes. This could replace the present expensive and tedious procedures used for extracting the hormone from the pancreas of pigs and cattle.

Defective or missing genes might be replaced with normal genes. Sickle-cell anaemia is caused by a gene for haemoglobin being defective. If this gene could be replaced in persons with sickle-cell anaemia much suffering could be avoided. For example, attempts are being made to remove the bone marrow cells, which form blood cells, from an animal, and insert into those cells the DNA segment coding for normal haemoglobin. If the added segment is incorporated into the cell, it can be transplanted back into the animal. This is a model for gene therapy of inherited blood diseases such as sickle-cell anaemia and thallasemias. The development of *in vitro* fertilisation has vastly increased the potentialities for human genetic manipulation. One can obtain in a test tube the earliest stages of the human embryo. By introducing DNA, or cells altered in the laboratory, into this early stage embryo, the possibility exists for introducing genetic change in the cells of the body including the sex cells. Thus the changes would be passed to subsequent generations. Gene therapy thus offers the possibility of reducing the deleterious effects of genetic disease.

Experiments are also in progress for transplanting genes from peas into carrots, that would enable the carrot to convert nitrogen from the air into nitrates, thus disposing of the need for nitrogenous fertilisers. If this could be done with wheat the saving of fertilisers would be enormous.

The possible advantages of genetic engineering appear to be great. Why then the outcry from both the public and scientists that curbs should be put on this sort of research? The first concern to be raised was the possibility that in experimenting with microorganisms a new microorganism could be made inadvertently that was pathogenic to humans or to the ecosystem. The escape of a new microorganism from the laboratory could be disastrous. In the case of a human pathogen it could spread rapidly since humans would not be immune. In the case of a microorganism that was pathogenic to an ecosystem (for example by destroying microorganisms involved in the nutrient cycle of plants) we would be confronted with a pollutant that reproduced itself!

The result of these concerns has been world-wide restrictions on the design of laboratories where this work is done to cut down the possibility of escape of microorganisms. Some people complain, perhaps justifiably, that strictures can never be strict enough. Allied to this is the concern that a malevolent person or group of persons might deliberately produce a pathogenic microorganism to terrorise a nation.

A second concern expressed by some geneticists is that if genetic engineering were applied to people it could lead to alteration of the human 'gene pool' in ways that could be deleterious. This concern springs from an uncertainty as to the role of genetic diversity in the future of the human species. The gene pool is already being altered by the practice of negative eugenics. Since this involves the removal of deleterious genes there is little objection. With positive eugenics there is greater objection since this, if widely practised, would reduce the incidence of genes that are not particularly deleterious, if at all. Genetic engineering, if widely practised, could affect the gene pool in both these ways and perhaps in others too.

A third objection is more general. It is the concern that genetic engineering opens up the way to making people according to specifications, with visions of Huxley's *Brave New World.* Some people have a somewhat unarticulated concern that we should accept what comes naturally and not interfere with nature, especially if some of the possibilities appear to be horrendous.

A fourth objection is that research and development in genetic engineering divert scarce resources away from areas where more people could benefit. Its medical emphasis is on treating symptoms and victims rather than on removing the primary causes. For example, King (1980, p. 271) remarks:

> Genetic engineering technology will focus attention on affected individuals and their genes. As a result there is a strong tendency to lose sight of the agents that caused the demage in the first place; mutagens, carcinogens, radiation, etc. Our problems are not in our genes, they are in recreating a society in which the genes of individuals are protected from unnecessary damage.

This is the general objection, already discussed, to the development of expensive medical technologies that are available only to the few and which deprive others of basic health care.

A fifth objection is that research and development in genetic engineering in microorganisms and plants for agriculture is ecologically inappropriate. As King (1980, p. 269) remarks 'the strains developed in the industrialised countries will be designed for capital intensive agriculture, requiring chemical fertiliser, pesticides and the destruction of any indigenous ecosystem. The most productive uses with respect to preservation of human and natural resources probably will in fact come from less manipulative technologies.' He went on to point out that existing bacterial strains are already being used in India, China and Pakistan for converting manure and waste to clean gas for use in cooking. The residue provides a good source of fertiliser. Similarly, advantage has been taken of existing strains of blue-green algae that fix nitrogen in rice paddies. Projects using such strains to provide necessary nitrogeneous nutrients for rice have been successful in India, Burma and Nepal.

The effect of these concerns has been a call on the part of some people for a moratorium on all recombinant DNA research until such time as assurance can be given as to the safety of the procedures. Others call for an outright ban on this research. We have reached the limit, they say, to which this research should be carried. Their wish is to prohibit further investigation on the grounds that accidents in the work are dangerous and/or that in the ultimate application of genetic engineering the disadvantages outweigh the advantages. It is as though research on atomic physics should have been stopped at that point when someone realised that an atomic bomb could be made. A great deal has been written about these problems of genetic engineering in which technical worries about health hazards are mixed up with the larger philosophical questions as to the use and abuse of knowledge (e.g. Goodfield, 1977; Rogers, 1977; Wade, 1977). Some people opposed to genetic engineering invoke safety as an argument of convenience in support of conclusions reached on ideological grounds.

We can focus on the critical ethical issues by asking the question – does genetic engineering have the potential for increasing richness of experience? Its promise of reducing human suffering by control of genetic disease, of providing a ready means of producing life-giving chemicals and of making plants that produce their own nitrogenous fertilisers could all be positive gains if they were to be achieved. Against these gains we must balance costs. The principle has already been stated that all creative advance is made with some cost. So the critical question becomes: how much should we be willing to pay for the promised advantages? To be able to do this we need to make judgments on all five costs already referred to. The assessment of risk and cost is very difficult. This should not be made by experts alone. As has been said, to ask the experts alone is like asking incendiaries to run the fire brigade. This is not to attribute any deliberate evil intent to experts to require that assessments are made by others as well. But they do have a vested interest. Furthermore, there is the principle that the people who are going to be affected by new advances should have some say in the matter. This is now becoming recognised. For example, the possible construction of a containment facility for research at the University of Michigan was debated before the Regents. It brought experts from across the country to debate the merits of the provisional guidelines. Concern about the research was voiced by the faculty, students, and members of the Ann Arbor community. After some delay, the Regents voted to allow the construction to proceed.

A similar hearing was held by the Committee on Research Policy at

Harvard University, where the construction of a containment facility in the Harvard Biological Laboratories was planned. Brought to the attention of Cambridge Mayor Alfred E. Vellucci by publicity surrounding this hearing, recombinant DNA research became an issue before the city government. After two crowded public hearings, which exposed the deep rift in the molecular biology community over the wisdom of pursuing recombinant DNA research in the city, the Council voted for a moratorium on those experiments judged to be most hazardous and established a citizen review board to examine the issue and recommend appropriate action. The Cambridge Experimentation Review Board was composed of eight city residents chosen by the City Councillors who were neither involved in the debate nor themselves molecular biologists. They recommended that the research be allowed to proceed within the city, but that much more stringent safety procedures should be followed than those provided by the National Institute of Health (NIH) guidelines. The Cambridge citizens' panel requested that the biological containment procedures and laboratory monitoring techniques for the research be strengthened. Amongst other recommendations they proposed that a Cambridge Biohazards Committee be set up to oversee all recombinant DNA research in the City of Cambridge (Loechler *et al.*, 1978). This is a nice example of the democratisation of knowledge. What many experts considered to be the private reserve of their expert elites became a shared experience of a much broader section of the citizens.

There are many technologies that are potentially dangerous, but it is not inevitable that dangerous technologies will be used. If the possible abuses of a new technology are anticipated then it is possible to build in restraints. But will this suffice?

We are confronted by a dilemma. Research in recombinant DNA has gone so far, is so widespread and promises so many gains that it is impossible to stop it. The best that can be done is to introduce and enforce safeguards. This is being done to some extent. If the high standards of safety that have been proposed and adopted in some places could be enforced everywhere and forever, it would be reasonable to consider that the remaining risks belong to the inevitable risks involved in all creative advance. Unfortunately, such universal and permanent standards are almost as unlikely to be achieved as would be a total moratorium on continuing research. Thus far commercial interests have prevented the United States from enacting regulations enforcing these standards. Should high standards be enforced throughout the industrial world, we would have to reckon with the movement of commercial research centres to third world countries where such regulation had not been developed. Nevertheless, there is at present no better response to the situation than to call for strict enforcement of rigorous precautions not only in industrialised nations but in the third world as well.

At the same time, it is important to emphasise again that the values and commitments that have led to this situation are not those that are encouraged by the ecological understanding of living things. In Easlea's (1973, p. 263) words, these are to make a more beautiful world, a world in which people are progressively able to become more self-aware, more able to determine their own lives and to become more human. The purpose of science, said Francis Bacon, is none other than the relief of man's estate. In the closing pages of his *The Great Instauration* he said:

> Lastly I would address one general admonition to all, that they consider what are the true ends of knowledge, and that they seek it neither for pleasure of mind, or for contention, or for superiority to others, or for profit, or fame, or power, or any of these inferior things; but for the benefit and use of life, and that they perfect and govern it in charity. For it was from lust of power that the angels fell, from lust of knowledge that men fell, but of charity there can be no excess, neither did angel or man ever come in danger by it (quoted by Ravetz, 1971, p. 436).

Conclusions

Human beings now live in a situation in which they are confronted with ever more difficult ethical decisions. Scientific and technological advances are making possible ever greater manipulations of life including human life. There is no doubt that some of these manipulations are life-enhancing. But the new knowledge also has vast potentialities for evil.

One of the worst features is that the new knowledge is likely to have benefits only for a few. The needs of the vast majority of the world's population are not addressed. The gap between the rich and the poor is heightened. The rich wonder what kinds of new pleasures to give themselves. The poor have dietary deficiencies which sap their energy and make them vulnerable to preventable diseases.

This situation forces a recognition that the energies of science and technology have been misdirected. This misdirection continues and even accelerates. Overwhelmingly and increasingly science and technology are controlled in the direction of their work by military and industrial interests which pay the bill. Even where this obvious control does not exist, scientists and technologists are often guided by the prestigious interest in highly complex questions which have little relevance to the urgent needs of masses of people. What can be more selfish than to turn one's talents to narrow tasks which will bring immediate reward to oneself and only the most remote and unknowable benefits to one's fellow man? In a day when science was a lonely investigation of a few passionate seekers after truth, such criticism was inappropriate. But in our day when it is one of our major social projects, the question cannot be avoided.

A biology more oriented to human need will also be a less manipulative and more ecologically sensitive biology. It will be a biology that studies the life processes in their complex interconnections in nature more than one that abstracts the living organism from its natural surroundings. In short, it will be a more ecological biology teaching human beings how to relate to other creatures in a way that brings greater health to the biosphere and to themselves.

But one cannot now wish away the actual results of the manipulative biology which has dominated research. The situation is like that with the motor car. When the motor car was introduced into the world no one asked about the possible cost in human lives that would result when cars became as common as they are in the developed world today. Now that motor cars abound, does the knowledge of the cost in human lives influence the having or the not having of them? It would seem not. People are evidently willing to pay the price in terms of human lives slaughtered and human bodies maimed each year on the roads. Everyone agrees that it would be ideal to prevent all road deaths. Everyone knows how to do it: chop up the roads and all the cars! But when someone says they would like to abolish road deaths, they mean at no cost. Given the economy, which has to continue, then people reckon that people are expendable. They reckon to kill a few thousand to keep the economy going and to retain their conveniences. The risk in having motor cars is known exactly. Every holiday week-end estimates are made with considerable accuracy of the number of people who will die on the roads. The cost now seems to us far too high to proceed as we are going. The time has come when freedom to drive a private vehicle may have to be severely restricted. We need also to envision new ways of organising our communities in which private motor cars will not be needed (see Chapter 10). With hindsight it now seems that the only sensible thing to do with new technologies is to estimate as far as possible the cost before embarking on them and to decide then what price we are prepared to pay. Meanwhile ethical problems involved in dealing with the present carnage on the highways have to be faced.

Similarly the ethical problems raised by our increasing ability to manipulate life for good and evil have to be faced. In doing so one can insist upon the values highlighted by the ecological model. One can insist that the freedom of individuals and of communities to choose be respected, even while recognising that the larger society has a stake in their decisions, and must sometimes insist upon policies contrary to the wishes of the individuals most closely involved. One can insist that the interconnectedness of people with one another and with other creatures be recognised, that the intrinsic value of experience of all living things be appreciated, that the full consequences of our actions be considered and not only the intended and desired ones. One can also insist that the energies and resources of science and technology be redirected into less manipulative channels more calculated to prevent disease and to improve the quality of human experience generally, without introducing greatly disturbing risks.

This does not mean that risks should never be taken. It is a case of where to draw the line. 'It is the business of the future to be dangerous,' said Whitehead (1926*a*, p. 207). The major advances in civilisation are processes that all but wreck the societies in which they occur. At the same time it is sheer foolishness to proceed in ignorance of the consequences. The great French physiologist Claude Bernard said 'Science teaches us to doubt and in ignorance to refrain.'

8

A just and sustainable world

> We need some quite new concept of reality, some revolutionary upheaval in past habits of thought ... this is the concept, made increasingly explicit by new methods and tools of scientific research, of the entirely inescapable physical interconnectedness of the planet which the human race must share if it is to survive.
>
> Barbara Ward (1979, p. 264)

A limited range of ethical issues was discussed in Chapter 7, namely those generated by the study of life itself and by the technology which that study has generated. Even there it was not possible to divorce the questions faced from the wider context of global injustice. That chapter called for science in general, and biology in particular, to redirect its energies toward dealing with basic global needs. When these needs are attended to from the perspective of ecological thinking it is immediately seen that they are not only problems of social justice. They are equally problems of ecological sustainability. Redistribution of wealth within countries and even between countries will be of little value if the ecological basis of life continues to distintegrate.

The discussion of values and ethics in Chapter 5 focused on individuals, although even there the limitation of that approach was recognised and with it the need to extend the approach to global concerns. In dealing with global problems the prime concern is still the enhancement of the richness of experience of individual human beings and other animals. And for human beings this primarily means the realisation of existing potentialities. But for these new purposes at the global level additional principles and guidelines become necessary.

The World Council of Churches has singled out three words to express the most important features of the human societies that need to be promoted. They should be just, participatory and sustainable (Abrecht, 1979). For simplicity's sake we select justice and sustainability as the two key terms, believing that participation can be contained within justice. The first two sections of this chapter discuss the meaning of justice and sustainability respectively.

The importance of this discussion cannot be appreciated unless we be aware of the remoteness of our present situation from these norms.

The third section, accordingly, presents a selection of the facts that illustrate how profoundly unsustainable, and how profoundly unjust, present practices are.

The appropriate response to this realisation includes the abandonment of the growth-ideal in the developed world. But this is not at all generally accepted. Indeed, an apparently logical alternative is proposed by others. They argue that with sufficient energy and innovative technology a future of universal prosperity is possible. Further, they argue that unlimited energy can also solve all environmental problems. These views are almost always associated with the call for nuclear energy, which in its fusion form is held to provide the unlimited energy we need. In the third section this ideology of unlimited growth is explained and rejected.

In this chapter and in Chapters 9 and 10 frequent use is made of the word development along with developed, developing and other adjectives to describe diverse societies and their goals. This does not mean that what has been meant by development is to be taken as normative. On the contrary, both the dominant models of development and the normative judgments derived from them are to be severely criticised. The advocacy of this chapter has more in common with the ideas of those who call for 'liberation' instead of 'development'. Nevertheless, it is difficult to avoid this language. The term development can have connotations which would justify its normative use. The delegates to the United Nations World Conference of the International Women's Year meeting in Mexico City in 1975 described the aim of development as: 'To bring about sustained improvement in the well-being of the individual and society and to bestow benefits on all' (Papanek, 1977, p. 14). This is an appropriate meaning of development when it is conceived in terms of an ecological model of human existence.

Justice

Justice and sustainability are sometimes presented as alternative human concerns. Those who emphasise justice sometimes see sustainability as the cry of those who benefit from present structures of society and want to distract attention from the abuses of power within it. However, more and more people are coming to realise that although these two principles may sometimes be in tension in the structures of present society, they require one another. There can be no justice without sustainability, and there can be no sustainability without justice. A society which abandons concern for justice breeds resentment which can only be controlled by force. This is a pattern which is inherently unstable. But it is equally true that a society which seeks only justice without regard to the consequences of its behaviour for the future cannot be just. It is unjust to many generations yet unborn. It denies to them opportunities for richness of experience at least equal to ours. If the notion of justice is expanded adequately to include justice in the future, sustainability can be subsumed in justice. Because sustainability has been so seriously neglected up until now, sustainability is retained as a separate theme.

Key terms are extremely difficult to define, and fortunately they can often be sufficiently understood for many purposes without exact definition. But it is necessary to offer some comments on their meaning and on the problems and possible misunderstandings associated with them.

The idea of justice can be synonymous with the idea of the good society. In this sense it implies the right ordering of all human affairs, leaving open for investigation the question of what ordering is good. Plato showed that a quite authoritarian and hierarchical society may be considered just. At the opposite extreme there are those for whom the meaning of justice is almost identified with equality. For them a society approaches justice as all differences in wealth, cultural achievements and power disappear. Equality is certainly an important norm. One could, however, imagine a society which was just in this sense in which there was little possibility of richness of experience. The concept of absolute equality is an abstraction. Justice does not require absolute equality, but as Rawls (1972) has argued, it does require that we share one another's fate and provide equal opportunity for each person to develop his or her talents.

To emphasise equality in this sense is quite different from insisting upon equal distribution of wealth among nations. This ideal has a commendable aspect, but in its application it can easily blind us to deeper questions of justice. This can be shown through a comparison of Iran and Sri Lanka. Iran under the Shah attained an annual *per capita* income of $1250, far above that of most developing countries, whereas Sri Lanka's *per capita* income remains around $130 a year (Brown, 1978*a*, pp. 302–3). From some points of view Iran's achievement could be commended as expressing a more just redistribution of international wealth, whereas Sri Lanka would appear an appropriate object of pity.

Because the evaluation of the success of development programmes in terms of *per capita* income has proven so inadequate, a Physical Quality of Life Index has been proposed by the US Overseas Development Council. This is based on infant mortality, life expectancy and literacy. Measured on this index Iran's score was only 38 on a scale of 0–100. Sri Lanka's was a remarkable 83! (Brown, 1978*a*, pp. 302–3). Iran had followed typical development policies, concentrating on sophisticated industrialisation. This led to increasing the maldistribution of income within the country and did little for the quality of life of the masses of the people. In Iran infant mortality remained high and life expectancy and literacy low. By contrast, Sri Lanka is a poor man's 'welfare state'. Mortality declined as a result of the spread of health care, control of malaria and improved nutrition. The infant mortality has fallen to 50 (in 1979) per thousand persons per annum which is a very low rate for a poor nation. The average for the developing countries is about 140. Life expectancy (in 1979) was 68 years in Sri Lanka which approaches the average for rich countries and greatly exceeds the average for the developing world. Education is not only free but freely available. There are minimum wage laws and old age security schemes. Small farmers have secure tenure for their land following land reform. Most striking of all has been the fall in the birth rate. In 1979 Sri Lanka's birth rate was 16 per 1000 having fallen from a high 38 per 1000 in 1950. The most reasonable explanation of this dramatic fall in birth rate is the socio-economic policies of the country.

Other countries which have made striking advances in the physical quality of life while their *per capita* annual incomes are still between $300 and $400 annually are the People's Republic of China and the Indian State of Kerala (Ward, 1979, pp. 212, 253). Those with a somewhat higher gross national product (GNP) *per capita* of about $1000 that have achieved a high physical quality of life are Taiwan and the Republic of Korea (1979 World Population Data Sheet). Political policies which led to these improvements mainly concerned land redistribution, taxation, labour and capital utilisation, food subsidies and the establishment of priorities in the fields of better and low cost delivery systems for education and health care. Besides achieving a higher quality of life, another benefit which has followed from the equalisation of opportunity is greatly reduced birth rates (Grant, 1977).

The concern for equality has been properly expressed by those countries which undertook to deal directly with the needs of the masses of their people instead of waiting until there would be a larger economic pie to divide. This does not guarantee that wealth will be equally shared within the country. It certainly does not assure that the country as a whole will have a 'fair share' of the global economy. But it deals with more basic human needs than those measured by economic indicators and thus expresses a keener sense of justice as equality. It also provides a context within which the people as a whole can profit from needed economic growth in the future.

But justice is not a matter of equality alone. Equality can be imposed by authoritarian governments on unwilling people. This is not yet justice. Justice entails that people will participate in decisions about their own destiny. Their own capacities to think and choose will be called into play rather than suppressed.

Self-reliant development involves the participation of people. Few countries may be self-sufficient in their productive efforts but all can be self-reliant in the sense that development provides scope for the fulfilment of the individual humans it serves. This can only happen if the individual participates fully in the process and his or her talents and the country's resources are developed hand in hand. The opposite of self-reliance is dependence which too readily becomes exploitation from outside. The Indian economist Parmar (1975, p. 5) wrote:

> If poverty and injustice are the main facts of economic life, the potentiality of the poor must be the main instrument for overcoming them. This would be possible if the people in developing countries discover a sense of dignity and identity within their socio-economic limitations. To assume that only when we have more, when we are nearer to the rich nations, we will have dignity and identity, is a new kind of enslavement to imitative values and structures; an enslavement that dehumanises. Many developing countries need liberation from such bondage into which the growth mania of the last two decades has led them. People's participation can be effective when people discover the power of their potentiality, one could even say, the power of powerlessness. The spirit engendered by that would be the most valuable capital for a society. It would set our feet firmly on the road to human development.

Just as equality and participation can be subsumed as essential elements within the concept of justice, so too can personal freedom. Personal freedom is closely connected with participation. Participation involves the freedom of people to dissent. For many people today there is no justice because they are subjects of political oppression. Moltmann (1979, p. 99) has said: 'So long as all men are not free, the people who believe themselves to be free are not truly free either.' Arbitrary arrests, torture, politically motivated disappearances, long term detention without trial, extrajudicial executions and political murders are tragically widespread in the modern world. The 1979 Annual Report of Amnesty International contains information on thousands of prisoners held without trial or charge in Argentina. Thousands have also disappeared there since the *coup d'état* in 1976. Bangladesh has over 3000 political prisoners in its jails. Indonesia still holds thousands arrested in the aftermath of the so-called 30 September Movement of 1965 as well as more recent detainees including students and opponents of the central government who seek greater provincial autonomy. Detention without trial is practised extensively in South Africa. Vietnam is holding 50 000 people in camps for political 're-education'. Amnesty International worked during 1979 for 220 prisoners of conscience in the German Democratic Republic, almost all of whom were imprisoned under laws restricting the non-violent exercise of human rights. It reported over 300 persons in the USSR sentenced to imprisonment, exile or confinement to psychiatric hospitals for non-violent exercise of human rights. Not a continent is free from violations of human rights.

Justice involves at least the three elements of personal freedom, equality and participation. By any one of these criteria huge numbers of people in the world today are condemned to suffer grave injustice.

Sustainability

The meaning of sustainability is somewhat easier to express: to be sustainable is to be capable of indefinite existence. Sustainability came into use in the global context in relation to the environmental crisis. In 1966 Kenneth Boulding contrasted the wasteful 'cowboy economy' with a 'space-ship economy'. In the cowboy economy resources are regarded as infinite and are exploited wastefully and extravagantly. In a space-ship economy resource use is geared to the finite amount available (Boulding, 1971). The same idea was presented by Meadows *et al.* (1972) in *The Limits to Growth*, Daly (1973) in *Towards a Steady State Economy* and later in *Steady State Economics* (Daly, 1977) and Henderson (1978) in *Creating Alternative Futures*. These authors refer to the steady-state economy in contrast to the ever-growing economy whose destiny is to collapse upon itself because the planet is finite. It was the lack of a positive connotation of the phrases 'steady-state', 'stationary-state' or 'equilibrium state' and their unacceptability to the third world that led to the phrase 'sustainable society' being coined at a meeting of the World Council of Churches on this subject in 1974 (World Council of Churches 1974, p. 12). This use of the word sustainability emphasised the necessity of sustaining the life-support systems of the earth and the resources on which they depend. It is an ecological sustainability. However, sustainability has a wider connotation when it indicates as well as ecological sustainability the sustainability of social structures and political systems. It could be argued, for example, that neither capitalism nor socialism, as at present practised, are sustainable political systems.

Sustainability in the real world is a relative matter, just as is justice. The call for a sustainable society in this chapter refers to the indefinite future not 'the infinite future'. We will do well indeed to envision social forms that can persist for even a few hundred years, although that is a short time from an evolutionary point of view.

The association of sustainability with a steady-state economy suggests to some people that only a static situation is sustainable. However, far from the sustainable state being static, only change is sustainable. As pointed out in Chapter 3, if change and movement ceased for even a moment all life would end. Nevertheless, out of an extremely dynamic basis, life achieves its stable forms. The relation of stability and change in the idea of sustainability is a complex one. Here nature is a good teacher.

For millions of years the thin envelope of life around the earth which we call the biosphere has sustained the resources necessary for its life in a most wonderful and complex way. The oxygen in the atmosphere comes from plants. The carbon dioxide in the air, soil and water comes from two sources, living organisms and volcanoes. All the molecules of oxygen, water and carbon dioxide are continually recycled by living organisms. Nature's global society is kept sustainable because the molecules keep moving.

Within this planetary system nature has produced many sustainable ecosystems such as rain forests and coral reefs. Here is a great diversity of plants and animals which, instead of exhausting the resources of the environment, sustain them. A little-appreciated fact about a natural community of plants and animals such as a rain forest community is that it recycles virtually all materials used as resources. Apart from the water and carbon dioxide the only resource that comes in from outside is the energy of the sun.

The trees and other plants in a rain forest take up minerals from the soil, with very great efficiency, and turn them into their own tissues. The leaves and branches fall to the ground or are eaten by animals which die and decompose and produce waste products, all of which go back to the forest floor to be decomposed by bacteria and fungi, and eventually turned again into inorganic compounds. These are taken up by the plants again, so rapidly and so effectively that the water that runs off from the undisturbed Amazonian rain forest is virtually the same as distilled water in composition. It contrasts with the water in the so-called white rivers which drain eroded areas and so are white with clay and other minerals. The mineral-free rivers are called black rivers, such for example as the great Rio Negro which joins the white rivers of the Amazon at Manaus. The black rivers are so called because their clear waters make them appear black. There is virtually no life in them (by contrast with the white rivers) because there are no inorganic compounds to support plant life. They have been removed by the roots of land plants before the water reaches the river.

The verb 'to grow', as Daly (1977, p. 99) says, means 'to spring up and develop to maturity'. The notion of growth includes some concept of maturity beyond which physical accumulation gives way to physical maintenance. Every plant, animal, population or community has a growth phase which is followed by a no-growth maturation phase. In their growth phase individual plants and animals use resources to build their tissues. They increase in size. In the maturation phase they no longer grow. Instead they use resources at a lower rate, not for growth but for the maintenance of the activities of the mature organism. The same principle holds with communities of plants and animals. This is clearly seen in a rain forest. Every rain forest started with seeds that grew into seedlings that grew into shrubs and trees. There was an increase in biomass. While this was going on, more and more energy was being trapped from sunlight for growth. Eventually the forest reached a mature phase. Energy was then used primarily to maintain the mature community. Of course lots of seeds were still produced, but in the struggle for existence most of them never saw the light of day. A mature forest we see today has reached the limits to growth. Yet it survives perfectly well, perhaps for millions of years, as a dynamic sustainable society.

Within the context of relatively enduring ecosystems millions of new forms of life have emerged. These too achieve relatively stable structures and patterns of behaviour, but, viewed as a whole, living species

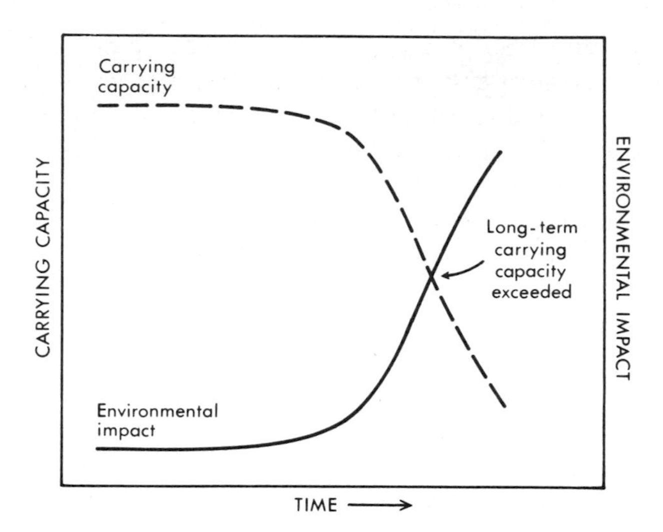

Fig. 8.01. The generalised relationship between economic growth and the carrying capacity of the earth. Growth results in environmental deterioration with consequent reduction in carrying capacity.

are remarkable for their dynamic adjustments to changing contexts and for their participation in the changing of their contexts. Where learning from adventure and experimentation assumes a larger role the rapidity of change becomes still more marked.

A sustainable society will respect the limits of the planet earth. The earth is finite in three aspects: it has a limited capacity to produce renewable resources such as timber, food and water, it has a limited amount of non-renewable resources such as fossil fuels and minerals, and thirdly it has a limited capacity for providing its free services for the maintenance of the life-systems such as its pollution absorption capacity.

These three limits determine the capacity to carry people. Growth in population coupled with economic growth reduces the carrying capacity of the earth because of the environmental impact of industrialisation (Fig. 8.01). Suppose that today the human demand on the environment is equivalent to only 5 per cent of the carrying capacity of the global ecosystem, which is surely a gross underestimate. Take into account the overall environmental impact, which, according to the American study SCEP (1970, p. 22), is growing at 5 per cent per annum. The environmental demand would reach the saturation point

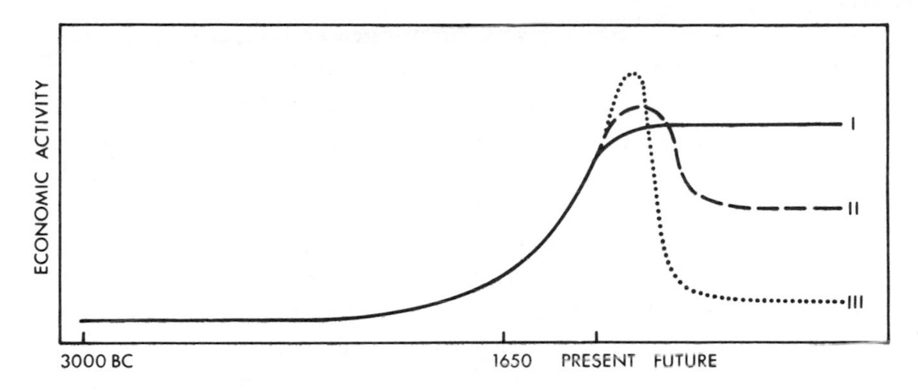

Fig. 8.02. The generalised ecological history of the world, past, present and future. I, direct transition to high level sustainable state; II, belated transition to somewhat lower-level sustainable state; III, reversion to a pre-modern agrarian way of life.

of 100 per cent by the year 2036, even if the carrying capacity were not itself reduced by growth.

The sheer momentum of growth and the lag in social response to deterioration of the environment predispose the world system to overshoot the level that would be sustainable over a long period. The inevitable consequence of overshoot is collapse. The basic options open to us are shown in Fig. 8.02. Throughout most of human history growth in economic activity has been very low. It began to sky-rocket after the industrial revolution and has continued ever since. The first policy option is an early and direct transition to a sustainable steady state. If this option is not taken, overshoot will occasion a fall to a lower sustainable state than could have been achieved by careful planning and timely action or even to a still lower level which could be tantamount to a reversion to a pre-agrarian way of life. Human beings stand at a crossroads of civilisation in which it is possible to begin now to create a just and sustainable society. The epoch that human beings could manoeuvre to enter could be a turning point in the ecological history of the human race of comparable importance to the neolithic revolution. But that choice has not yet been made.

The history of ecocatastrophes is the history of the human race ignoring limits and replacing sustainable societies with unsustainable ones. The singularity of this generation is that it is doing this on a global scale.

The world can be visualised as a hollow vessel with an inlet and an outlet. For population the inlet is birth-rate. The outlet is death-rate. For a steady state no-growth population, birth-rate and death-rate must be equal. Both should also be as low as possible, perhaps around 11 per thousand people per annum. There is no advantage in having high death-rates and high birth-rates, unless we are interested in manufacturing suffering. Similarly, with goods or material wealth in the world. The inlet is the production of goods. The outlet is the consumption of goods. One should equal the other, giving a constant level of material wealth. The level at which people and wealth stabilise is very largely an ecological problem. Traditional economic systems that encourage growth aim to maximise the throughput of goods and materials, whereas in a sustainable state it would be minimised.

Not everything is held constant in the sustainable society. A sustainable society develops, but its development is not dependent upon quantitative growth. A steady state economy is not prevented from growing. It differs from a growth economy, not in rejecting growth but in aiming at maturity. The economies of the developing world have not yet reached maturity. Hence growth is important. But this growth should aim at meeting real needs and then ceasing rather than becoming a built-in requirement of the system.

Even within a mature economy which has no further reason to increase the quantity of its output, there need be no cessation of creative change. John Stuart Mill long ago saw this when he wrote:

> It is scarcely necessary to remark that a stationary condition of capital and population implies no stationary state of human improvement. There would be as much scope as ever for all kinds of mental culture, and moral and social progress; as much room for improving the Art of Living and much more likelihood of its being improved, when minds cease to be engrossed by the art of getting on. Even the industrial arts might be as earnestly and as successfully cultivated, with this sole difference, that instead of serving no purpose but the increase of wealth industrial improvements would produce their legitimate effect, that of abridging labor (Mill, 1857, p. 326).

From these considerations some conclusions can be drawn about the characteristics a sustainable society will have.

1. The population will be well within the carrying capacity of the planet. What that population would be depends on the economic habits and social organisation of the society.

2. The need for food, water, timber and all other renewable resources will be well within the global capacity to supply them.
3. The rate of emission of pollutants will be below the capacity of the ecosystem to absorb them.
4. The rate of use of non-renewable resources such as minerals and fossil fuels will not outrun the increase in resources made available through technological innovation.
5. Manufactured goods will be built to last; durability will replace inbuilt obsolescence. Wherever possible materials will be recycled.
6. Social stability requires that there be an equable distribution of what is in scarce supply and that there be common opportunity to participate in social decisions.
7. The emphasis will be on life not things, on growth in quality not quantity, on services not material goods.

The unsustainable and unjust world

A farm which is overstocked with sheep or cattle is unsustainable. A sustainable farm is one in which the population of animals does not exceed the carrying capacity of the farm. Overstock the farm and it deteriorates. It may end up as a dust bowl. The earth as a whole has a carrying capacity for the living organisms on it, including humans. Sometimes rabbits or locusts multiply to the point at which they exceed the carrying capacity of a region. That region may be decimated in consequence. But it is only humans whose numbers have continued to grow virtually all over the earth without the major periodic set backs that characterise growth in numbers of animals such as locusts and rabbits.

No one can say what the carrying capacity of the planet is with respect to human beings. It depends on the way society is organised and the demands made by individual human beings upon their environment. Lester Brown (1979, p. 38) warns that we have already passed some critical points. 'Avoiding a decline in the *per capita* fish catch would have been possible if world population growth had come to a halt at 3.6 billion in 1970. If population had stopped growing at 3.8 billion, grain output per person might still be increasing.' Brown recognises 'the great difficulty in bringing population growth to a halt quickly,' but he argues that 'this must be weighed against the social cost of failing to do so in time to avoid the collapse of the earth's major ecological systems' (p. 39). He proposes a programme to level off population growth at six billion early in the next century. Even if this ambitious goal is attained, it will be very difficult to develop a just and sustainable society for that many people.

Not only have human numbers grown. So too has the human appetite. Human beings are consumers of the resources of the earth. Ever since the agricultural revolution, appetites for consuming products have increased. There was a huge increase in consumption in the world following the industrial revolution. This was because people learned how to get access to vast quantities of non-renewable resources such as coal and iron and how to use them industrially. As a consequence incomes rose in the industrial world. People were then able to buy more and so to consume more. Meadows (1975, p. 60) has estimated that one American now requires in a lifetime:

98 million litres of water
28 tonnes of iron and steel
1200 barrels of petroleum
29 000 kilogrammes of paper
10 tonnes of food
$10 000 in public expenses.

One American throws out:

10 000 no-return bottles
17 500 cans
27 000 bottle caps
2.3 automobiles
35 rubber tyres
126 tonnes of garbage
9.8 tonnes of particulate air pollution

Some arithmetic with these figures answers the question as to whether the earth could support its 1980 population of 4300 million all living at that standard of material consumption. The result is startling. If all the people of the world were to consume resources at the rate of Americans, the total known reserves of petroleum would be used up in six years and the annual consumption of timber, copper, sulphur, iron and water would exceed available known reserves of these resources. The high rate of consumption of resources by Americans would be impossible for all the people of the world. This has been called the 'impossibility theorem' and as yet it has not been refuted (Daly, 1980, p. 361).

This poses a very obvious problem of justice. Can Americans and others who belong to the rich world morally justify living in a way, or consuming at a rate, which it would be impossible for the rest to enjoy? In a sustainable world with a more just distribution of wealth, a nation is overdeveloped when its citizens consume resources and pollute at a rate which is greater than would be possible for all the people of the world. The Swedish economist G. Adler-Karlsson has called for an 'inverted utilitarianism'. This would mean 'organizing our society in such a way as to minimise suffering'. He maintains that 'increasing the material standard of the already affluent does not have any value so long as suffering still is widespread ... Nobody should increase his affluence unless everybody has got his essentials.' This makes 'less demands on the material resources of the globe but more on our moral resources' (quoted by Omo-Fadaka, 1977, pp. 21–32).

This is a radical challenge to societies which have encouraged anticipation of ever greater personal affluence and based the meaning of personal life largely on these expectations. But this does not make the challenge irrelevant. It only shows that the changes required will not be made by a sufficient number of individuals acting alone. Society will have to change.

If the carrying capacity of the earth would be exceeded by a world population all as rich as Americans are now, then either a small proportion of the world can be rich and a large proportion will remain poor (as at present) or else the rich will reduce their rates of consumption in order that the poor may increase theirs. Yet economic systems the world over operate as though the problems of injustice would be solved by letting everyone, rich and poor, grow in wealth together. To fail to recognise that the carrying capacity of the earth sets limits is to live in a fool's paradise. It is true that technology has in the past tended to push the limits a bit further away. But technology cannot be relied upon to remove all limits. Indeed, there is now much evidence that modern technology itself, through its polluting activities, reduces still further the limits of the earth for carrying life. The technological fix becomes the technological trap.

The human race has joined forces with entropy against life. Human beings are fast depleting the world's accessible reserves of minerals and fossil fuels and fresh water. According to the report by Leontief, Carter & Petri (1977) to the United Nations on 'The Future of the World Economy', in the remaining years of this century the world will consume from three to four times as much minerals as have been consumed during the whole of previous history. The report anticipated that after the year 2000 absolute shortages of particular minerals would become critical. At accelerating rates the energy stored in fossil fuels is being converted into atmospheric heat and carbon dioxide, while the effect this will have on life in the future is as yet unknown.

And what of the earth's renewable resources? Human beings are utterly dependent on the products of the earth's four basic biological systems; grasslands, croplands, forests and fisheries. These four systems of renewable resources supply not only all our food but, with the important exception of minerals and petrochemicals, all the raw materials for industry as well. There is now evidence that the global *per capita* productivity of each of the four natural systems has peaked and is now declining. This is particularly the case with fisheries in which the total world catch peaked in 1970 and in forestry where the area of forests in the world shrinks by 20 hectares a minute, which in a year is an area the size of Cuba (Brown, 1979).

In pursuit of more water for growing crops great rivers all over the world have been dammed. In so doing a Pandora's box of problems has been created. Because silt is no longer deposited each year on the plains of the Nile Valley, the farmers there now use artificial fertiliser instead, at an annual cost of $100 million. Because of rising water tables following irrigation, salt accumulates in their otherwise fertile soil. To alleviate this problem Egypt is now engaged in the world's most expensive drainage project costing $180 million (Omo-Fadaka, 1977).

Complex self-sustaining ecosystems have been replaced by various forms of intensive agriculture, much of which is probably unsustainable in the long run. According to the United Nations Environment Programme (UNEP, 1977, p. 20), 'The present rates of soil loss through erosion may be as high as 2500 million metric tons per year – over half a ton of soil for every man, woman and child on the planet.'

Life-supporting ecological cycles of growth and decomposition in lakes, rivers, estuaries as well as on land have been blocked through pollution. And no one really knows how pollution of the oceans is affecting the health of life in the oceans. There is some suggestion that recent reduced fish yields are due to increasing pollution of the seas as well as over-fishing.

If little is known about the effects of pollution upon nature's goods, even less is known about its impacts on nature's services. These services include the degradation of organic waste, the fixation of solar energy, the maintenance of atmospheric gas balances and the cycling of nutrients. The magnitude of these services is only just beginning to be investigated. For example, every hectare of freeway built through San Bernadino pastureland in California not only will increase the production of carbon monoxide from automobiles but also will reduce the natural carbon monoxide removal capacity by 440 kilograms per annum (Westman, 1977). In the south-eastern United States, a detailed calculation of the value of wetlands as tertiary waste-water treatment facilities and as fisheries concluded that each hectare cleansed as much water as could be handled by $200 000 worth of modern pollution-control equipment. To the extent that pollution inhibits these natural functions, society suffers a very real loss (Hayes, 1979, p. 7).

The greatest threat to sustainability is war and the preparations for war. Or as Rifkin (1980, p. 163) says: 'In the end, warfare, and its preparation, are the most highly entropic form of human activity. After all, there are only two things you can do with a missile – use it for destruction, or store it until it becomes obsolete and has to be scrapped.' In 1979 the world had stockpiled 50 000 nuclear weapons. The US nuclear arsenal exceeded 6000 megatons in 1980. Some 200 megatons of this total, if delivered to the 200 largest cities in the USSR would, it is estimated, kill one-fifth of the population of the USSR and destroy two-thirds of its industry (Lewis, 1979). Likewise about 10 per cent of the Soviet nuclear arsenal aimed at 70 of the largest metropolitan areas in the US would kill half the US population within 30 days, injure tens of millions, destroy two-thirds of the nation's industrial activity and 98 per cent of its key industries, such as energy and transport (Knox, 1980). It is clear that the USA and USSR have enough nuclear weapons to destroy each other several times over.

Even a fraction of the physical and human resources spent on war, if diverted to the wise support of life, could solve many human problems. The annual military bill in 1980 approached 450 billion dollars, which is about one million dollars a minute (Brandt, 1980, p. 11). The military expenditure of only half a day would suffice to finance the whole malaria eradication programme of the World Health Organisation. For the price of one tank (one million dollars) storage facilities could be provided for 100 000 tonnes of rice. The same amount of money could provide 1000 classrooms for 30 000 children. For the price of one jet fighter (20 million dollars) 40 000 village pharmacies could be set up (Brandt, 1980 p. 14). There is no simple solution to the arms race. It will probably not be solved at the level of national governments and international negotiations, although they must be encouraged. Deep attitudinal changes will have to occur throughout the world before drastic reduction of military expenditure is likely. A world organised more along the lines suggested in this chapter and the two chapters that follow would involve attitudinal changes of the sort that would reduce fear and suspicion among nations sufficiently to make progressive disarmament more likely. To continue the present trends of the global society is obviously to continue on an unsustainable path.

Nor is the existing global society just. Statistics about inequities have become deadeningly familiar so that they have almost lost their power to move us. But they do represent realities that are all too deadening for the hundreds of millions of victims of the injustice they represent. In 1979 the GNP *per capita* in the United States was $8640 and in Australia $7340. In Bangladesh it was $90 (Population Reference Bureau 1979 data sheet). Some 1300 million people live in countries with a mean GNP less than $300 *per capita*, 800 million in countries with a GNP of $300 to $2000 *per capita* and 700 million in countries with a GNP of $2000 or more *per capita* (1975 figures); the mean GNP *per capita* of developed countries is 13 times that of developing countries (Grant, 1979). In terms of industrialisation, the industrialised countries of Europe, USSR, North America, Japan and Australia, whose people constitute 30 per cent of the world's population, control 93 per cent of all industrial production, 95 per cent of all exports, 85 per cent of all armament production and 98 per cent of all research and scientific activity. These countries consume 87 per cent of the world's energy production, 78 per cent of all fertilisers, 94 per cent of aluminium production and 94 per cent of all copper production (UN/UNCTAD Handbook of International Trade and Development Statistics Supplement, 1973).

In terms of health, the life expectancy at birth in 1979 in the USA was 73 years, in Bangladesh 46 years; the infant mortality rate (annual deaths of infants under one year per 1000 live births) in 1978 in the USA was 14, in Bangladesh 153 (Population Reference Bureau Data Sheet for 1979). According to the World Food Council (UN World Food Council, report of executive director, April 1979) over one thousand million people may not get enough to eat to meet their joule requirements, over 450 million of these are estimated to suffer from serious under-nutrition. Over 60 per cent of these undernourished people live in Asia and most of them in rural areas. Every year 15 million children die from malnutrition and disease in developing countries as compared with half a million in developed countries. In the poorest countries, of those children who survive, between one-quarter and one-half suffer from severe or moderate protein and energy malnutrition. Vitamin A deficiency causes at least 250 000 children to go blind every year. And all this in spite of the fact that if all the food that is produced in the world each year were divided equally between all the people on earth, all would have an adequate diet (Abbott, 1972). That is to say all would have the requisite number of joules and the requisite amount of protein. There is enough for everyone's need but not enough for everyone's greed.

Distribution of resources within countries is also skewed. The super-rich have incomes thousands of times greater than the most destitute. A useful measure of social justice is the comparison of the average income of the wealthiest fifth of the nation's population and that of the poorest fifth. The ratio in Ecuador is an appalling 29 to one. In the USA, Great Britain and Australia, it is around five to one. Some Eastern European countries have achieved ratios of three to one (Brown, 1978*a*, pp. 221–2).

With some notable exceptions (e.g. China, Sri Lanka and Kerala) development programmes have not reduced the profound injustice in the world. Indeed, the gap has widened in the last decade. The poorest of the poor are not only relatively worse off than a generation ago but they are probably more impoverished absolutely than ever (Grant, 1979).

This horrendous injustice is a threat to the world which should scandalise us as injustice scandalised the prophets of old. For them, as Heschel (1962, pp. 3–4) remarks:

> The world is a proud place, full of beauty, but the prophets are scandalized, and rave as if the world were a slum... To us a single act of injustice – cheating in business, exploitation of the poor – is slight; to the prophets, a disaster. To us injustice is injurious to the welfare of the people; to the prophets it is a deathblow to existence; to us, an episode, to them, a catastrophe, a threat to the world.

If there is to be ecological sustainability and social justice on this earth there can be no question but that a new social and economic order is required. This is recognised by the United Nations in its call for a new international economic order. But there is much question as to the direction in which it is to be sought. Amid all the intricacies of the debate, two basic directions can be discerned. The one toward which the whole perspective of this book leads is the ecological view. It recognises the finiteness of the planet and the intricacies and vulnerability of its life-support systems. It learns from the way in which life has evolved and persisted how life might further be enabled to flourish on the earth. It would develop economic and political systems appropriate to these considerations that would also enable full participation of individuals in shaping their own diverse destinies. It would reverse the long process of exploitation of the resources of the earth and would seek a level of use of resources that could be sustained indefinitely. It gives preference to renewable resources such as solar energy and to low energy-use technology and to a use of soil that sustains its fertility indefinitely.

There is, however, a second view of the future which Ehrlich, Ehrlich & Holdren (1977) call the cornucopian view. It is the view that still governs most of the affairs of men and nations. This approach considers that all problems of ecological sustainability can be solved through technology and that some measure of justice will be achieved through unlimited growth and hence by the rejection of all limits. It is the view advocated by the Hudson Institute of New York in the book *The Next 200 Years: A scenario for America and the world* (Kahn, Brown & Martel, 1976) and by others such as Beckerman (1975) in his *Two Cheers for the Affluent Society*. In their classic analysis of the two views Ehrlich *et al.* (1977, p. 6) wrote:

> Which view of civilization's prospects is more accurate is a question that deserves everyone's scrutiny. It cannot be decided merely by counting the 'experts' who speak out on either side and then weighing their credentials. Rather, the arguments must be considered in detail – examined, dissected, subjected to the test of comparison with the evidence around us. This is an ambitious task, for the issues encompass elements of physics, chemistry, biology, demography, economics, politics, agriculture, and a good many other fields as well. One must grapple with the arithmetic of growth, the machinery of important environmental processes, the geology of mineral resources, the potential and limitations of technology, and the sociology of change. It is necessary to ponder the benefits and shortcomings of proposed alternatives as well as those of the *status quo*; and it is important to ask where burdens of proof should lie.
>
> Does civilization risk more if the cornucopians prevail and they are wrong? or if the pessimists prevail and *they* are wrong? Could an intermediate position be correct, or are perhaps even the pessimists too optimistic? What is the most prudent course in the face of uncertainty?

The ideology of unlimited growth

Industrial society has been built on the energy stored in fossil fuels; firstly coal, then oil and now back to coal again – while it lasts. Since all these fuels are exhaustible and are being exhausted at an accelerating rate, there is no possibility of unlimited economic growth based on fossil fuels. The cornucopians present other possibilities. Theirs is a vision of a continuous supply of cheap energy from nuclear fission and when that runs out from nuclear fusion energy (the techniques of which have yet to be invented). With vast supplies of energy, the huge stores of minerals in low concentration in sea water and common rocks will permit technology to produce more of everything and to do it cheaply, enabling the poor to become prosperous and the rich to become richer. All this is to be accomplished even in the face of continued population growth. The benefits of expanded technology are seen as outweighing the environmental and social costs which are considered to have been greatly exaggerated anyway. The cornucopians with their Promethean strategy hold that industrial society is very much on the right track and that more of the same – continued economic growth – is all that is needed to usher in the technological golden age.

Of course energy and minerals are not the only needs of the industrial society. Food, wood and also water are required. But with adequate energy these too can become superabundant or, if not, then suitable substitutes can be found. As agricultural lands disappear under factories and cities, chemicals can replace soil. With sufficient energy, chemicals can be produced in unlimited quantities. With sufficient energy (and capital!) desalination plants will produce fresh water from the sea. When wood runs out new chemical compounds will serve the purposes once served by wood. With sufficient energy the problems of waste disposal and pollution can be solved. Hence, in the cornucopian view, nuclear energy combined with great technological ingenuity can produce unlimited quantities of all the things that human beings may desire.

But is this really possible? Assume for the moment that sufficient fuel can be obtained from nuclear power. It still leaves nuclear power as a dangerous source of energy for a number of reasons. In order of increasing importance these are*: (i) The disposal of solid radioactive

* This order of priority was reached in discussion with Dr John P. Holdren, Energy and Resources Group, University of California, Berkeley.

wastes. As yet there is no known safe way of disposing of solid wastes for thousands of years (AIRAC, 1979). However, this problem may be more amenable of solution than other problems of nuclear power. (ii) The routine release of gaseous radioactive wastes, which problem may have been underestimated in the past. (iii) The chance of a catastrophic accident in a nuclear power plant. (iv) The proliferation of nuclear weapons as a consequence of the multiplication of nuclear power plants.

This latter problem is the most serious objection to nuclear power. There are those who have argued that the peaceful use of nuclear power can be promoted whilst at the same time restricting the rise of nuclear armaments. The argument has depended upon the supposed difficulty of developing weapon-grade fissionable material from fuel used in civilian plants for the purpose of constructing a bomb. The theory has been that a nation or group which attempted to do this could be detected early enough to stop the process. However, this now seems unlikely. Lovins, Lovins & Ross (1980) argue convincingly that there is no way to stop military proliferation in a world dependent more and more on nuclear power. They argue that 'all concentrated fissionable materials are potentially explosive' (p. 1139), and that 'both bomb-making and reprocessing are relatively easy' (p. 1143) once the fissionable material is abundant. A 'timely warning system can be provided neither for separated plutonium *nor* for spent fuel, so that *all* nuclear fission will be unsafeguardable in principle' (p. 1143).

But this does not mean that we must simply reconcile ourselves to a world in which scores of nations have nuclear armaments. There is nothing inevitable about the continued growth of nuclear power plants as Lovins *et al.* (1980) point out. On economic grounds alone nuclear power is much less attractive than once thought (p. 1149). The disappearance of civilian use of nuclear energy would not in itself ensure that nuclear weapons would not proliferate. But the situation would be far more hopeful. 'Detection and deterrence of bomb making require that it be unambiguously identifiable; and for that, phasing out of nuclear power and the supporting services it justifies would be both a necessary and sufficient condition' (p. 1148).

Proponents of the nuclear solution who reluctantly acknowledge the limitations of fission point to fusion as the final answer. No one knows today whether the enormous problems surrounding the use of fusion for production of usable energy can be solved. Even if they are, it is quite likely that the first fusion plants will be built without all the necessary safety features because of enormous cost. Furthermore, there is the same objection to fusion as to fission, namely, that it adds to the problem of proliferation of nuclear weapons (John P. Holdren, personal communication).

Suppose that fusion could become a safe source of unlimited energy and that this energy together with advances in technology could accomplish all that is proposed. Suppose, for purposes of discussion, that physically speaking unlimited growth in the production of goods is possible. Suppose further that the difficulties in the transition from our present world to the world of fusion can be surmounted. The question is whether such a world would be ecologically sustainable and socially just. Is nuclear fusion the best hope for humanity? Are the critics of this direction faithless traitors to the human cause? Or does our society now err in devoting its major research resources to the development of this kind of technology? If the fusion society would be the only truly just and sustainable one, then we should put our doubts aside, face the difficulties, and make every effort to realise that goal. But if that society is inherently undesirable, then the endless debates about its possibility are pointless. The decision is finally more ethical or religious, than scientific and technological.

What, then, would a society of unlimited growth based on fusion be like? For the most part it would be a projection of recent trends; for it is precisely for the sake of continuing such trends that this society is proposed. One of these trends is toward larger and more complex technologies. The theory is that people grow wealthier as they become individually more productive, and they become individually more productive as they control more and more energy. Hence new technologies substitute energy for human labour. Some economists are now reluctantly conceding that this can be a cause of unemployment. But there are other problems associated with high technology which have been pointed out in the report of the British Council for Science and Society in 1976 entitled 'Superstar Technologies'.

1. Large systems are more vulnerable to failure than small ones. The more components a system has in it the more vulnerable it is to breakdown and the more time has to be spent in repairs. In communication and control engineering, such as used in Apollo spacecraft, systems are triplicated so that a back-up system comes into operation when one system fails. But heavy electrical and mechanical systems cannot be protected by redundancy, a steam turbine or electric generator is bound to be put out of action for a long time if part of a rotor

breaks off. Since the possibility of accidental failure of a component cannot be entirely discounted, a 'fail-safe' system may be designed as in nuclear reactors. This demands, in its turn, an extraordinary level of imagination and ingenuity, hence the 'inevitability of the unexpected' which was demonstrated, for example, at the Brown's Ferry reactor accident in Alabama, where a candle being used to test for air leaks from the reactor containment set alight cables controlling the reactor and most of its essential safety systems.

2. Large systems have dangerous, often cumulative, side effects. Any technology has hazards. The recognition of slow-acting side effects entails expensive preventive measures, but sometimes the side effects are not discovered until the damage is done as, for example, in the notorious case of poisoning in Minamata Bay in Japan. The industrial pollution of the bay started in 1930 with the production of vinyl chloride. Spent catalysts containing mercury were discharged into the bay. Microorganisms converted this to methyl mercury which is soluble. This poisonous compound got into fish and shellfish. People ate the fish with disastrous consequences. The first account of the disease was reported 15 years after the first release of mercury into the bay. It was not diagnosed until three years later. In some cases experimental tests of side effects are possible in principle but these would require an enormous investment. To demonstrate, for example, the carcinogenic effects of low-level radiation would require several billion mice! It has not been done.

3. Large systems have unsatisfactory social aspects. The quality of life of many people who work on assembly lines is dull and monotonous. It compares unfavourably with participative creativity in small enterprises.

4. Large systems have disasters. Windscale, Flixborough, Ronan Point, Séveso, Torrey Canyon, Three Mile Island, are now familiar words in our vocabulary. They epitomise major technological disasters. A disaster concentrates attention. Remedial action often follows. But our society is much less responsive to the slowly accumulating toll of the unspectacular disaster such as the mounting road death toll, a direct result of advanced technology combined with human fallibility. 'If we are to devise institutional procedures for bringing superstar technologies under social control, we must keep in mind the way people *are*, not how they ought to be.' So says the British Council for Science and Society.

Superstar technology is often inappropriate technology not only because of its violence to human beings but also because it is violent to the rest of nature. The point is illustrated in the story (Watts, 1976, p. 48) of a mythical king of ancient India who, oppressed by the roughness of the earth upon soft human feet, proposed that his whole territory be carpeted with skins. However, one of his wise men pointed out that the same result could be achieved far more simply by taking a few skins and cutting off small patches to bind beneath the feet. In this case human beings adapt themselves to nature rather than adapting nature to their wants. In the parable the technically skilled solution is the one in which human beings adapt themselves to nature. Unfortunately, growth-economics and superstar technology have a strong tendency to be violent to nature and force it at great expense to be adapted to us. The inflated GNPs of the developed countries reflect this violence. They do not provide the model or norm for the undeveloped countries. Fukuoka (1978, pp. 82–4) gives a modern example of the choice between superstar technology and appropriate technology in Japan. Chemical fertilisers are used in large amounts in much Japanese agriculture. Most of what is added to the land leaches into streams and where these flow into contained areas such as the Inland Sea, algae multiply to massive amounts causing a red-tide to appear. One proposal to clean up the Inland Sea is to cut a channel through Shikoku Island to let the clean waters of the Pacific Ocean flow into it. But Fukuoka also asks: where is the electric power to come from to run the factory which will manufacture the steel pipe? How about the power required to pump the water up? A nuclear power plant is proposed to provide the needed power. But this sows the seeds for second and third generation pollution problems which will be more difficult to solve than the initial ones. Fukuoka's own modest proposal is to eliminate the source of the problem by spreading straw on the fields and growing clover so that less fertilizer is used.

A decision that gives preference to 'appropriate' technology is still a viable option. In the case of energy, for example, to proceed into a nuclear future is to eliminate that option. The natural resources that are accessible to even present-day superstar technologies will be rapidly exhausted. New technologies that dwarf these will then have to be employed. A continuation of exponential growth will cause this day to come upon humanity sooner than is yet widely realised.

Those who see coal as an almost inexhaustible alternative, the massive mining of which could provide a transition to the nuclear society, need to consider the effects of exponential depletion rates on coal supplies. At present rates of consumption, the world's coal reserves might last 5000 years. But if the rate of consumption were to rise by just 4 per cent each year, all would be gone in 135 years (Maglen, 1977). Furthermore the burning of coal at these rates will accelerate the increase of carbon dioxide in the atmosphere, with the likely raising of global temperatures causing massive dislocations of weather patterns (National Academy of Sciences, 1977). Meanwhile, since we are already using wood supplies globally far more rapidly than they are replaced, exponential demand will rapidly destroy the remaining forests and force a shift to other materials. The plastics now used as wood substitutes, however, will not be available for this purpose since they are oil based. Whole new technologies and industries must come into being to satisfy the continuing exponential demands.

Not only would there be no road back from superstar technologies, there would be no road back to feeding people from the soil. An 'all systems go' approach to feeding the world will rapidly reduce the capacity of the planet to produce food by natural means while the need for food soars. New technologies of factory food production and hydroponics will be the only options.

That these and other needed marvels of technology are actually possible is by no means evident. That the present economic system which is projected to work its way forward through them can do so is extremely doubtful.

The ultimate limit to energy use is imposed not by human ability to find more energy but by ecological considerations that have mostly to do with weather and the maintenance of life-support systems. There is a limit to the amount of energy that can be used because all energy used is converted into heat. Provided the heat is produced at a rate no greater than the rate of escape of heat to outer space, the earth will not heat up. But once this critical rate is exceeded the earth heats up and with it the atmosphere. If the present rate of energy growth of about 5 per cent per annum were continued for possibly less than a century the heat produced would cause an intolerable global average increase in temperature of 11.5 °C or 21 °F (Ehrlich *et al.*, 1977, p. 679).

In addition to thermal pollution an ever-growing industrialisation presents a threat to the ozone layer of the upper atmosphere. This is the layer which protects life on earth from being burned up by ultraviolet radiation. Every society would have to eschew those activities such as supersonic flights and fluorocarbon aerosols that may damage the ozone layer. Alternatively, some new technology would have to be devised either to replenish the ozone layer to an appropriate degree or else to protect people on the planet from the damage which it now wards off. To believe that ecological sustainability is possible on these terms is to have a faith in human capacity for management and a stable social order which is quite without grounds in human history.

Suppose that humanity might overcome these gargantuan environmental threats, would a society of this sort be a just one? Since continued exponential growth is the god of this economic system, it must be served at all costs. As mistakes become ever more costly, they simply could not be allowed. Their avoidance would require ever increasing centralisation of power. Economic power and political power would merge. Opposition parties, public dissent and free experimentation would not fit with such an economy. Yet human alienation from work would be likely to increase. The greater the restlessness, the greater would be the need to restrain it – by force if necessary.

The need for centralisation of power and control of citizens would be accentuated by the exponential multiplication of nuclear power plants whether fission or fusion. Since the basis of the economy would be a form of energy capable of enormous destruction, these plants must be protected against enemy agents, subversives, insanity and even carelessness. A class of managers would be required whose coolness of nerve and loyalty to the government was beyond question, and complete security would have to be maintained to prevent others from entering the plants. This system would have to be defended and rapidly extended without mistakes, not for a generation, but for thousands of years. The prospects of success are vanishingly small, but even with success, the system would be repulsive and replete with injustices.

Dependence upon the chemical industry would be enormous. In combination with energy technology it would be relied upon to come up at an exponential rate with new materials to replace old and to meet new needs. It would, of course, have to be much more careful than it has been in the past, or chemical poisoning would soon destroy human life. As more and more dangerous products were produced, the factories would require the same kind of quasi-military protection as the power plants. Society simply could not take chances. In short, the world would be filled with more and more deadly substances from all of which human beings would have to be protected. The assumption would have to be made that despite the psychological strains of such a society, no one in a key position would crack over a period of millenia.

Another necessary assumption is that there would be no war. Since nuclear technology would be commonplace, the chances that it would be used by anyone threatened with catastrophic defeat are considerable. The balance of terror would have to work. No adventurer could appear, and no revolutionaires succeed. As Rossel (1980, p. 260) says:

> There are insidious but fundamental risks inherent in the very nature of the nuclear industry, dedicated as it is to gigantism and centralization and serving as the driving force in a society dominated by a technology of ever-expanding proportions. The intellectual equilibrium of the human community and its spiritual foundations are at stake in this dramatic evolution of a society which is perverting itself and turning its back upon the values on which it was built.

Proponents of unlimited growth who recognise the seriousness of some of these problems with nuclear energy might appeal instead to the use of solar energy. This would be ironic, since solar energy has been widely belittled by growth advocates as unable to make a major contribution. However, a shift seems to be taking place as enthusiasts for superstar technologies discover the potentialities of superstar solar technologies. Some of them are now advocating, for example, huge solar collectors in outer space which would beam down energy which would not otherwise reach the planet.

This source of energy would have some advantages over nuclear energy in that it would probably be less dangerous and in need of somewhat less police protection. On the other hand, outer space collection of solar energy would have the same effect of heating up the earth's atmosphere as would the production of any other form of energy.

If, to avoid that, solar technologies were restricted to the utilisation of energy already reaching the earth from the sun, a growth economy based on solar energy would certainly have real advantages over any other type. There would still be environmental effects from covering larger and larger areas of the earth's surface with solar collectors, but the main objections would be that if this energy were used to fuel unlimited industrial growth, the many other problems noted would still apply. A society based on exponential growth, even if it were fuelled with solar energy collected on the surface of the earth, would be unsustainable and unjust.

Even if it were possible, the goal of limitless economic growth is inherently absurd and few probably really believe in it. It would require either infinite increase of human population or infinite increase of individual consumption or some combination thereof. Both are ridiculous.

Other advocates of the growth direction admit that growth will end at some point but they believe that point is still distant. They may argue that global production needs to be doubled several more times before it will be at a level that satisfies legitimate human desires. To achieve such growth, they argue, nuclear energy is essential. Against such a more moderate position, the scenarios above would require modification. Nevertheless, their basic thrust remains persuasive. Nuclear energy as now developed introduces into our world elements of unsustainability and injustice we cannot afford. The necessity of preventing its destructive use hampers the movement to a freer society. The focus on quantitative growth as the major global need directs our efforts in the wrong direction.

Sometimes it is argued, somewhat hypocritically, that to oppose growth is to oppose the development of the poor nations and hence to condemn many of the world's poor to perpetual poverty. Growth in the poor world is necessary. But it is to be hoped it will be a different sort of growth from that which the rich world has pursued. Very often those in the rich world who advocate all-round economic growth want the poor world to embrace their own unsuitable technologies. Nuclear energy is an example. It is peculiarly unadapted to the needs of the poor. It involves centralisation of economy and political power and disfranchises the masses of people. It makes the developing country dependent on outside economic and political power. Moreover, it produces electricity only, whereas a prime need of the poor world is for liquid fuels.

Not all growth is to be opposed. In much of the world production needs to be doubled, probably more than once. But this will not best be achieved by a global doubling and quadrupling of production. Such increases are simply not needed in much of the industrial world.

Lester Brown points to the needlessness of such growth in many instances with the following example:

> Using the official economic-growth projections of the Swedish economy, Backstrand and Ingelstrom examined the consumption patterns for the year 2000. The projected growth rate of 4 to 6 percent per year yields an overall level of output four times greater than at present. This expansion in the gross national product includes a sevenfold increase in steel output and a tenfold growth in the production of chemicals. With Sweden's population on the verge

> of stabilizing, the authors ask what the country could possibly do with such volumes of steel and chemicals, considering that consumption levels are already quite satisfactory (Brown, 1978*a*, pp. 261–2).

The realisation of the urgent need for economic growth in the developing nations combined with the recognition of the reality of global limits to growth has one obvious and revolutionary corollary. The only way the poor world can grow is for the rich world to curb its own growth. Only when the rich world has redirected its attention from the question, 'how can we grow?' to the question, 'how can we best adjust to the cessation of growth?' will promising answers to social problems begin to appear.

Conclusions

To those who are sensitive to fundamental features of living things the fact that there are limits to economic growth is so evident that it is difficult to appreciate the sincerity and conviction with which others deny that there are such limits. Two structures of experience, two faiths, confront each other here. The first sees humanity as a part of the natural world, a very remarkable part in which the element of transcending, present in all life, is extremely prominent. For them human beings transcend most successfully as they respect most sensitively the conditions of all life. They can learn to use the limited resources of the world more efficiently and more effectively. By attentiveness to the strategies of life, they can learn what true growth is, they can develop appropriate goals and find deep satisfaction in their attainment. This is a religion of life. It is to place faith in Life.

The other sees humanity as outside of nature, having nature as its object, its possession, to be used for human ends. Its adherents recognise little difference between living things other than human beings and inanimate materials. Both are resources for human management, manipulation and consumption. Human desires are, and rightly so, unlimited. And human ingenuity is similarly unlimited. To place confidence in human intelligence, and in the science and technology in which it is expressed, will enable growth in numbers and in consumption for as long as these are desired. Proponents of this faith believe that humanity can move forward to a Golden Age! The danger is that human beings falter. They lack trust in Reason. They turn their backs upon the promise because of problems whose resolutions at the moment they cannot see.

From the perspective of the religion of Life, these 'problems' are not challenges to be overcome but limits to be respected. The Technical Reason in which we are asked to put our faith seems something abstracted, separated from Life and devoid of wisdom. It wants to analyse each situation in terms of isolated problems which are to be solved in abstraction from all the other interrelated factors. It does not grapple with concrete reality, but prefers the world of mathematical formulae. Indeed, it seems to enjoy creating new problems to solve more than it enjoys restructuring a situation so that such dangerous problems do not arise. From the perspective of the religion of Life this faith is misdirected toward a lifeless and disembodied form of reason which has placed itself in the service of human greed. To accept it, as so many people have, is to accept the choice of death.

This is stated baldly. More than rational argument based on factual information is involved here. Human beings have an enormous capacity to select that information that is supportive of their basic convictions and life-orientations and to neglect what does not fit. Even when they are forced to accept information that is uncomfortable, they can adapt it subtly in such a way that it requires of them only moderate adjustments. It is remarkable how much new wine we can in fact put in old wineskins! Nevertheless, the time does come for some when the old wineskin bursts. The old paradigm collapses under the weight of information that strains it too far. A new paradigm emerges that for a time organises all the information, opens the way to new and convincing perceptions, and orients life with new wholeness and freshness. The ecological model of living things and its accompanying religion of Life can serve in this way and hasten the erosion of the old paradigm and its religion of Technical Reason.

This chapter has not gone far toward saying what is concretely entailed by the religion of Life. It has said that it calls for a just and sustainable society and that this requires the abandonment of the dream, or nightmare, of unlimited growth in population and consumption. The ideology of growth has been allowed to set the terms of the discussion. But from the perspective of the religion of Life it is obvious that the ideology of growth misunderstands the true nature of growth and aims at a monstrosity that will destroy life.

The question is not really whether human beings will accommodate themselves to the earth's capacity to support them, but how. It is not whether they reduce their consumption of oil. They will. It is not whether the use of oceanic fisheries and whales be limited. It will. It is not whether the existing manufacturing industry give way to something else. It will. The question is whether limits are reached in a catastrophic way or whether human beings engineer a smooth transition, recognising the limits, to a just and sustainable future. Chapters 9 and 10 propose directions in which such a transition could be pursued.

9

Economic development in ecological perspective

> It is very arguable that the science of political economy as studied in its first period after the death of Adam Smith (1790), did more harm than good ... it riveted on men a certain set of abstractions which were disastrous in their influence on modern mentality. It dehumanized industry ... It fixes attention on a definite group of abstractions, neglects everything else, and elicits every scrap of information and theory which is relevant to what it has retained. This method is triumphant, provided that the abstractions are judicious. But, however triumphant, the triumph is within limits. The neglect of these limits leads to disastrous oversights.
>
> Alfred North Whitehead (1926*a*, pp. 280–1)

> The ideas of economists and political philosophers, both when they are right and when they are wrong, are more powerful than is commonly understood. Indeed the world is ruled by little else. Practical men who believe themselves to be quite exempt from any intellectual influences are usually the slaves of some defunct economist.
>
> John Maynard Keynes (1936, p. 383)

In Chapter 8 it was argued that justice and sustainability necessitate the acceptance of limits to economic growth. The implications of this are radical. They run counter to the dominant patterns of economic thinking that guide global development. Economics is a study of enormous importance, as Keynes points out in the quotation that heads this chapter. The authors of this book are laymen with respect to economics. Nevertheless, we think that we are able to identify defunct economics and its influence on practical affairs and that it is incumbent upon us to criticise the model of human existence which underlies most economic thinking and to indicate the different results that follow from the adoption of an ecological model of human beings. Furthermore, we cannot avoid consideration of the implications of the ecological model of living things for economic thinking. The commitment of so much economic theory to continuing and unlimited growth in production and consumption of goods is bound up with the basic image of human beings which grounds the theory.

From the perspective of the ecological model, economic thinking appears to have many of the same limitations as the mechanistic model that has played so important a role in biology. The economist Georgescu-Roegen says:

> A curious event in the history of economic thought, is that, years after the mechanistic dogma had lost its supremacy in physics and its grip on the philosophical world, the founders of the mechanical school set out to erect an economic science after the pattern of mechanics – in the words of Jevons, as 'the mechanics of utility and self-interest'. And while economics has made great strides since, nothing has happened to deviate economic thought from the mechanistic epistemology of the forefathers of standard economics (Georgescu-Roegen, 1980, p. 49).

And further on in the same article (p. 50), he criticises the 'self-sustained mechanical analogue' of economics on the grounds that it virtually ignores the mutual influence between the material environment and the economic process. It was this influence which was discussed in Chapter 8 in the context of physical limits to economic growth.

The acceptance of physical limits to economic growth does not mean favouring a fixed and unchanging economic world to be blueprinted in advance. The goal is to plan for a future in which people will have the maximum opportunity to shape and reshape their own destiny, not to live out our particular hopes. Blueprinting the future at any point in the past would have been disastrous to life. Lewis Thomas (1979, pp. 28–9) questions whether molecular biologists could have achieved what nature achieved in creating the DNA molecules if they had flown in from another planet to start life on earth.

> We would have made one fatal mistake: our molecule would have been perfect. Given enough time, we would have figured out how to do this, nucleotides, enzymes, and all to make flawless, exact copies, but it would never have occurred to us, thinking as we do, that the thing had to be able to make errors. The capacity to blunder slightly is the real marvel of DNA. Without this special attribute, we would still be anaerobic bacteria.

So it is with the life of the sustainable and just society. Models are necessary but they should be models that have an inbuilt capacity for being remodelled as circumstances change.

The first section of this chapter presents the dominant economic model in terms of which most thinking about global development has been done. Just as the actual practice of biologists has led to work and conclusions which are more appropriate to the ecological model, so the reflections of economists on the results of following the dominant paradigm have already modified programmes in an ecological direction. Nevertheless, the change is not sufficient, and one main reason is that the basic paradigm has not been radically reconsidered.

The second section describes the ecological model in its relevance to economic theory and practice. This relevance is not spelt out in technical terms, but something is said about the different implications to be drawn with respect to development policies especially in the developing world. The acceptance of this model would accentuate shifts in thinking and implementation that are already going on and, in many instances, would reverse the priorities that have governed development in the past.

The choice of an economic system is often prescribed as a choice between socialism and the market economy or between communism and capitalism. These alternatives are examined in the third section. It turns out that the most important decisions that are now needed are not between these systems but rather cut across them. Neither has been based on an ecological understanding of human life.

The dominant model

The paradigm which has dominated reflection on public policy in the West has both political and economic components, favouring human rights and democratic processes on the one hand and a market economy on the other. This chapter focuses more on the latter, but begins with an overview of the model that recognises the political component as well. A fine statement by a proponent of the dominant model is found in *Equality and Efficiency: The big tradeoff* by Arthur M. Okun (1975). After this model is discussed it is contrasted with the ecological model.

In the title of Okun's book, *equality* represents the political goal and *efficiency*, the economic goal. Okun believes in both and urges us to accept the tensions that result from pursuing both. The market is geared to encourage efficiency at the cost of great inequalities. The role of government is to guarantee inalienable rights equally held by all citizens and to undertake programmes to reduce the inequalities generated by the market. Okun believes that income taxation is the most important and useful means of reducing the inequalities generated by the market without unduly reducing the efficiency. His conclusion is that 'the market needs a place, and the market needs to be kept in its place' (p. 119).

Okun considers that the efficiency to which the market is dedicated 'implies that more is better, insofar as the "more" consists of items that people want to buy' (p. 2). He recognises that 'some warn that the economic growth that generates more output today may plunder the earth of its resources and make for lower standards of living in the future' (p. 3). This position has been affirmed in Chapter 8. But without argument Okun announces that he is quite assured 'in essentially ignoring' this 'type of criticism of the "more is better" concept of efficiency' (p. 3). Indeed, it is striking that the dominant model is so well entrenched that the challenge to the ideal of infinite growth is not regarded as requiring a reply!

Okun takes more seriously a second criticism of the dominant model. He writes:

> In relying on the verdicts of consumers as indications of what they want, I, like other economists, accept people's choices as reasonably rational expressions of what makes them better off. To be sure, by a different set of criteria, it is appropriate to ask sceptically whether people are made better off (and thus whether society really becomes more efficient) through the production of more whisky, more cigarettes, and more big cars. That inquiry raises several intriguing further questions. Why do people want the things they buy? How are their choices influenced by education, advertising, and the like? Are there criteria by which welfare can be appraised that are superior to the observation of the choices people make? (Okun, 1975, pp. 2–3).

However, having noted these significant questions about the assumptions of the dominant model, he continues: 'Without defence and without apology, let me simply state that I will not explore those issues despite their importance. That merely reflects my choices, and I hope they are accepted as reasonably rational' (p. 3).

There is little doubt that those choices are accepted as reasonably rational in the community of economists and political thinkers to whom Okun's book is chiefly addressed. They are the choices that reflect and underlie the dominant model. But from the perspective of the ecological model matters appear quite different. The decisions of people whose lives are shaped by a certain kind of education and advertising and a certain way of thinking of themselves and their society, all deeply influenced by the dominant model, do not appear reasonably rational when one is convinced that the model is wrong and that the resulting activity is destructive.

In the dominant model, so far as economics is concerned, human beings are regarded as individuals who have appetites for acquiring and consuming goods. The goal is to increase the possession and consumption of goods. It is the task of the economic system to organise the resources of the state so as to satisfy the wants of its citizens. It is the economic system which determines the nature of the productive system. In the ecological model the relationships are the other way around. The basic needs of the community as a whole determine what is and is not produced. It is the task of the economic system to order the productive system in relation to the capacity of the ecological system to deliver the needed goods whilst at the same time maintaining it as a sustainable system. And whereas in the dominant model it is the economic system which is constantly monitored for its state of 'health' (i.e. growth), in the ecological model it is the ecological system which would be monitored for its state of 'health' (i.e. sustainability).

But even measured by its own criteria of health, society based on the dominant model is anything but healthy. It has now reached the point where the wealth-creating activities of industry and commerce are generating such great social costs, and the interrelations between industry, finance, government, trade unions and the public services have become so intertwined, that the workings of the system are grinding to a halt. The American economist, Hazel Henderson (1976, p. 9), describes this as:

> the entropy state [which she says is] a society in which technological complexities and interdependencies have reached unmodelable and therefore unmanageable proportions. It begins to generate an increasing backlog of unanticipated environmental and social costs – not only in pollution, waste, and resource depletion caused by current patterns of production and consumption, but in the costs of caring for the social drop-outs and human casualties of vast, incomprehensible technologies and organisations, and the costly requirements for government co-ordination and mediating the resultant conflicts.

Her analogy is a good one, for life, as has been shown in chapter 1, is a fight against 'the entropy state'.

In the dominant model pollution is treated as one might treat symptoms of a disease, ignoring the causes. Most pollution control is unecological because it is merely shifting of environmental impact. Raise the chimneys of factories in Britain and produce acid rain in Scandinavia! At best it is either a reduction of pollution at the source or a mopping up after pollution has occurred. Motor cars are made with after-burners and other devices to reduce exhaust gases. In contrast the ecological approach is to reduce the need for motorised transportation and to replace the internal combustion engine as it exists today with a non-polluting engine.

The dominant model gives priority to the efficiency of production with secondary attention to the lives of the producers and the quality of the environment. Feelings, beliefs or patterns of social life that interfere with maximisation of production and appropriate distribution are often viewed as superstition or sentimentality. This applies to the feelings of the small farmers for their land and to many of the customs and social relations of peasant villages everywhere. Even the breakup of families is not too high a price to pay for 'progress'. The need, on the dominant paradigm, is to uproot such bonds so as to achieve rational industrialisation, apply business methods to agriculture, and achieve economies of scale. To do this, advanced technologies are transferred to developing countries thus binding them into patterns of global trade which work to the benefit of industrialised nations and a few unusually resource-rich developing ones. Once developing countries are in this system, their growth depends on growth in the developing world in such a way that 'it is in the interest of the poor countries that the rich grow as fast as possible' (Lewis, 1978, p. 68). It is regrettable that the Brandt (1980) report on the future of the developed and developing countries could find no better proposal than this conclusion of the dominant economic model.

In contrast to all this the ecological model gives priority to exploring ways of meeting the full range of human needs while maintaining or restoring healthy living in a healthy environment. This environment is both social and natural. Increase of production is often desirable, but only when it can be sustained and when it fosters improved relationships. Generally it is better to help relatively self-sustaining communities to meet their recognised needs more adequately than, for the sake of more efficient production, to make heretofore independent communities dependent on international economic systems that they are powerless to influence. Indeed, once the negative effects on the future of further economic growth in the developed world is recognised, tying the growth of the poor to that of the rich can no longer be supported. Fortunately, in fact, as Lewis (1978, p. 71) says, 'their development does not in the long run depend on the existence of the developed countries'. The less developed countries 'have within themselves all that is required for growth'. Further (p. 74) the primary 'engine of growth should be technological change rather than trade'. Many of the poorer countries can make use of their customs and social patterns while introducing appropriate technology to build up their existing communities in ways that are ecologically sound.

In the dominant paradigm wealth is viewed as possession of goods and especially of the means for acquiring and producing them. Robertson (1978, p. 82) suggests that we should learn to reconceive wealth. Concerning the 'new wealth' he gives a telling quotation from Tom Burke, director of Friends of the Earth, London.

> The new wealth might count as affluent the person who possessed the necessary equipment to make the best use of natural energy flows to heat a home or warm water – the use which accounts for the bulk of an individual's energy demand. The symbols of this kind of wealth would not be new cars, TVs or whatever, although they would be just as tangible and just as visible. They would be solar panels, insulated walls or a heat pump.
>
> The poor would be those who remained dependent on centralised energy distribution services, vulnerable to interruption by strike, malfunction or sabotage, and even more vulnerable to rising tariffs set by inaccessible technocrats themselves the victims of market forces beyond their control. The new rich would boast not of how new their television was but of how long it was expected to last and how easy it would be to repair.
>
> Wealth might take the form of ownership of, or at least access to, enough land to grow a proportion of one's food. This would reduce the need to earn ever larger income in order to pay for increasingly expensive food. Wealth would consist in having access to most goods and services within easy walking or cycling distance of home thus reducing the need to spend more time and earning more money to pay for more expensive transport services. A high income would be less a sign of wealth than of poverty since it would indicate dependence on the provision by someone else of a job and a workplace in order to earn the income to rent services. Wealth would consist in having more control over the decisions that affected wellbeing and in having the time to exercise that control.

Adherents of the dominant model may rightly object that they are not oblivious to the concerns to which ecologically sensitive people attend. The dominant model recognises that there are many needs besides efficiency in the increase of production. Okun emphasises equality as such a consideration. Environmental quality and sustainability can also be considered by those who work with the dominant paradigm. They are viewed as 'externalities', and economists working within the dominant approach can help determine how such social goals can be met at the least expense to efficiency.

Often ecologically sensitive people accept the definition of the problem as formulated within the dominant paradigm. They emphasise externalities to which public policy is not yet sufficiently attentive, and they share in formulating proposals as to how these can be better taken into account. This participation in the dominant context is important, for otherwise what is said by ecologically concerned people is likely to be perceived as irrelevant. But the full meaning of the ecological model cannot appear in this way. Some of what can appear only as externalities for the dominant model constitutes the heart and core of the ecological model.

Fortunately, some leaders in global development policy in the sixties have been moving in directions appropriate to the ecological model. Morawetz (1977, p. 7) notes that whereas in the 1950s and 1960s *per capita* increase in GNP was sought singlemindedly as the goal of development, 'since the early 1970s there has been a sharp shift in these objectives ... Other aims related to poverty reduction need to be considered as well: improving income distribution, increasing employment, fulfilling "basic needs".' World Bank policy reflects this shift with work in urban slums and the direction of increasing resources to rural development programmes calculated to assist small farmers rather than, as sometimes in the past, to displace them with large scale commercial ventures (cf. *World Bank 1979 Annual Report*, Washington, DC). A World Bank staff paper summed up things this way. '... economic growth appears to have done very little for the poorer of the Third World's rapidly growing populations ... the economic emphasis has tended to lose sight of the ultimate purpose of the policies... The demand now is to put man and his needs at the centre of development' (Cleveland & Wilson, 1979, p. 17). The 'Cocoyoc Declaration' adopted by participants in the symposium of the United Nations Environmental Programme in Cocoyoc, Mexico, in 1974 proclaimed that 'our first concern is to redefine the whole purpose of development. This should not be to develop things but to develop man. Human beings have basic needs: food, shelter, clothing, health, education. Any process of growth that does not lead to their fulfilment – or even worse disrupts them – is a travesty of the idea of development' (Cocoyoc Declaration, 1974).

Whereas economist W. Arthur Lewis in 1955 began *The Theory of Economic Growth* by stating: 'First it should be noted that our subject matter is growth, and not distribution' (Lewis, 1955, p. 9), and this focus set the tone; by 1965 Lewis (1965, p. 12) expressed concern at the unanticipated rising levels of unemployment that sometimes accompanied economic growth. Still later, in 1978, reflecting back over the period, Lewis (1978, p. 75) recognised that development programmes which did not lead to improved food production have been unsuccessful. He concluded (p. 75) with this statement: 'The most important item on the agenda of development is to transform the food sector, create agricultural surpluses to feed the urban population, and thereby create the domestic basis for industry and modern services. If we can make this domestic change, we shall automatically have a new international economic order' (p. 78).

This gradual recognition of the inadequacy of the policies inspired by the dominant paradigm in the fifties and sixties does not go far enough. Much depends, for example, on whether the transformation of the food sector aims at short-term increases in production or whether it is guided by ecological sensitivity and commitment to sustainability. Gradual adjustment of the old model to more and more externalities does not suffice. A new model is needed.

The ecological model

In contrast with the dominant model, the ecological model of living things is bound up with the awareness and acceptance of limits. Since Chapter 8 was devoted to this concept it is not re-emphasised here. In addition, the ecological model forbids the formulation of a static blueprint for a future Utopia (see Chapter 6). In this chapter six other features of the ecological model, relevant to policy formation in the years ahead are discussed. None of them is wholly unrecognised in other approaches to public policy, but taken together in their interconnectedness, they point in fresh directions.

The first feature is the emphasis upon the intrinsic value of every person. This is measured in terms of richness of experience. It is precisely and only in individuals that intrinsic value is to be found and any policy which is not directed toward the enrichment of the experience of individuals is misdirected.

The second feature is the emphasis on relationality. Individuals are constituted by their relations. Richness of experience is richness of relations and depends upon the richness of what is experienced. This means that individuals exist in community and are constituted by the community in which they exist.

The third feature is the emphasis on transcending. Although individuals are constituted by relations with others, they also transcend the others to which they are related. No one is merely the product of relationships. Each one creates a creative synthesis from relationships. The fullness of the relationships and the richness of the experience depend upon creative freedom of the new individual.

The fourth feature is the emphasis on the limits of transcending. Scientific thinking and political planning are engaged in by human beings who are conditioned and therefore constrained by their historical situation and their interests. It is as important to acknowledge the limits of transcending as to recognise its reality.

The fifth feature is the emphasis on the continuity between human beings and the rest of the natural world. Intrinsic value is not limited to human beings. The relationships that constitute human beings are not only relations to other human beings. Human beings are impoverished by the decay of their non-human environment, and furthermore, what happens to the non-human world is important in itself. As Midgley says (1978, p. 363): 'humanity can neither be understood nor saved alone'.

The sixth feature is the emphasis on the possible symbiosis of desirable goals. It is not necessary to organise society as a zero-sum game. Gains in sustainability need not be at the expense of justice. Full employment need not depend on destructive growth. Goals which appear competitive in the light of the dominant model can become mutually supportive.

These six emphases are now considered in turn as a means of comparing the implications of the ecological model with the dominant model.

The emphasis upon the intrinsic value of every person. The two models agree on the intrinsic value of individual persons. But the focus of attention in the dominant model is not on the richness of their experience but on their possession and consumption of goods and services. Of course, it may rightly be argued that in many cases possession and consumption are correlated with richness of experience. But it must also be recognised that often they are not. The Physical Quality of Life Index discussed in Chapter 8 is a better measure of the condition of a people than is *per capita* wealth or consumption. But this measure, too, is obviously very crude.

It can be argued by defenders of the dominant model that since the richness of another's experience cannot be measured, the thing to do is to create a context within which people are individually free to choose what they want. The free market is a means of maximising such freedom of choice. But the model of individuals with desires for material things is too abstract to guide policy adequately. Also, when policy is based on such a model it tends to encourage people to accentuate the aspects of their lives that conform to it. People begin to think of themselves as individuals with private desires for material things and to suppose that happiness consists in such possessions.

A sad example of the 'success' of development based on this model is given by Hill (1979, pp. 23–7).

> In Nauru, where the providence of defaecating birds generated the highest *per capita* income in the world the country went on a spree. No food is now produced on the island; supermarkets are stocked with packaged foods, particularly imported sugar. Every adult member of the population has a car in which to drive around the twenty-mile circuit of the island ... an epidemic of chronic non-communicable diseases has broken out, as it has, with modernization, in many of the major Pacific centres; dramatic increases in obesity, diabetes, hypertension and heart disease have occurred in Nauru as well as across Polynesian and Melanesian cultures.

Human beings should be considered in their role as producers as well as consumers. In this role, too, the dominant model recongises the subjectivity of people in so far as it proposes that they should be free to offer their labour wherever they find it most advantageous to do so. But the model directs attention to using labour as efficiently as possible for production of goods and services for consumers. Consideration of the worker's enjoyment of work or satisfaction in creative activity is subordinated to questions of quantity of output.

Frederick Taylor, the father of modern 'scientific management', expresses this characteristic of the dominant model in a quite unqualified way. Scientific management involves 'the analysis of work into its simplest elements and the systematic improvement of the worker's performance of each of these elements' (in Braverman, 1974, p. 88). The goal is maximum control of the worker's activity, leaving as few decisions as possible to the worker. 'The key principle of scientific management... is separation of thought from action, of conception from performance. The management becomes the mind and the workers the body' (Rifkin & Howard, 1979, p. 189).

Of course, many industries and businesses are finding that boredom and alienation on the part of workers interfere with productivity. Also much of the work which scientific management taught could be done best by machine-like workers is now done by machines. Hence there are movements away from the dehumanising use of workers.

Nevertheless, the changes allowed within the dominant paradigm do not go far enough. They are mere concessions to the fact that when people are treated without regard to their enjoyment they are not efficient producers. What is needed is to see that the richness of experience of people as producers is inherently as important as the richness of their experience as consumers. The ecological model calls for an economic system in which methods of production would be judged by their contribution to the enjoyment of the labourer as well as the consumer. This would entail maximum participation of workers in the decisions that govern their work. It might involve worker participation in ownership or management of business and industry. Sweden and Yugoslavia have led the way in experimentation along these lines.

The emphasis on relationality. Consideration of the intrinsic value of individuals in this chapter has already led to a discussion of relationality and community, the second emphasis of the ecological model. Like the ecological model, the dominant model expresses the recognition that the world is not composed simply of individuals and that goals are set at other levels of social organisation. But here, too, there are differences. For the dominant model the setting of goals is the task of politics, and the most important political unit is the nation state. The state establishes goals, and it is the task of economists to assist the government in realising many of these goals. In addition to seeking an increase of goods, these goals usually include an increase of power both in relation to other states and in relation to the people who are governed. An enormous portion of the state budget of most nations is devoted to military and police purposes.

The ecological model cannot ignore these realities. But it can point toward other definitions of community, encourage an increase of decision-making at the level of these communities, and encourage development policies that share in this shift. Usually the family is the most important community, followed in much of the world by the village and the tribe. Existing political boundaries are often disruptive of these communities. All over the world ethnic, linguistic, religious and cultural groups are calling for more recognition over against political boundaries which are for them artificial. From the point of view of the dominant model these are irrational disturbances which interfere with progress, but from the point of view of the ecological model they represent a legitimate protest of real communities over against artificially defined groupings.

Of course, real communities can be hostile to one another, and a major justification for maintaining larger political boundaries and centralised power within them is that in this way tribal warfare can be prevented. However, the ecological model rejects tribalism without affirming nationalism. Our interdependence does not end with either the tribe or the nation, or even the whole of humanity. The ecological perspective calls for recognition of a truly global community of communities.

The proper identification of communities is even more complex. For the ecological model the community is that group of people and other creatures who most deeply affect one another, whose lives are most richly intertwined. In predominantly agricultural societies these intimate relations are usually based on proximity of residence and kinship patterns. In industrialised societies class divisions, common economic and professional interests, participation in voluntary associations, or shared ideals create many other communities which criss-cross one another in confusing ways rich with both destructive and constructive potential. Some of these communities fall within nations, and sometimes a nation succeeds in constituting itself as such a community. Others of these communities transcend national boundaries. The point of appreciating this is that the focus of the dominant model on individuals and political units leads to very abstract thinking. When these abstractions guide policy formation for the actual world, the consequences are often disastrous.

When individuals are thought of primarily in terms of their unlimited desire for material goods and when it is recognised that most such goods are in limited supply, then competition for scarce resources becomes the basic human situation. Viewed in this way it follows that as some satisfy more of their desires, others can satisfy fewer. This is the dominant model.

The ecological model, in contrast, asserts that the well-being of others contributes directly to the well-being of oneself. A son is not benefited by having more food at the expense of his mother whose poor health then affects the total climate of the home. Families who attain wealth at the expense of the impoverishment of their communities do not thereby attain true well-being. A village which supplies its current needs by deforesting all accessible hills is not well-off as it passes prospects of misery on to its children. We are members one of another and our individual happiness is bound up with the happiness of others. The economic goal is the enhancement of the sustained well-being of communities by the most appropriate use of those things which the community needs. This entails that the community attain its own well-being in ways that allow and enable other communities to attain theirs.

The emphasis on transcending. The third emphasis of the ecological model is on transcending. According to this model all individuals transcend the world to which they are so intimately related. In every moment there is the possibility of creative novelty. No one has to think and act just as they have been taught to think and act.

The dominant model also recognises the importance of novelty, especially technological innovations. It is expected that improvements in technology can enable us to increase consumable goods from our resource base and to substitute new resources as old ones are exhausted. Faith in the unlimited possibilities of such technological improvement has encouraged the view that other considerations should be subordinated to the promotion of this development.

The ecological model does not deny the importance of improved technology, but it measures improvement in a different way. The best technology is that which aids the production of truly needed goods with the least use of resources and without disrupting the life-sustaining character of the natural environment and of human communities. This does not mean that the present form of human community and its relationship with the natural environment should remain unchanged. On the contrary, the ecological model puts primary emphasis on overcoming destructive forms of human interaction and abuse of the environment. It is precisely to encourage the healthy transformation of present patterns that the emphasis on transcending is so urgent.

The difference is that when people view themselves as constituted by their relations to one another and to their fellow creatures they will use their abilities to transcend so as to imagine and to implement ways of improving these relations through improving the conditions of all. In this context new technology becomes very important. Indeed not only are new technical instruments needed but also new ways of conceiving the relationship between technology and social life (see Chapter 10).

Governed by the dominant paradigm, agricultural technology has been developed for the purpose of increasing productivity per worker and maximising profit. As farming practices have been adjusted to the new technology, a high cost has been paid in the quality of rural community life and of the soil. If agricultural technology were developed according to the ecological model, the first consideration would be a healthy rural life and the maintenance and renewal of the soil. A few examples of what this might mean are given in the first section of Chapter 10. They are sufficient to show that what is needed is far more creative novelty than that which is exhibited in further technological development along present lines.

It is also because people are constantly transcending what is given to them in their relationships that they can often transform apparent tradeoffs into new solutions which satisfy a multiplicity of needs. It is usually assumed that if, for the sake of the environment, people avoid using large quantities of chemical fertilisers and insecticides, they will have to accept greatly reduced yields. But Life functions to enable people to transcend such either–or alternatives (see Chapter 6). There is need to envision new types of agriculture which can be as productive as the old without destroying the life in the soil or poisoning our fellow creatures and ourselves. Enough is already being done in this direction to assure us that if energies were redirected in this way on a massive scale, the remaining problems could be overcome. But this will require a high order of imaginative freedom.

The ecological model, by accentuating the importance of individual transcending or freedom, has implications also for the political arena. Some of these are shared by the dominant model. But the accent is somewhat different. In the dominant model the freedom that is emphasised is the freedom to choose between given goods. Individuals should be allowed to move freely in quest of the jobs they prefer, they should be free to spend their earnings as they wish and they should be free to share in selecting those who will rule them politically.

The ecological model supports the freedom to choose representatives in the political arena, and goods and jobs in the economic sphere. But what is most important about freedom is not the ability to choose between existing goods. It is first the ability to envision new goods and to shape life so as to attain them. A healthy society is one in which members are encouraged to transcend the given pattern. Since society is nothing more than the patterns of relationships through which its members are constituted, a healthy society as a whole is always in the process of transcending itself.

The second element in freedom that is of particular importance for the ecological model is a transcending of given boundaries of concern. The human tendency to organise the world into 'us' and 'them' is a very powerful one (see Chapter 4). By emphasising the importance of intimate relations the ecological model may seem to give support to this practice. But in fact it does not. So far as individuals are not self-transcending, so far as their freedom consists only in the ability to express their given desires, the we–they distinction will govern their lives. But in so far as our individuality consists finally in our transcending the given, the power of this dichotomising of the world is broken. Our horizons of concern are not limited to the range of interconnections which most importantly constitute our being. We can have concern for human beings and other creatures disproportionate to their importance for us. In short, we can be moral. It is out of their own moral concern that people can encourage the appreciation and strengthening of communities as the prior and primary context within which increase in production can be appropriately sought.

Favouring more local autonomy rather than a rationalised world is not to abandon global concerns but to express them. The most intense and hostile expressions of local and ethnic interests come from people who have been denied the opportunity to develop their own communities in some autonomy from others. There should be no tradeoff exchanging global vision for community autonomy. On the contrary, healthy communities will be better able to see their interests in interconnection with others and will also be able to some degree to transcend their interests in the light of the needs of larger communities. This expresses the conviction that the degree of freedom is highly correlated with the richness of relations and especially of intimate relations.

The emphasis on the limits of transcending. It is as important to stress the limitations of transcending as to emphasise its reality and value.

Sometimes these limitations are not recognised. Some people suppose there is a transcendent realm, for example, the realm of reason. They think that they can operate in that realm of reason. They think that they can operate in that realm quite apart from their concrete situation in this world. They may even suppose that persons in political power can plan and govern on the basis of such a transcendent rationality. They may think that scientific work is not affected by the social, economic and political situation of the scientist. This way of thinking about political and even scientific reason is sharply challenged by the ecological model. Science is not, and never has been, carried out in a vacuum. It is subtly influenced by its social context. The sort of science pursued and practised by society is not an untarnished reflection of the objective world. It is a reflection that mirrors the society in which it is practised. The Newtonian apprehension of nature was conditioned by history and culture. Even the purest of pure science, quantum physics, is, according to Paul Forman (1971), a product of a reaction to a hostile intellectual environment of the Weimar culture in Germany after the First World War. Likewise the various revolutions in biology can be interpreted as reactions to particular social contexts. Mendelsohn (1977) makes this point in his analyses of the origin of the reductionist theory of the cell in the early part of the nineteenth century. And Charles Darwin's revolutionary ideas about evolution came out of the context of a stuffy Victorian culture deeply influenced by a rigid theology of nature.

The dominant model tends to alternate between a view of individuals as completely self-interested and governed by their private desires, on the one hand, and of the possibility of an objective science and disinterested planning for the good of all, on the other. The policies actually proposed tend to be worked out as compromises between these two views. Some things are to be left to the market place, which operates on the basis of purely interested decisions by individual citizens. Other matters are to be decided by a governmental planning which is to express someone's disinterested transcendent perspective. It is not surprising that there are problems and that one of them is a growing cynicism about 'the system'.

The ecological model stresses both the universal presence of transcending and the high degree to which each act of transcending is conditioned by the concrete situation in which it arises. There is no transcendent sphere into which we can move for purposes of science or political planning. People think scientifically and engage in planning as the particular people they are and have been formed to be by their social experience. They can transcend their conditioning by recognising that they are conditioned and by becoming critical of their own ideas and habits of minds. But such transcending is always very limited. People should never suppose they have attained a neutral and unconditioned perspective, nor should they expect such an attainment from anyone else. But people should not cynically suppose that those in power are merely behaving in self-interested ways any more than are those who put them in power or those who are powerless victims of social, political and economic structures. Elements of transcending occur in all people. They should be encouraged. But this is done neither by belittling them nor by expecting too much of them.

Thus the ecological model helps to show why policies and patterns based on the dominant model have difficulties. It helps in the acceptance of the limits of any 'rational' approach to politics and economics. It helps in the recognition that no one political or economic system is appropriate to all kinds and conditions of human beings. It helps in not expecting too much from any of them and yet not becoming cynical about human beings and their capacities for social and political organisation.

The emphasis on the continuity between human beings and the rest of the natural world. The fifth emphasis implicit in the ecological model is the continuity between human beings and other creatures. This is in marked opposition to the dominant model. If human beings are continuous with the remainder of nature, then the deep-seated habit of viewing other things as having value only as they are valued by human beings is erroneous, and the economic and political theories that are based on this principle require change. In the ecological model people are guests on this planet, guests of each other and all other creatures (see Chapter 5).

The dominant model divides the world between the agents who make decisions and the objects which they compete to possess. This model has had consequences which are not intended by its advocates. They think in terms of all human beings as agents, but the actual systems of society strongly tend to restrict the agency of many. Until recently most civilisations have been based on slavery. Through much of history women have been treated as possessions of fathers and husbands rather than as agents with equal rights with men. Even today we find that practical development policy largely ignores the agency of women and assumes that only men are the true agents of development.

Sensitive women call attention to the extent to which in the male vision women and nature are treated in the same objectifying way.

Similar patterns can be found in the relation of ruling classes to those over whom they exercise power. Whole classes of human beings are assimilated to that nature which is to be possessed, managed and exploited for the good of those who exercise power. Theoretical commitment to a model which treats all human beings as having intrinsic value while regarding everything else as valuable only in the service of human beings does not protect most human beings from being valued only for their use to the dominant classes.

There is, of course, no assurance that abandonment of this dualistic model will put an end to the exploitation of the powerless by the powerful. But it may help. Instead of the extension of the sphere of objects or resources to include much of the human species, the ecological model proposes the extension of the sphere of subjects and agents to include all fellow creatures. The community to which human beings belong contains these creatures as well, and their welfare contributes to human well-being.

There is nothing in the dominant model, as such, that requires the subordination of agriculture to industry and of the rural to the urban sector. But one can hardly doubt that the policies generated by the dominant model have attended to trade and manufacturing more than to rural life. It seems that once the human is separated from the natural, attention tends to focus on that aspect of humanity which is most removed from nature. Progress is supposed to consist in transforming nature into forms that are imposed by human beings, and this is most fully accomplished in factories and in cities. The peasant village is too close to nature and participates too much in the rhythms of nature to be taken seriously as part of the truly human world. Accordingly programmes of national development focus on urban and industrial growth. The agricultural sector is seen as providing raw materials for industry, food for industrial workers and products for export in exchange for machinery and for oil to operate the machinery.

Those who recognise the continuity of humanity with all other creatures cannot be so blind to the importance of the community of living things. Those human activities which are most closely bound up with other living things, such as the production of food from the soil, cannot be dismissed to the periphery of thinking.

The emphasis on the possible symbiosis of desirable goals. The human failure to be concerned with the welfare of other creatures has, as has been argued in Chapter 5, a negative impact upon the prospects for humanity as well. On the whole, those policies that are sensitive to the right of other species to flourish in their native habitats are also policies which promise a healthy future for humanity.

This widespread symbiotic character of relational existence has been stressed in many ways in dealing with the first five emphases. It needs extension into basic thinking about human goals. The goals of justice and sustainability, far from being in conflict with each other, require each other. This is the proposition developed in Chapter 8. In the first section of this chapter the critique of the dominant model presupposed that its goals of satisfying legitimate human wants could be met in ways that did not have such high social and environmental costs. Chapter 10 makes more specific suggestions about how agriculture could be so developed as to feed humanity adequately without continuing to degrade the environment and destroy existing communities. It shows how justice to women can also help to solve the problems generated by unacceptable increases in the human population. It also shows that there are policies in the area of energy, transportation and urban planning that reject the zero-sum mentality for one that envisions liberating life symbiotically on many fronts together.

The dominant model encourages people to think that their legitimate goals are mutually competitive. People are told that to fight inflation they must accept high interest rates, recession and unemployment. Similarly, if a community reduces the rate of destruction of the environment, it is told that it must accept increased unemployment and inflation. Gains in one area are traded off for losses in the other. This idea of tradeoffs is a fundamental aspect of the dominant model. Since present social structures and business practices are taken as the fixed context for reflection, little thought is given to ways of achieving full employment in an ecologically responsible society. The political process is expected to be the arena in which compromises between competing concerns are worked out. And we know that when the crunch comes, it is the environmental controls that are relaxed.

In the ecological model environmental quality, employment and reduced inflation are not tradeoffs. They have to be considered together from the beginning in a responsible way. This is the perspective of modern movements that work for conservation and full employment together.

Of course the difference between the two models on the subject of tradeoffs is not absolute. Those operating with the dominant model know that there are times when it is possible to increase two desirable variables together rather than being forced to reduce one in order to increase the other. In Okun's book the subtitle, 'The Big Tradeoff', points to his conviction that at many points equality is a tradeoff with efficiency, but he also considers how increase of equality of opportunity can increase efficiency as well.

On the other side, those operating with the ecological model know that the image of the tradeoff is sometimes applicable. A time may come when global food supplies will be so limited in relation to population that some people can be fed only at the cost of starving others. But when confronted by apparently conflicting goods, the ecological approach is to think first about the kinds of changes that would be required so as to make possible the attainment of both. With respect to equality and efficiency, this would involve changing the concept of efficiency by changing the goals in terms of which efficiency is measured. If people ask what systems lead to the increase of richness of experience and measure efficiency in these terms, then they may expect equality and efficiency to support one another to a much greater degree than when efficiency is measured only in relation to quantity of production of desired goods and services. Relative equality will become part of the goal in terms of which efficiency is understood.

Socialist and market economies

The question of economics is sometimes posed as the choice between capitalism and socialism or between the free market and the centrally planned economy. There are merits in these alternative systems of organising the economy, and it will be well to recognise this. But neither system is appropriate to the world's present needs.

The free enterprise system or the market economy, when great centralisation of economic wealth and power is avoided, encourages individual initiative. Even within socialist countries it has been discovered that farmers produce more efficiently on small privately controlled plots of land than on the great collectives (Weitz, 1971, p. 34). Where individuals find themselves rewarded for initiative and innovation, they are more likely to display these virtues to the gain of society as a whole.

Also a free market economy is more likely to employ resources efficiently than is a controlled economy. For example, in the United States, if each family had been required to pay the actual cost of supplying it with utilities, the suburban sprawl which is so wasteful of energy would have occurred much less. It is government regulations which have required city dwellers to subsidise their suburban fellow-citizens. Also government policies have kept the costs of established types of energy unrealistically low, thus encouraging its prodigal use and discouraging the development of new types, whereas a truly free market would, before now, have rewarded conservation, efficiency and inventiveness. Indeed, if the free market had worked ideally and Americans had met their energy needs in the most economical manner possible during the sixties and seventies, by 1978 they would have used 28 per cent less oil, 34 per cent less coal and 43 per cent less electricity (Sant, 1979).

Nevertheless, it is generally recognised that there are many social needs which the free market system handles poorly. Products dangerous to health and destructive of the environment are not discouraged by this system. The system does not prevent abuse of the commons. Also, the free enterprise system allows, if it does not encourage, vast concentrations of economic power in a few hands, concentrations which in fact subject the market to manipulations that violate its principles. Finally, the system does not solve the problem of poverty. Structures of political power are required to intervene and regulate, but these in their turn are subject to undue influence from those who wield economic power. Both the economic and the political structures tend toward increasing centralisation, bureaucratisation, destruction of individual initiative and massive alienation.

Socialism rightly teaches that the function of the economic structures is to serve the needs of all people. And socialist countries have on the whole reduced the differences between rich and poor more successfully than have those which relied on free enterprise. Further, the removal of a class of people who own the means of production and exploit the labour of others for their own profit can be a step toward a more just society. Socialist systems allow for long-term planning for the conservation of scarce resources and for the immediate prevention of the production of unhealthful and polluting goods, even if they have been slow to make use of these strengths. They can avoid the generation of false 'needs' through advertising.

But socialism, too, has its severe limitations. There is, as yet, no instance of a thoroughly socialist country in which wide diversities of opinion and theoretical dissent have been tolerated over a long period of time. Intellectual and cultural life, as a result, is at a lower ebb there.

The full and free development of Marxist thought, for example, is to be found in bourgeois countries rather than in those committed to Marxism. The Yugoslav Marxist philosopher, Mihailo Marcovič, wrote shortly before being expelled from Yugoslavia:

> private ownership of the means of production has not been transcended by really social ownership but has been modified into state and group property. Professional division of labour still largely exists, and work is as long, monotonous, stultifying, and wasteful as in capitalism. The market is no longer the exclusive regulator of production; it has been supplemented by state planning. But this latter way of regulating production is still far from being very rational and democratic, and it still preserves a good deal of profit motivation. The bourgeois state has not been transcended by a network of self management organs but has only been modified into a bureaucratic state which allows a greater (in Yugoslavia) or a lesser (in Russia) degree of participatory democracy in the atomic units of social organization. The party (as a typically bourgeois type of political organization) tends to be perpetuated. True, the social composition of the rank and file membership of the Communist Party shows a shift toward the working class, but the organization is even more authoritarian and ideological indoctrination even more drastic. The fact that there is only one such organisation which monopolizes all political power is hardly an advantage over bourgeois pluralism (Marcovič, 1976, pp. 87–8).

The choice between capitalism and socialism is a false one on three grounds. Firstly, both the free enterprise system and socialism are based, from the perspective spelled out earlier in this chapter, on an inadequate understanding of living human beings. Human beings are internally related to one another and to other creatures. This makes each individual unique and actually constituted by a network of relationships. Classical liberal economics abstracts 'economic man' from this reality and treats him as a system of infinite wants which it is the duty of the system to satisfy as far as possible. These wants establish what is truly desirable, so that the desire of the rich for luxuries is on a par with the desire of the poor for food. Indeed, since the market regulates not according to wants but according to ability to buy, it satisfies the desire for luxuries on the part of the rich in preference to the need of food on the part of the poor. Marxist thought is keenly sensitive to these evils of the capitalist system, and it is to its credit that it perceives human individuals as social beings. Nevertheless, 'social' easily becomes 'collective' in a depersonalising sense. People are treated as members of a class and as producers rather than in terms of their actual complex, personal individuality based on relational existence.

Socialist governments have at times been prepared to sacrifice the welfare of masses of people in one generation for the sake of realisation of the socialist ideal in future generations. It is often claimed as a merit of the market economy that it defends personal freedom against these abuses. To some extent this is true. But there are too many examples of nations which have adopted the market economy without the commitment of the political order to guarantee elements of equality and to keep the market in its place. In some countries political and military power is devoted to defending the market from the resentment of those who suffer from the extreme inequalities it maintains and increases. In these countries too, the masses of people in one generation are sacrificed. It is not clear whether they are sacrificed for the benefits of their descendants or simply for the benefit of the wealthy few for whom the free market works so well. But at least the moral justification that is offered is similar to that in socialist dictatorships.

According to Berger (1977, p. 176), Brazil's economic 'miracle', bought at the expense of the oppression of many of the Brazilian people, was justified by Brazil's Minister of Finance in these terms: 'True many people in Brazil today are suffering. But the government plans are designed to elevate Brazil to the status of a fully developed society by the end of the century. When this goal has been achieved, there will be a sharp and general alleviation of misery.' In other words, the suffering of this generation will contribute directly to the happiness of the next. In elementary human terms parents are made to sacrifice themselves to the future of their children. Whether this forced sacrifice of masses of people is carried out in the name of socialism or capitalism, it is a violation of human beings and is based on a false model of living things.

The Yugoslav Marxists of the *Praxis* group, to which Marcovič belongs, display a rare understanding of the relational and transcending character of human existence. The quotation above continues as follows: 'Real supersession of political alienation will materialize only when all monopolies of power are dismantled, when authoritarian and hierarchical organizations such as the State and Party are gradually transcended and replaced by self-governing associations of producers and citizens at all social levels' (Marcovič, 1976, p.88).

This vision is eminently appropriate from the point of view of the ecological model, and the fact that it comes from Marxists may be an indication that Marxism has more to offer. However, it is significant that it comes from a Marxist who was driven out of the most liberal of socialist countries. In some countries, such as Sweden, privately owned companies are experimenting with more worker participation in decisions about production, and there seems to be no inherent reason that market economies could not move toward much more worker ownership of businesses. These possibilities are well outlined by Okun (1975, p. 64). But he also notes lack of enthusiasm for moving in this direction on the part of both business and labour. It seems that new ways of understanding the situation and new structures of power will have to emerge in both socialist and capitalist countries before either can evolve in the needed direction. A process of decentralisation is required which is in contrast to past development in both. From the ecological perspective success in developing this kind of participatory economy is more important than the choice between capitalism and socialism.

In the developing world there is a second, related reason why the choice between capitalism and socialism is a false one. Both of these systems are designed for industrial economies. The real choice in many developing countries is between a programme that begins with urban-industrial development and one which is based on the rural-agricultural sector in which most people still live.

It is clear by now that the former choice often leads to massive injustice and to unsustainable forms of society. It creates an urban, Westernised élite dependent on expensive imports in the midst of an impoverished peasantry and a largely unemployed urban proletariat. The latter choice can lead to a just and sustainable society in which urban industrial life can develop gradually. In countries where there are large estates, land reform is essential both for justice and for adequate food production. In countries where rural communities have been neglected, these communities need help in taking responsibility for their destinies and in adopting appropriate technologies. They need elementary medical care and education. They need systems of security for old age so that people will not depend for their survival upon their children. They need help in family planning, in reforestation and in the development of cottage industries. With such assistance they can constitute the sound political and economic base on which urban life can be built. They can absorb more of their increase in population rather than export it to swell the roles of urban unemployed.

Mahatma Gandhi was fully aware of the importance of the village as the basis for Indian life. He chose the home spinning wheel as the symbol of his movement for independence from Britain. After his death, his successors viewed his economic ideas as sentimental. They invested heavily in huge urban industry. The result was the decline of village life and rising unemployment in the cities. The wisdom of Gandhi's vision was vindicated, and in November 1977 the ruling Janata Party pledged to dismantle the textile, shoemaking and soap-making industries and to return the production of these goods to the villages (Norman, 1978, p. 14).

Such a society will be relatively self-sufficient. That by no means forbids trade. It does mean that the country's basic policies and programmes should not be geared to the exigencies of world trade. The feeding of its own people should take precedence over using its agricultural resources for growing crops for export. Nations should avoid as far as possible making themselves dependent on imports for the basic needs of their people. They should also avoid importing goods that benefit only a tiny minority of the people. In short they should aim at a high level of self-reliance.

Mahbub ul Haq, who was one of Pakistan's principal planners during the late fifties and sixties, later recognised the need for this direction of development. 'We were taught to take care of our GNP, as this will take care of poverty,' he said, 'but let us reverse this and take care of poverty, as this will take care of the GNP' (Brown, 1978*a*, p. 224).

A third reason why the choice between capitalism and socialism is a false one is that the need is for economic systems that are not fundamentally oriented to growth. Both capitalism and socialism are economic systems committed to such growth. As Daly (1977, p. 8) says:

> The appeal of growth is that it is the basis of national power and that it is an alternative to sharing as the means of combating poverty. It offers the prospect of more for all with sacrifice by none . . . If we are serious about helping the poor, we shall have to face up to the moral issue of redistribution and stop sweeping it under the rug of aggregate growth.

People are addicted to growth, says Daly, because they are addicted to large inequalities in income and wealth.

It may be that a system based on state planning is better able to adjust to a steady-state economy than is a system based on the free market. This has not yet been established since no socialist government has accepted a steady-state economy as its goal. In any case, the requirement of a steady-state, sustainable economy cuts across the choice between capitalism and socialism. It stands as a challenge to both.

Surprisingly little attention has been given to the conditions necessary to achieve a healthy non-growing industrial sector. An exception is the work of Randers (1977) in Scandinavia. He has established alternative models for the Scandinavian timber industry which has now reached the limits of sustainable production. It is not possible to cut trees down any faster than is being done at present, without making the industry unsustainable. One of his models is exemplified by the existing forest products industry in Finland. In response to reaching the sustainable limit in the mid-1960s this industry formed a central agency to oversee the expansion plans of individual companies to ensure that total capacity did not exceed available wood supplies. This agency has been successful in limiting the expansion of the forest products industry. In large part, says Randers, this is probably due to the understanding of everyone interested that violations would threaten the whole industry.

The significant fact is, that after more than a decade of such regulation, the no-growth Finnish forest products industry is still alive and well. Randers suggests that in an attempt to establish an equilibrium in production far-sighted entrepreneurs in particular industries could gather around them people of the same bent and start no-growth co-operatives. National authorities could support such developments by gradually introducing subsidies to industries that limited expansion. The challenge lies in developing viable new institutions to govern economic and social relations in the material steady state. An important function of these institutions will be to run retraining schemes for people whose employment is threatened and to phase them into alternative creative employment.

In many countries of the industrialised world a satisfactory steady-state economy could be at a considerably lower level of consumption than the present one. Present production is more than sufficient to feed, clothe, house and provide basic health care and education. There are problems of poor distribution of goods and of social organisations that turn luxuries into necessities, and these must be addressed in a steady-state economy. But the major problem is that our whole economy is based on continuing growth whether its products are needed or not. It has encouraged the development of a society that continuously redefines its needs upwards. And it exhausts the resources of the planet while polluting it with wastes.

This growth economy is now faltering. Growth rates in the seventies have declined along with rising unemployment and inflation. Many countries have cut their future projections of growth in the socialist world as well. We seem to be entering a no-growth period (Brown, 1978*a*, pp. 189–90).

Economists worry that without growth, unemployment cannot be avoided. The growth economy requires it. But as Daly (1977, p. 126) says:

> A condition of non-growth can come about in two ways: as a failure of a growth economy, or as the success of a steady-state economy. The two cases are as different as night and day. No one denies that the failure of a growth economy to grow brings unemployment and suffering. It is precisely to avoid the suffering of a failed growth economy (we know growth cannot continue) that we advocate a steady-state economy. The fact that an airplane falls to the ground if it tries to remain stationary in the air simply reflects the fact that airplanes are designed for forward motion. It certainly does not imply that a helicopter cannot remain stationary.

A growth economy and a steady-state economy are as different as an airplane and a helicopter.

One feature of growth economies has been particularly absurd. They have encouraged the exploitation of new resources rather than conservation by reuse or recycling. The United States still has depletion allowances to encourage the consumption of virgin materials and freight rates that are higher for secondary materials. Obviously advantageous proposals for a system of returnable bottles still meet such powerful opposition that it is not politically expedient to press for them (Hayes, 1978, pp. 33–4). If, instead of a system of incentives for waste, we adopted an opposite policy, Hayes believes that 'at least two-thirds of the material resources that we now waste could be reused without important changes in our life-styles. With products designed for durability and for ease of recycling, the waste streams of the industrial world could be reduced to small trickles. And with an intelligent materials policy, the portion of our resources that is irretrievably dissipated could eventually be reduced to almost zero' (p. 5). Further, by reusing materials a large amount of energy could be saved in comparison with processing virgin ore. The percentage of such saving varies from 8 per cent in the case of glass containers to 97 per cent in the case of copper (p. 17).

Even if growth economies were sustainable, we would have to challenge their justice. The Mobil Oil Company has published such advertisements as the following: 'Growth is the only way America will ever reduce poverty ... While the relative share of income that poor people get seems to be frozen, their incomes do keep pace with the economy ... Even allowing for inflation, the average income of the bottom tenth of the population has increased about 55 per cent since 1950' (Johnson, 1975, p. 168). Mobil Oil's facts may be accurate, and the advertisement accurately reflects widespread thinking of the dominant economists. But the fallacy of this approach is being recognised more widely today. First, students are reluctantly coming to the recognition that there can be great growth of GNP in developing nations accompanied by tragic consequences for the poorest people. According to Morawetz (1977, p. 68) many of the poorest people in the poorest countries 'improved their economic situation very little during the twenty-five years; although the issue is in dispute, significant numbers of them may even have become worse off in absolute terms'. Others who view the situation in more concrete and personal terms rather than economic ones have little doubt about the suffering that has accompanied development!

Furthermore, even where the poor have followed the rich in their proportion of economic improvement, the problem of poverty remains critical. Morawetz (1977, p. 38) recognises that 'relative, and not only absolute poverty is indeed an important concept even for the very poorest people'. He further notes that in countries following dominant development policies 'income inequality increases in the early stages of development'. He thinks it diminishes when the country becomes more prosperous, but

> a society that begins with an unequal income distribution is quite likely to remain unequal or become more so, whereas one in which initial disparities are small may be able to avoid significant increase in inequality ... Once growth is taking place and incomes are being earned, it seems to be difficult to redistribute income by means of taxes, public employment, and the like ... The historical evidence suggests that it simply may not be possible to 'grow first and redistribute later' (p. 41).

Finally, those who support the trickle down theory by attempting to show that the poor also gain by general increase in wealth, neglect the pattern of changes in living requirements that often accompany such growth. A poor person may have 55 per cent more purchasing power than in 1950, but the structure of the society may make ownership of a car necessary now whereas much less expensive transportation may have sufficed then. If so, much of the apparent gain vanishes.

A steady state economy in the developed world might well move toward a smaller GNP. A steady state economy in the developing world would move toward a larger GNP. The enormous disparities now existing between rich and poor nations should be greatly reduced. But justice does not require an equal ratio of population to GNP in all countries. Indeed, any way we set out to achieve complete equality could hardly fail to induce enormous suffering and loss of freedom.

One reason for the pointlessness of seeking equality in this sense is that the GNP is an extremely poor indicator of the welfare of the people. We have already seen that the people of Sri Lanka have been much better off than the people of Iran under the Shah, despite the fact that the *per capita* income in Iran was ten times as high. The goal of justice is the opportunity of a good life for all the people of the world, and this would indicate that more improvement is needed in Iran than in Sri Lanka! The ideal is that the good life is to be made possible by the least production and consumption necessary rather than by its maximum increase. Accounts of justice and sustainability in relation to agriculture, the role of women, energy, transportation and the urban habitat which show that improvement of life can be attained by means that do not require increased consumption are given in Chapter 10.

Even if an accurate measure of the real extent to which countries offer opportunity for the good life to their people were found, equality could function only in a very vague way as an ideal. A healthy international order would be one in which individual countries and communities were encouraged to experiment freely, reaping inevitable rewards from success and penalties from failure. The hoped-for world would be one in which cultural diversity would flourish, and such diversity counts against abstract notions of equality. An exaggerated stress upon equality almost inevitably leads to a static order. But the fact that equality in this sense is one ideal among others, and one that should be subordinated to others, in no way suggests that the gross disparities now existing are tolerable.

Conclusions

If economic growth cannot be continued indefinitely, economic theories geared to growth need to be re-examined. The model in terms of which traditional economics has understood human beings is in conflict with the ecological model. Present economic theory cannot simply be kept as it is with the addition of a principle of limits to growth. A new economic theory built on the ecological model of human life is needed. In some respects the model of economic man is very much like the mechanistic model in biology. But it differs in that human wants are taken seriously in the economic model. Economics aims at the satisfaction of those wants and in this sense it considers human beings as ends. However, people with their individual wants are conceived as essentially self-contained units which can be viewed in abstraction from their relations to others. These relations are seen as incidental to essential human life and satisfaction. The call of the ecological model to see living things, and especially human beings, as constituted by their mutual relations has just as radical consequences for economics as for biology.

Neither socialism nor the free market economy in any existing formulation is based on an ecological model of human existence. There is no mystery as to the broad outlines which the needed new economic theory must take. Some of the practical implications of the needed paradigm shift can be seen even when the detailed theory is still lacking.

Economics and politics are intimately interrelated. Accordingly, much in this chapter is political as well as economic. Nevertheless, the political problems that would have to be faced in implementing the paradigm shift called for have not been considered. In particular there is the troubling fact that one main reason for the political commitment to economic growth is that this is seen as enhancing political, and especially military, power.

10

Rural and urban development in ecological perspective

> Meaningful action is whatever increases the confidence, the autonomy, the initiative, the participation, the solidarity, the egalitarian tendencies and the self-activity of the masses, and whatever acts in the demystification. Sterile and harmful action is whatever reinforces the passivity of the masses, their apathy, their cynicism, their differentiation through hierarchy, their alienation, their reliance on others to do things for them and the degree to which they can be manipulated by others – even by those allegedly acting on their behalf.
>
> Paul Cardan (1974, p. 2)

Human beings live in societies that are, to varying degrees, restrictive, oppressive and corrupt. For many people there is little hope of any real fulfilment of life because of the structures of the societies in which they live. Mahatma Gandhi said: 'Some people are so poor that God can only appear to them in the form of bread.' So whilst the primary task before us is a spiritual one that has to do with a reorientation of human life to new values, meanings and purposes, there will be no such reorientation for many until the structures of society are changed. Whilst the options for a full life are denied, there is no real liberation of life.

The ecological model of development poses a challenge both for the transformation of the meanings and purposes people hold and for the transformation of the structures of society such that these purposes can be carried out. One does not come after the other. Both go together.

Liberation of life takes place today in the struggle for faith in Life against apathy and despondency; in the struggle for human freedom and human rights against political oppressors; in the struggle for the rights of non-human life against its exploiters; in the struggle against racism, nationalism and sexism; in the struggle for economic justice and in the struggle for ecological sustainability in both urban and rural development.

This chapter is primarily concerned with the struggle for justice and sustainability in rural and urban development. In this context the rights of women require special attention.

People with ecological sensitivity are often accused of being prophets of doom and insensitive to injustice. Many are indeed profoundly troubled and fearful that humanity is headed for a precipice. But the ecological perspective is not in principle pessimistic about the possibilities of the future. It is pessimistic only about the consequences of continuing along the lines of thought and action of the dominant model of development. Nor is the ecological perspective indifferent to justice; on the contrary, it calls for the rich world to curb its use of resources in order that the poor world may share more fairly. One of the characteristics of the dominant model of development is that it works from the top down. Development policies are for nations. Land is viewed as a resource for national economic development. Individuals are viewed abstractly as producers and consumers, many of whose individual differences and personal needs can be ignored. The ecological model on the other hand begins with organisms which cannot be abstracted from their environment. It stresses the need for a transformation of values in individuals and small groups rather than a coming to power of new elites. It points to the relation of human beings to the land and to the community of organisms the land supports and from which people derive their sustenance. As has already been indicated in Chapter 8, all the resources needed in the modern world, with the exception of petroleum based products and minerals, derive from four basic biological systems; croplands, pastures, forests and fisheries. All four systems are being overstressed in way they are managed at present. The first section of this chapter deals with what is happening to these systems, focusing on modern agriculture. It points to some hopeful examples of a just and ecologically appropriate agriculture.

The second section deals with the role of women in the just and sustainable society. As Rosemary Ruether (1975) shows in *New Woman – New Earth* and as Altman (1980, p. 120) argues, male attitudes of dominance over nature and over women are closely related. Development programmes have been planned and implemented by men. Although women constitute more than half the world's population, their role in the economy has largely been ignored. As a result of neglect, the condition of women has often worsened, even where developers suppose themselves to have had some success. Efforts to deal with the critical questions of excessive population growth without addressing the legitimate claims of women for justice are doomed to fail.

The third section deals with justice and sustainability in the production and use of energy. Nuclear energy was discussed in Chapter 8 where the hope was expressed that it would play a diminishing role in the future rather than be appealed to as the basis for an indefinite continuation of economic growth. But even if limits are accepted and there is a move toward a steady-state economy, enormous quantities of energy will still be needed to continue fuelling the world's industry. Meanwhile much of the poor world is already facing an acute crisis in the growing scarcity of firewood – one of its major sources of energy.

The fourth section is concerned with transportation and the urban habitat. One of the major pressures on the environment is caused by present transportation systems, especially the private automobile. In the developing world public policy oriented to the bicycle, supplemented by buses, is more just and sustainable than the building of highways for the few who can afford motor cars. In the developed world, cities could be built in such a way that dependence on private cars and even buses would be greatly reduced. Building cities in that way would reduce the need for energy for other uses as well. It would provide a new and more favourable context for dealing with many of the acute social problems which are accentuated in existing cities.

No one can be an expert in all these areas. But it is the right and even the duty of non-experts to challenge experts to think in new ways. Most experts, whether they know it or not, operate from a world view and models established by the recent history of their disciplines. The task is to foster a hearing for those who view their work from a different perspective that is conducive to the aims of the ecological model. A community of such thinkers is growing, but it is still on the fringe of the establishment. Nevertheless, through them is emerging a vision of a more livable future in which life could be liberated as never before.

A just and sustainable agriculture

No society is sustainable that does not have sustainable sources of food. The main sources of food now and for the foreseeable future are agriculture and fisheries. Overfishing and pollution are the two threats to a sustainable world fisheries. The fact that the total global fish yield reached a peak in 1970 and the collapse of a number of major fish stocks, such for example as the anchovy of the Peru current and the herring of the north-east Atlantic, suggest that present practice may not be sustainable. One problem is that fisheries management has been based on knowledge of stocks of single species only, whereas no species lives in isolation. A more ecological approach is now called for in which the abundance of food of the stock and its predators and all other components of environment that might affect numbers are taken into account. The time is overdue for a more realistic ecological approach to fisheries management (Gulland, 1976, p. 302 *et seq.*; Robinson, 1980, p. 20).

Some forms of agriculture are sustainable. Others are not. The contrast between the pastoral husbandry of temperate northern Europe and the herding and cultivation of the arid Mediterranean Europe is a contrast between a sustainable agriculture and one that was not. In northern Europe, even though forests were cut down there was emphasis on their replacement with grasses and legumes to provide food for stock and to maintain a soil cover. The original soils of western Europe, except for the Po valley and some parts of France, were in general very poor in quality. They are now highly productive and have been for centuries. In arid Mediterranean Europe over large areas the vegetation was overgrazed and the soils eroded to the point where neither plants nor animals could be sustained. Even today that deterioration continues. For example, a report of the Organisation for Economic Co-operation and Development (OECD) described the extensive abandonment of farmland in Italy thus: 'it is generally agreed that in Italy two million hectares have been abandoned in the last ten years ... the farming methods used in the marginal land have led to deterioration of the soil so that the land was consumed in the literal sense of the term' (Brown, 1978*b*, p. 18).

There are three main problems associated with the sustainability of modern agriculture; loss of soil and mineral nutrients through erosion, excessive use of pesticides and the form of energy input.

Soil erosion and loss of mineral nutrients occur in practically all climatic conditions. In Java, as a result of overpopulation, hillsides have been deforested and then misused by land-hungry farmers. 'Erosion is laying waste to land at an alarming rate, much faster than present reclamation programmes can restore it' (Brown, 1978*b*, p. 18). Similar stories can be told from the wet regions of countries such as Brazil and Venezuela. In arid and semi-arid regions it is the world's grazing lands that suffer most from 'desertification'. In 1977 a UN Conference on desertification considered that nearly one-fifth of the world's cropland was being degraded and had already suffered an average 25 per cent reduction in productivity. This generally results from overgrazing which itself often springs from the introduction of modern technology such as sinking of wells. The increased supply of water results in more stock being added which overgraze, especially in drought years. This happened in the Sahel countries in the 1960s and 1970s. 'As the number of people who rely on the pastures and crop-lands of the arid zones climb, once sustainable social patterns and production techniques begin to undermine the biological systems on which life depends' (Eckholm & Brown, 1977, p. 7). Eckholm & Brown estimate that 650 000 square kilometres of land once suited for agriculture and grazing has been forfeited to the Sahara over the past 50 years along its southern fringe. In the Sudan the desert boundary has moved south by an average of 90–100 km in the past 17 years. Desertification is also a major problem along the northern edge of the Sahara in Morocco, Algeria, Tunisia and Libya. Overgrazing and the extension of grain farming into lands that cannot sustain it has led, according to one estimate, to the loss of more than one hundred thousand hectares of land to desert each year (Eckholm & Brown, 1977, p. 11). The problems of desertification are by no means confined to Africa. Deserts are creeping outwards in Asia and Latin America as well. And ill-managed range lands in Australia and the United States have lost much top soil and vegetation.

The loss of soil by erosion is also a major problem in productive cropland outside the tropics. Techniques designed to raise productivity in the near term lead to excessive loss of soil, such for example as the continuous cropping of corn in the American Midwest. A report from one of the world's most productive farming areas says: 'The 200 million tons of soil lost from Iowa cropland each year simply cannot be replaced within our life time or that of our children' (Brown, 1978*b*, p. 23). President Carter said in his message to the Congress on Environmental Priorities and Programs, August 1979: 'Over the past half-century we have invested $20 billion of federal funds in efforts to conserve soil ... Yet in that same half-century wind and water erosion have removed half the fertile topsoil from nearly one-third of the nation's potentially usable croplands' (Council on Environmental Quality, 1979, p. 743).

According to a 1977 report by the California State Water Resources Control Board, there were critical problems with 95 000 000 hectares in that state. Annual loss of soil ranged from 0.5 to 2.4 tonnes per hectare

(Flannery, 1979, p. 19). The United States Secretary of Agriculture, Bob Bergland, explained what happened:

> When grain prices went up so high, as they did in '73 and '74, a lot of marginal land was brought into production – shelter belts bulldozed over, terraces torn up, farm land that was put in good conservation use was put into a policy of mining during that period, and some of those bad habits have held over. We are on a collision course with disaster ... Our water supplies are being depleted, and we have mined our soil, in fact the erosion of America's farm land, today, is probably at a record rate, and this simply cannot go on (Face the Nation radio broadcast, 26 November 1978).

Much farmland needs to be protected against over-intensive use by inclusion of fallow periods or alternating crops with pastures. In 1973 the US Department of Agriculture abolished the Soil Bank – a programme that paid farmers to keep part of their land out of production. The poorest land was kept fallow most of the time and was thus protected against erosion. Abolition of the Soil Bank led farmers to continuous cropping over larger areas of their farms. With the neglect of soil conservation practices soil erosion rates rose to equal those of the 1930s Dust Bowl era. In addition to the loss of soil with loss of productivity this erosion results in the pollution of waterways and silting of reservoirs (Lockeretz, 1978; Risser, 1978).

Irrigation is a sustainable form of agriculture only provided there is adequate underground drainage or an effective flow of surface water out of the fields. The annual flooding of the Nile in Egypt flushed accumulated salts out of the soil each year. For thousands of years the valley of the Nile was one of the world's most productive areas. Irrigation projects in Egypt outside the Nile valley flood plain have problems of salt accumulation. So does the Nile Valley itself, now that the Aswan dam has eliminated flooding and the natural soil desalination. Waterlogging and salinity on both new and old farmlands in Egypt are problems that have yet to be overcome if agriculture is to become sustainable again (Eckholm, 1976, p. 118).

A United Nations report in 1977 indicated that some 21 million hectares of irrigated land were waterlogged, one-tenth of the total irrigated area. The productivity of this land had fallen by 20 per cent. An estimated 20 million hectares was affected by salination, with again about a 20 per cent reduction in productivity (Brown, 1978*b*, p. 19).

There are some success stories. Much of Israel's Negev desert which suffered thousands of years of overgrazing and deforestation is now productive and prosperous as a result of innovative irrigation practices, improved dryland farming and controlled grazing. China too has halted deterioration and boosted productivity in many of its huge desert areas (Brown, 1978*b*, p. 34).

A second threat to sustainable agriculture, especially where crops are grown as huge monocultures, is the excessive use of pesticides. The chemical-based system of controlling pests and crop diseases is now widely recognised as an ecological disaster which also fails to solve the pest problem (van den Bosch, 1978). Pests have become resistant to the pesticides. More and more toxic pesticides have been developed that now constitute a threat to humans. In addition valuable predators of pests have been selectively killed by the new pesticides so that the last state is often worse than the first. In the long run the pests win out unless other means of controlling them are initiated. van den Bosch (1979) wrote of pesticide usage in the United States: 'thirty years ago, at the outset of the synthetic-insecticide era, when the nation used roughly 50 million pounds of insecticides, the insects destroyed about 7 per cent of our pre-harvest crops; today, under a 600-million-pound insecticide load, we are losing 13 per cent of our preharvest yield to the rampaging insects'.

The basic principle of so-called integrated control of pests is to combine a limited use of pesticides geared to life cycles with the use of predators and ecological changes in farming practices such as mixed crops instead of monocultures. For example in Indonesia combinations of corn and rice have been shown to be both more resistant to pests and more responsive to the application of nitrogenous fertilisers (IUCN, 1980, p. 14).

The energy input of modern agriculture constitutes a third threat to sustainability. Fertilisers require huge amounts of energy in their manufacture, as does the operation of farm machinery. In addition, in developed countries, there is a huge expenditure of energy in the processing, distribution and preparation of food. The cost of energy most affects poor countries that import petroleum and have inadequate foreign exchange such as Bangladesh and India. An appropriate technology in these countries would be labour intensive rather than capital intensive and would use organic rather than chemical fertilisers.

In terms of petrol the equivalent of 728 litres is used in the United States to raise one hectare of corn. For every joule of energy input the yield is only 2.8 joules. In rice paddies in China one joule input gives

50 joules of crop yield. The cost of moving from 'primitive' agriculture to modern agriculture is a huge 'energy subsidy' (Brown, 1974, Chapter 8).

In the United States about 5 per cent of the total national energy budget goes to producing food on the farm with an additional 7 per cent to processing, distribution and preparation (Ehrlich *et al.*, 1977, pp. 350–1). On a world basis, Campbell (1979, p. 48) says that consumption of energy on the farm constitutes about 3 to 4 per cent of the national energy budget.

If the world food system used the amount of energy to produce, transport and process food that the United States uses the total energy used would constitute about 40 per cent of the world's commercial fuel consumption in 1972. The likelihood that such an energy use could be sustained is remote (Ehrlich *et al.*, 1977, pp. 350–1).

Bearing in mind these three threats to a sustainable agriculture, consideration is now given to some of the special problems in developing and developed countries and how these might be faced realistically.

In developing countries the long term needs for food will have to be met in the main by food grown within these countries, not by imports. Rising costs of food and transport make mass movements of food across great distances prohibitive, except in emergencies. At present only 4 per cent of food produced in the world crosses a national boundary. To increase this percentage substantially would involve a massive increase in transport and in the use of energy. It is better to produce food where it is needed for other reasons as well. For example, one additional tonne of appropriate fertiliser will produce about twice as much additional rice or wheat in South East Asia as in Australia or North America. However, until developing countries become self-sufficient in food there will be an ever-present need for emergency food supplies from abroad. The main component of this programme would be an international 'food bank' controlled by a transnational agency. The reserve would act as a buffer against food shortages during drought and other emergencies. The United States and Australia and other countries which produce far more than they themselves consume would be suppliers to such a bank.

Great hopes were placed in the 'Green Revolution' as a major component of a sustainable food supply for developing countries. However, the Green Revolution depends upon high energy technology and this imposes huge costs with rising fuel prices. To reduce energy imports developing countries must look towards substituting manpower, currently replaced by machines, to organic fertilisers rather than to chemical fertilisers and to practices which reduce the need for pesticides such as multiple crops in single fields. Much of the Green Revolution was promoted and run by 'agribusiness' in the form of huge farms. Apart from the problem of the sustainability of these methods in developing countries agribusiness has been singularly unsuccessful in channelling the food produced to where it is most needed. Some have claimed that more food from agribusiness has meant more hunger. The food goes to those who can afford to buy it, not to those whose need is greatest, namely the poor (Lappé & Collins, 1976). As a result of these experiences the World Bank has now moved its support away from agribusiness to the small farmer. The object here is to enable the small farmers to get going by means of loans and technical advice in the hope that they will first become self-sufficient in food for their immediate families and then later for a wider community.

The substitution of capital intensive and energy intensive farming for traditional methods may have been a quick way of increasing agricultural production but its long term consequences could be disastrous. Bede N. Ikigbo, Assistant Director of the Farming Systems Programme, International Institute of Tropical Agriculture, Ibadan, Nigeria, pointed out at the World Food Conference of 1976 that the agricultural methods of the developed world are inapplicable to millions of small farmers who produce the food for the developing world. He recognised that some chemical fertiliser is needed, but held that this should be minimised by the use of biological methods. Similarly, integrated pest management should be encouraged rather than dependence on general use of chemical insecticides. He concluded as follows:

> It should always be borne in mind that: (1) improved agricultural technology can best function as integral components of improved rural development programs, and (2) all efforts devoted to increase food production should be based on the philosophy that we are mainly buying time until various other measures such as education and general improvement of the welfare of the world's rural majority can facilitate integrated approaches to the solution of the population and food supply problem. All available resources of all governments and private bodies should be directed towards this end (*Proceedings*, 1977).

The key to a just and sustainable society for much of the world's population lies here rather than in the more visible and highly touted industrial development.

Ikigbo is himself working on more radical changes in tropical agriculture. He practices what the forest teaches. With deep litter, or mulching, degrees of shade and sunlight from trees, he keeps 'forty to sixty plant varieties per acre under cultivation at all times'. The forest teaches diversity, the constant cover of the soil, and sustainability. He is applying these lessons and achieving a natural control of pests and freedom from fertilisers. He accomplishes this with little or no tillage. This looks like a sustainable agriculture in tropical Africa.

Other important experiments in zero tillage agriculture are taking place in Japan. For many centuries Japan had a sustainable system of wet rice production, but after World War II Japanese farmers were persuaded to shift to American methods of farming. The result has been deterioration of the quality of the soil. To show the possibility of reversing this trend, Masanobu Fukuoka took land that had been worn out by abuse and experimented successfully with a new style of agriculture influenced by traditional Oriental philosophy. Without irrigation, tillage, mechanical equipment, chemical fertiliser or insecticides, his 'do-nothing' method of co-operating with nature has brought the land back to fertility while producing crops comparable to the best of the ordinary farms in Japan. His methods are summarised by Larry Kohn in his introduction to Fukuoka's book (Fukuoka, 1978, p. xxi):

> In the fall Mr Fukuoka sows the seeds of rice, white clover, and winter grain onto the same fields and covers them with a thick layer of rice straw. The barley or rye and clover sprout up right away; the rice seeds lie dormant until spring . . . The rye and barley are harvested in May and spread to dry on the field for a week or ten days. They are then threshed, winnowed and put into sacks for storage. All of the straw is scattered unshredded across the field as mulch. Water is then held in the field for a short time during the monsoon rains in June to weaken the clover and weeds and to give the rice a chance to sprout through the ground cover. Once the field is drained, the clover recovers and spreads beneath the growing rice plants.

Thomas Odhiambo of the International Centre of Insect Physiology and Ecology in Nairobi, Kenya, has recently pointed to the widespread rejection of 'the erstwhile clarion call for the application of existing scientific knowledge to LDC problems' (Odhiambo, 1980). In some instances traditional agricultural methods are far more sustainable than those introduced by developers from the temperate zone. For example, the Wakara were not able to shift agriculture from one field to another because the land on their island is too limited and must support one person on every two arable hectares. They have devised a successful and sustainable solution which could suggest better solutions to problems elsewhere than are offered by supposedly scientific proposals introduced from a different context. Odhiambo describes the system as follows:

> The Wakara have also devised a 3-shift rotation system. In the first shift, bullrush millet is sown soon after farmyard manure has been applied to the land. After the millet has germinated, a slow-growing leguminous crop for making green manure is interplanted with the millet, which allows the latter to be harvested long before the leguminosae are dug in as green manure several months later. The leguminosae are dug in at the beginning of the second year, thus allowing the second shift of crops (millet, interplanted with groundnuts) to be sown. In the third year, millet is again sown, after farmyard manure has once again been applied, and this is later interplanted with sorghum and often also with cassava.

Another proposal for a sustainable agriculture in tropical Africa is to shift from domesticated animals to indigenous animals such as deer. Whereas the grazing and watering habits of domestic cattle are highly destructive of the vegetation and the soil in much of semi-arid Africa, native mammals have lived in large numbers in these areas without this negative impact. Efforts along these lines have been under way for at least 15 years. As yet they have received little support from agricultural experts, economists and government leaders. The overwhelming investment in research and development continues to be in cattle. Appraising the concept of cropping wild animals in preference to domesticated animals (Caughley, 1976, p. 235) has this to say:

> Two fair comments can be made on this theory. First, it is a theoretically sound and eminently reasonable expectation. Secondly, attempts to demonstrate its validity over the past 15 years have been largely unsuccessful. That paradox cannot be resolved easily. On one hand the theory squares with modern concepts of ecology: on the other, I know of not one example where a sustained yield of game was shown to be more valuable than a sustained yield of live

> stock in a comparable area. Some projects lost money, others barely broke even; and others, which at first glance appeared to support the theory, proved on second glance to have extracted a short term profit by capital reduction instead of by sustained yield. The initial failure of game cropping in Africa may have several causes. The most likely is the short period over which it has been tried. The failures might reflect little more than teething problems. Certainly, 15 years is not enough to achieve an optimization that farmers reached through thousands of years of trial and error.

Caughley concludes that harvesting game is at present economically unviable as compared with harvesting domesticated animals. However, what we really want to know is if it is more sustainable in the long run. Caughley (personal communication) doubts this on the grounds that domesticated cattle have been farmed in Africa for thousands of years. Still we do know that much of the African bush is being degraded by overgrazing by cattle. One could imagine, that with better supplies of wild game and the cultivation of the market, returns to the game farmer might improve. Many worthwhile developments are possible in this field.

David Hopcraft has pursued some of them. He wrote his thesis at Cornell University on 'The production of Thomson's Gazelle in East Africa'. On the semi-arid rangelands of his birthplace, where rainfall is between seven and fourteen inches a year, Hopcraft currently reports harvests of 300 per cent more meat per acre with gazelle than from the best managed Boron cattle. In addition gazelle meat has no fat deposits. Most important, no environmental damage can be detected. Gazelles do not track to water holes, as cows do, and they browse in ways which preserve the variety of indigenous plants. In the context of an economic system geared to cattle production and with more opposition than help from national and international leaders, problems continue. But a sustainable world requires that these problems be solved, not that the project be abandoned because it does not fit into the patterns of the past.

Half the world's population lives in the rural areas of developing countries (IUCN, 1980, p. 14), yet throughout the developing world many governments have concentrated their resources on urban development and have continued colonial policies of exploiting the agricultural sector for export crops. The results have been widespread misery for the rural poor and food deficits for most nations. A 1978 FAO/WHO study reports that 73 per cent of the Iban people on

Sarawak on Borneo suffer seriously from malnutrition. The agriculture there revolves around rubber and pepper for export. The farmers do not earn enough to buy adequate food, which must be brought from a distance. Agricultural research in Sarawak continues to be directed to increasing production of these export crops introduced by British colonialists in the nineteenth century.

In its statement on a World Conservation strategy the International Union for the Conservation of Nature (IUCN, 1980, p. 14) considered that the most serious conservation problem faced by developing countries was the lack of rural development. In their struggle for food and fuel, growing numbers of desperately poor people find themselves with little choice but to strip large areas of vegetation until the soil is wasted or blown away. An essential component of any strategy for a just and sustainable society in the developing world is to help hundreds of thousands of villages in which the mass of the farmers live to become more effective in their use of their resources and more self-sustaining.

In the developed countries also, agriculture faces problems of sustainability most of which have already been referred to. There remain two that have not yet been considered. In the developed world at least 300 square kilometres of prime agricultural land are submerged every year in urban sprawl (IUCN, 1980, p. 2). Only a new approach to urban living can help to halt this threat to agriculture. Secondly, the replacement of family farming with agribusiness is considered by some to be a major contributory cause to unsustainability in the long run. According to Brown (1978*a*, p. 286), 'By far the most successful agricultural systems are those built around the family farm.' Prior to World War II the United States was a land of family farmers. For generations these farm families had constituted the backbone of the American spirit of initiative and independence. During the years since World War II the number of farm owners has been cut in half by economic pressures caused by the mechanisation of farming. As farms became increasingly mechanised most of the farmhands moved to the cities, often to swell the ranks of the unemployed. Yet in spite of mechanisation and energy-intensive systems, farm income failed to keep pace with personal income generally. The heavy costs of capital outlay and fertiliser prevented the expected rise in profits (Commoner, 1976, pp. 159–69).

That the shift from family farms of modest size to agribusiness plantations has been socially detrimental, has long been established. The social cost was documented by Gaylord (1947) in his classic study of three Californian towns. But it has long been supposed that, contrary to Lester Brown's claim, agribusiness is required for maximum efficiency. This is now in question, in spite of government policies which encourage agribusiness. The economic limitations of agribusiness are sometimes recognised by its own leaders. For example, when Tenneco pulled out of agriculture, its November 1975 *Agricultural Report* stated: 'Agriculture is a high-risk business and typically shows little if any profit, especially for the large corporations ... There is no substitute for the small- to medium-sized independent grower who lives on or near his farmlands, who ... has a deep personal involvement in the outcome of his efforts' (Perelman, 1977).

There is hope that the shift away from family farming in the United States is slowing down. A shift back to the farm may be encouraged by the possible enforcement of a long-ignored legislation limiting water rights on federally irrigated land to family-sized holdings. As capital, machines and energy become more expensive, agribusiness may have increasing difficulties, and family farmers may hire more farmhands to assist them with their chores. This should lead to some reduction of unemployment. Agricultural schools and government agents may begin to shift their attention from ways of making agribusiness more profitable to ways of making family farms sustainable as well as profitable.

But more is needed. Although family farms are likely to abuse the land less than agribusiness, most of them have been forced to become heavily energy and insecticide dependent. We anticipate that in the interests of sustainability agricultural development will move in the direction of reducing dependence on oil and chemicals, and experimenting with farming methods that take maximum advantage of the sun and of the nitrogen fixing capacity of legumes and the use of organic fertilisers (Oelhaf, 1979). Another possibility is the cultivation of crops with little or no tillage and a switch from annuals to perennials (Jackson, 1978). We need radical experimentation and visionary thinking of a sort that characterised some of the earlier history of scientific agriculture.

A just and sustainable role for women

The roles of men and women in a just and sustainable society will be different from those many societies assign to the sexes. In almost all societies women have been subordinated to men and denied equal access to opportunities, jobs and positions of public authority. This is so despite the fact that in much of the world the woman is not only wife and mother, and manager of the home, but also the farmer who produces the food for the family. The subordination of women implies the domination of men. The consequence is that in many societies to be male is to be aggressive, dominating and lacking the more tender human qualities. Men need to be liberated as well as women. Ruether (1975, p. 74) diagnoses sexism as being 'based on misappropriated dualisms. The dialectics of human existence: mind/body, spirituality/carnality, being/becoming, truth/appearance, life/death – these dualisms are identified as male and female and are socially projected upon men and women as their "natures".' She further makes the interesting proposition that since women in Western culture have been traditionally identified with nature, and nature in turn has been seen as an object of domination by men: 'it would seem almost a truism that the mentality that regarded the natural environment as an object of domination drew upon images and attitudes based on male domination of women' (Ruether, 1975, p. 186). She thus sees, as does another feminist writer (Gray, 1979, Chapter 13), a close relationship between the liberation of women and the sustainable natural world of the future. There will be no new earth without a new woman.

The extent of discrimination varies between societies and between types of societies. Buvinič (1976, p. 234) asserts that 'Women have experienced least sexual inequality in simple and stable hunting and gathering societies and in horticultural and early farming societies and the most sexual inequality in more advanced agricultural (peasant) societies.'

One would like to think that the impact of European cultures upon agricultural societies elsewhere would have led to some improvement in the lot of women. There are some respects in which it has. But in general the condition of women was worsened by colonialism, and post-colonial development policies have continued to undermine both family life and the place of women. For example, women have been widely discriminated against in education. Because it was assumed that men were the leaders in the public world and that education was designed for such leadership, most of the resources for education have gone into the education of men. Even today massive inequalities continue with respect to the number of men and women who receive education and the length of the schooling. The differential here is reflected in differential earning power.

Advocates of development, says Tinker (1976, p. 22), have proposed: 'create a modern infrastructure and the economy will take off, providing a better life for everyone ... yet in virtually all countries and among all classes, women have lost ground relative to men; development, by widening the gap between incomes of men and women, has not helped improve women's lives, but rather has had an adverse affect upon them'. The deterioration of the condition of women dates from colonial times (Tinker, 1976, pp. 25–6). Whereas tribal custom had given women farmers the rights to the land, colonial regimes substituted private ownership for communal land-tenure rights. It was men rather than women who came to own the land. The women, who continued to do the work of farming, became dependent upon their men. Also, when colonial regimes introduced cash crops, men were employed to raise them. Agricultural development neglected the subsistence farming in which the women were employed, for the export crops grown by the men.

In order to get people to work on plantations or to grow cash crops themselves, colonial governments introduced taxes which could be paid only by those who entered the modern money economy. It was the men who could leave home and take paying jobs. Since the men now controlled the money, the women were made even more dependent upon them.

Even today the problem continues. As John D. Rockefeller (1978, p. 515) has said:

> In most developing countries women are crucial to the economic standing of the family because they are responsible for food crops, they supplement family income through home-based arts and crafts, they are often the head of the household. But unfortunately women are often discriminated against in education and employment or excluded from such opportunities altogether. And, in all too many cases, women have been virtually neglected in development plans and programs.

This neglect is closely connected with the neglect of food production and food processing for home consumption. Few resources have been devoted to strategies to reduce the incredible burden of women's work in developing countries. In Senegal, for example, women still spend four hours daily converting five kilograms of wheat into cous-

cous to feed the family. A simple grinder would do the job in minutes (Mink, 1980). This neglect is reflected in this statement by a Kenyan woman:

> During the anticolonial campaigns we were told that development would mean better living conditions. Several years have gone by, and all we see are people coming from the capital to write about us. For me the hoe and the water pot which served my grandmother still remain my source of livelihood. When I work on the land and fetch water from the river, I know I can eat. But this development which you talk about has yet to be seen in this village (quoted by Pala, 1977).

The neglect of women in development programmes has been recognised by the World Bank which now reviews its projects systematically to assure an understanding of the impact each project will have on women (Mink, 1980).

Not only has subsistence farming and food processing through which women feed their families been almost entirely neglected in dominant patterns of development, but also on big properties 'mechanisation has eliminated many of the supplementary sources of employment on which poor women depended' (Safa, 1977, p. 22). This form of development has not replaced the eliminated jobs with new ones to which women have access while continuing child care and subsistence farming. Hence the economic conditions and social status of women are both degraded.

In many cases the problem is not merely that the woman suffers economic deprivation and relative loss of status. It is also that family life is disrupted. Often the man leaves the family in order to earn wages in the city. The most extreme instance of this is the exploitation of black labour in South Africa, where economic policies deprive men of any means of livelihood other than factory labour, and women are not even allowed to come with their men to the barracks in which they are housed. But the situation is also serious when the motivation is a more normal economic one. In much of the Third World, development policies, rather than helping to create a just and sustainable family, are destroying the family while neglecting the women who produce most of the food. The reversal of priorities, from produce for the international market to necessities for home consumption, would lead to a totally different development programme, one geared primarily through very simple improvements, such as better farming tools, wheelbarrows and better seeds, to easing the lot of the woman farmer and improving her productivity. The result would go far toward recovering some measure of justice for women, while also reducing hunger and malnutrition and leading toward far more sustainable societies.

If such justice is to be attained, there will have to be drastic changes of deep-seated attitudes, especially on the part of men. Present attitudes distort perception of the real situation and hence lead to destructive policies. For example, Tinker refers to a statement issued by the US Department of Labour saying that 'in Africa only 5 per cent of the women work. This clearly is an absurd assertion about a continent where women are reported to be doing 60–80 per cent of the work in the fields and working up to 16 hours a day during the planting season' (Tinker, 1976, p. 23).

Tinker also shows how Western stereotypes as to the gender of farmers have led to failures in efforts to improve agriculture. She gives the example of a programme to encourage wet-rice cultivation in Liberia in 1974. A team of Taiwanese farmers was brought in to instruct the Liberians. The government paid people to come and observe the planting. As a result many men came, since they were not otherwise employed. But the women, who were the farmers, continued working in their fields (Tinker, 1976, p. 26).

Even if the distortions of First World planners could be overcome, there would remain the prejudices of the men in developing countries. Males in many societies resist gains by women. According to a Tanzanian report:

> it was considered important not to isolate the women too much for the purpose of learning new skills, and so create the possible impression of imparting to them an exclusive mystique. Otherwise, as past experience in rural areas had shown, husbands sometimes grew suspicious that the fearful prospect of female emancipation was being subtly introduced in order to undermine their traditional masculine authority (Boserup, 1970, p. 222).

Injustice to women goes even deeper. Families discriminate against them when food is scarce. Field studies by the Indian Council of Medical Research showed that more girls than boys suffer from malnutrition and specifically from kwashiorkor. Yet more boys than girls are brought to hospitals for treatment (Newland, 1979, p. 15).

Perhaps the most fundamental right of any human being is the right over one's own body. But this right has been widely denied to women. Women have been subjected to a total system of social expectations with respect to their role which many of them have not yet been able to question. Leghorn & Roodkowsky quote the comment of an Ethiopian woman: 'A woman has no choice really. The number of children she has is many – nine or eleven – and this is a social necessity she cannot question' (Leghorn & Roodkowsky, 1977, p. 31). Even where attitudes are not quite so fatalistic, the pressures to produce children are tremendous. Leghorn & Roodkowsky quote a Bolivian woman: 'Women who have few children usually fail. People gossip when there are just one or two. My son doesn't have any brothers and sisters, so my husband wants to kill me' (p. 30). In some cultures childless women or even women who do not bear sons are divorced for no other reason than this.

The great injustice which is perpetuated in this denial to women of their rights over their own bodies is also a major source of unsustainable population growth on a global scale. There is, of course, no guarantee that if women were entirely free to have the number of children they chose the result would be a stable population for the whole planet. But there are many indications that justice to women would greatly aid in dealing with this problem. For example, in a study of women in Thailand, 'women were asked what they consider the ideal number of children. In every case, they would have preferred fewer children' (Leghorn & Roodkowsky, p. 31). As women gain also the right to education, the results are hopeful. Lester Brown (1978*a*, p. 275) considers that 'the social indicator that correlates most closely with declines in fertility appears to be education for women, particularly the attainment of literacy'.

As the impossibility of supporting an ever-increasing human population becomes more and more evident, governments across the world are adopting programmes to encourage reduced family size. China has made use of its totalitarian structures to reduce its growth rate to near one per cent and is committed to still further reduction. New population control measures appear to be aimed at creating a nation of predominantly one-child families. This reduction in family size is facilitated by the Chinese view of both men and women largely as units of production of goods. Its slowed population growth is achieved more by changing the role of women than by direct pressure against having children. This is even more true in less authoritarian countries. In the seven countries of Europe which have actually attained population stability, government population policy was not a factor (in some of them, e.g. East Germany, the government encourages more births –

largely unsuccessfully). In these European countries 'the decline in fertility was a result of major social changes, such as improved access to family planning services, the liberalization of abortion laws, high levels of education and literacy, and expanded employment opportunities for women' (Brown, 1979, p. 41). These are changes which lead to greater justice for women.

If there were a serious commitment to justice for women, the problem of population stability would largely take care of itself. Unfortunately, most of the discussion of the population problem takes place at a macro-economic and global scale from which conclusions are drawn as to what women should do. The proposal in this chapter is that the ecological approach seeks justice and sustainability by working from the bottom up.

Justice and sustainability coincide. There can be no sustainable world unless population stability is achieved. There can be no justice for the 53 per cent of the world's population that is female until women have greater control over their own bodies, better education and better opportunities for employment. These social changes together with a system of social security for the aged of both sexes lead to reduced birth-rates. No other approach to reducing population growth is either just or sustainable.

The focus on the developing world in this discussion could leave the misleading impression that the problem of the family and of justice for women has been happily solved in the developed countries. This is certainly not the case. Although social changes favourable to women have brought cessation in population growth in most developed countries within sight, much injustice remains. In the industrialised world as well as in the developing world women are still largely excluded from the public sphere or at least from leadership positions. The current women's movement has vigorously called our attention to this injustice, and some changes are being effected. Yet the pace at which the earning gap between women and men is being narrowed in the United States is so slow that at this rate of change, women's salaries will converge with men's in about 2500 years.

Once again this pattern of injustice has its negative consequences for global sustainability as well. The pressure on our environment is a function not only of our numbers but also of our *per capita* consumption. Whereas in the developing world women are perceived very largely as producers of children, in the industrialised world they are perceived more as sex objects. They are treated as part of a consumer society and encouraged to contribute to consumption rather than to the shaping of public policy. With reduced roles as parent and home-maker, and denied equal opportunity in the earning of money and the satisfaction of professional life, they are encouraged to devote themselves to beautification of self and homes according to carefully manipulated norms. Men are similarly led to identify their masculinity with the driving of heavy, inefficient cars and to attract women through large expenditures on unnecessary objects. As women and men relate to one another more fully as persons, one major prop of the wasteful consumer society in which we now live will be removed.

Although consumerism is the special sin of the industrialised world, it is being imposed upon the developing world as well, and once again the direct victims are often women. A notorious instance is that of the successful promotion of substitute milk formulas for babies. 'In Kenya the annual loss of breast milk, calculated to be worth $11.5 million, equals two-thirds of the annual health budget' (Leghorn & Roodkowsky, 1977, p. 28). Whilst the formulas have saved some babies whose mothers could not produce adequate milk, they are less healthful than mother's milk. Because of ignorance of proper use, particularly the proper use of boiled water, they have led to the death of many babies. They have been sold successfully because they have been made to appear 'modern'.

Even more important than the misdirection of women's energies by their exclusion from public life is the loss to public life itself. Since women constitute more than half of the global population, their neglect by planners is a clear indication that male eyes do not see the whole picture. Women have already called attention to this. But the leadership in developing its implications and in responding to the neglected needs is still very much in their hands. To far too great an extent women are still forced to utter their protests from the periphery of the controlling institutions in society. And in spite of some learning that men have done about women's issues, it seems that concern for women will remain peripheral until women constitute a much larger part of the leadership. Only when women attain their just share of the decision-making roles of society will the decisions that are made be just to women.

But women's leadership is needed not only so that the needs of women will be more adequately addressed. It is needed because the kinds of changes this chapter proposes as essential to a just and sustainable world are more likely to be made through the leadership of women than through that of men. Males have long viewed women and nature alike as fields of conquest and control – with disastrous results. Even when men are individually enlightened at the conscious level, they are not fully freed from these destructive attitudes. Women on the whole have been more sensitive to the evils of hierarchical authoritarianism, impersonal power, the ravaging of the environment and war. They have been less prone to aggression and to the sacrifice of human life on the altar of glory and power. They have been more aware of the interconnectedness of human life with the rest of nature. In short, they are in better touch with the primal forces of life. Their sensitivities are those needed by society if it is to reverse the directions of destructive 'development' and to attain a justice and sustainability.

It is sometimes objected that where women have been in positions of power there have been no striking changes for the better. This is largely true. Many societies have allowed some women, as exceptions, to occupy positions of power. But to attain these positions these women have had to cultivate the habits of mind and behaviour of the dominant male society (see Maher, 1980). Some of them have succeeded brilliantly in the male-dominated game. But what justice and sustainability now require is something quite different. They require a society in which women are able to participate fully in leadership, not on men's terms but on their own. Only in this way can they bring the sensitivities and wisdom nurtured by their distinctive history and experience to the shaping of the new world.

Full participation of women in the public sphere along with men will make possible full participation of men in the family sphere along with women. That will require a quite new way of conceiving employment and of gaining status. It may bring an end to the 40 hour week in favour of a distribution of paid work to all who want it. Women who have suffered most from the separation of paid work from unpaid labour will be best able to design quite new arrangements in which the necessary productive work is done in an equitable way. They may be able to usher in forms of marriage that are free partnerships of free people in a free society. In such a world there is a chance that a stable population, consuming moderately, within the carrying capacity of the planet, can be attained.

This section has taken its cue from the literature of women's liberation. The most urgent need is that of justice. Justice for women will lead to sustainability both because of its effects on excessive population growth and because women can help develop the new attitudes toward the non-human world which are needed. Of the six emphases of the ecological model (Chapter 9), this discussion has given expression to the intrinsic value of women, to the importance of transcending, to the continuity between human beings and the remainder of nature, and to the symbiotic relation of diverse goals. Less has been said that reflects the importance of relationships in constituting individual existence and the consequent importance of community.

Too often the emphasis on community has justified the subordination of women to men. For the sake of community, women have been required to play secondary roles geared chiefly to home and children. Hence, in addressing the question of the role of women it is important to speak of liberation before relationality and community. Nevertheless, relationality and community are important too, and many women are among the first to stress the need for an understanding of all human life in those terms.

A basic community is the family. It is especially in this community that individuals receive their identity. It may be that in time the institution of the family will disappear and children will be raised in other communities than the traditional family (see for example Ramey, 1976). But in all of the confusion that now besets family life there are some models emerging of free persons freely relating in families, traditional or otherwise, that can be both just and sustainable. Because of all that women have suffered through unequal marital arrangements, they are particularly sensitive to the requirements of a just marriage. So in this area their leadership is particularly important.

Just and sustainable energy

For many people in the developing world there is a critical energy crisis. It is the shortage of firewood. Distances travelled to attain wood are increasing. The cost of wood is rising. Dung which used to be used as fertiliser is being burned. In some places villagers are no longer able to cook at all.

This energy crisis is directly related to population increases. Customs of gathering and using wood which were developed when populations were well within the carrying capacity of the land, have led growing populations to use up all available wood in larger and larger areas. The resulting deforestation of hillsides has caused erosion and floods. This energy crisis is not 30 or 50 years in the future. It is already here!

Clearly massive reforestation is essential. But except under very authoritarian regimes, reforestation programmes have had limited success. Villagers who need wood now find it hard to allow trees to grow for 20 years. What is required in tens of thousands of villages is something like the 'conscientising' process that Paulo Freire developed among Brazilian peasants, before being driven out by a military government. In the process of learning to read, Brazilian peasants also learned to name the reality in which they lived. They ceased to be mere victims and became agents who identified their needs and took steps to deal with them. The focus was on justice. But there can be no justice without restoring and maintaining the resources on which life depends. Justice and sustainability are inseparable.

A programme of integrated village development which helps villagers take responsibility for their own future can accomplish much. Simple devices can improve the efficiency of cooking with wood up to 70 per cent. Even if the improvement were only 50 per cent effective, a 10 hectare tree plantation could supply fuel for a 1000 family village in rural India (UNEP, 1977, p. 29).

The developing world is also suffering from a shortage of energy for agriculture because of the high cost of oil. This crisis results from the introduction into tropical agriculture of methods dependent upon oil based fertilisers and motorised equipment. It too has a solution, namely the return to methods and crops which rely on free and renewable solar energy. The same styles of farming which preserve and develop the soil are those which are free from dependence on imported energy.

Often the most practical means of employing the sun's energy is in the form of biogas made from plant materials. It is claimed that in the province of Szechuan, China, 2 800 000 families are equiped with tiny biogas installations suitable for home and farm use (Pflug, 1978). Production of liquid fuel from plants for automotive use is also possible.

The use of solar energy through the burning of plants can help deal with much of the energy crisis in the developing world. It is not a panacea. One of the limits of the earth's carrying capacity is its limited capacity to absorb the carbon dioxide produced by burning both coal and plant material. Fortunately there are other uses of solar energy that do not have such limits. It can, of course, be used to heat space and water in thoroughly sustainable ways, and in the tropical world needs for this kind of heat can easily be met almost entirely by solar energy.

Wind is also a sustainable form of solar energy, and there are many places where windmills can be used to pump water and perform other forms of useful work. Running water is a form of solar energy which played a large role in earlier periods of European history.

Electricity is one of the least efficient forms of energy. Impressive civilisations existed without it. The developing world will be well-advised to become less dependent upon it. Nevertheless, there are many modern amenities for which electricity is the most convenient and pleasant type of energy. Some electricity can be produced in villages by the use of windmills. Hydroelectric power, although damaging to the environment and not permanently sustainable, is already widely used in the developing world and can ease the transition to sustainable forms of solar energy. New solar cell technologies promise possibilities for converting solar energy into electricity in ways suitable for both the developed and the developing world.

Most of the developing world was almost entirely dependent upon solar energy until quite recent times. Unfortunately the ideals of development encouraged by the industrialised world entailed a shift away from solar energy to non-renewable forms. These forms are associated with centralised control and dependence on sources outside the country. The results have been neither just nor sustainable. The ideal we propose is not simply to return to the earlier situation. More energy is needed to enjoy a richer life and to participate in global cultural advantages. The ideal is to gain that new energy through improved use of the same renewable energy which made a sustainable life possible earlier. The technologies for such improved use of solar energy are already available. They can be employed and controlled at the local level. Such an energy policy can free developing countries from manipulation by external political and economic forces. It can encourage freedom, self-government and self-respect. It will lead to a more just and more sustainable society.

The energy crisis of the industrialised world is quite different. Here societies are addicted to consumption of vast quantities of fossil fuels by all sectors of the economy. Fossil fuels still exist in large but not inexhaustible amounts. Coal cannot be substituted for oil without increasing the possibility of critical damage to the environment. Economic and political problems are generated by the dependence of many countries on importing an increasingly large portion of their energy supply. The crisis expressed itself recently in waiting lines at petrol stations rather than in inability to find fuel with which to cook. Nevertheless, it is real.

Future energy requirements are often estimated by prediction, as if they were beyond human control. Daly (1978) has pointed out that such predictions have, in practice, become implicit planning. Yet it does not have to be that way. People predict events that are beyond their control but plan events subject to their control. Trend is not destiny. Until quite recently standard projections of future energy needs in industrial countries were based on exponential growth. That is, it was assumed that our needs for energy would increase by a certain percentage each year in order to fuel a growing economy. Like all projections of exponential growth, these led to predictions of doubling energy requirements every 10–15 years. Especially since enormous amounts of energy and capital are required to increase energy production, it was virtually impossible to conceive a scenario for such rapid increase of energy production even if no consideration whatever was given to environmental consequences. Yet the suggestions of environmentalists that society could get along without further growth in energy consumption was ridiculed as a proposal to return to the Dark Ages.

The course of actual energy consumption in the 70s and into the 80s has led to drastic revisions on the part of planners. It turns out that the rate of annual increase of energy is rapidly declining. The idea of Zero Energy Growth is no longer ridiculed. Indeed, projections of the International Institute for Environment and Development Energy Project indicate that there will be an actual decline of energy consumption in Great Britain in this century (Leach, 1979).

The important point to note is that this projection of stable or declining use of energy is not associated with stable or reduced economic production. On the contrary, it presupposes continued economic growth. The study shows that in the United Kingdom,

> if the gross domestic product (GDP) roughly trebles in round terms by 2025 . . . a series of simple, known technical fixes could keep energy (and electricity) demand more or less constant from here on. If GDP doubles, oil consumption could be roughly halved, coal output could remain constant. Nuclear output could fall steadily from 1990 (when the stations now under construction or planned are built) and primary energy consumption could be down about 7 per cent on today's levels by 2000 and 20–25 per cent by 2025 (Leach, 1979, pp. 68–9).

A study by the Harvard Business School is equally encouraging about the possibilities of more efficient use of energy in the United States.

> The US can use 30 or 40 per cent less energy than it does, with virtually no penalty for the way Americans live – save that billions of dollars will be spared, save that the environment will be less strained, the air less polluted, the dollar under less pressure, save that the growing and alarming dependence on OPEC oil will be reduced, and Western society will be less likely to suffer internal and international tension (Stobaugh & Yergin, 1979, p. 182).

These studies indicate that the developed world has a choice between continuing economic growth without increased use of energy and maintaining present economic levels with a marked reduction of energy consumption. Either of these scenarios greatly reduces the dimension of the energy crisis, and neither provides justification for proceeding with nuclear energy or crash programmes of increased use of oil-shale and coal. Neither addresses adequately the long-term problem of sustainability, but both indicate that there is time for development of sustainable forms of solar energy to replace oil as this is exhausted. Especially since there are many other limits besides energy, the appropriate choice for the industrialised world is for reducing energy use in a steady-state economy.

When all other reasons for building expensive new facilities for producing energy fail, we are usually told that we should do so in order to spur employment. The fallacy of this argument is pointed out succinctly by Amory Lovins (1977, p. 9).

> In fact, every quintillion joules/year of primary energy fed into new power stations *loses* the US economy some 71 000 net jobs, because power stations produce fewer jobs per dollar, directly and indirectly, than virtually any other major investment in the whole economy ... It is the conservation, solar, environmental, and related social programmes, not the refineries and reactors, that yield most energy, jobs, and monetary returns per dollar invested. Indeed, energy conservation programmes and shifts of investment from energy-wasting to social programmes create anywhere from tens of thousands to nearly a million net jobs per quintillion joules/year saved – lasting jobs that use widespread or readily learned skills and need personal initiative and responsibility, not transient jobs requiring exotic skills that are already in short supply.

To choose a high energy future is also to choose a hard technology answer. To choose a low energy future gives us the option of choosing a soft technology answer. Lovins (1977, pp. 38–9) identifies five characteristics of soft energy technologies:

1. They rely on renewable energy sources that are always there, whether we use them or not. Examples are sun, wind, vegetation. They rely on energy income not on energy capital.
2. They are diverse so that a national soft energy supply is an aggregate of many small contributions each designed for maximum effectiveness in particular circumstances.
3. They are flexible, and relatively low technology, that can be used without esoteric skills.
4. They are matched in scale and in geographic distribution to their ultimate use.
5. They are matched in quality to ultimate use.

The last point means, for example, that the use of electricity for heating is inappropriate when natural gas or solar heating is available. Indeed, if electricity were confined to its appropriate uses in the United States, it would constitute only about 5 per cent of energy used as compared with the present 13 per cent. 'Whereupon we could probably cover all those needs with present US hydroelectric capacity plus the cogeneration capacity available in the mid to late 1980s' (Lovins, 1977, p.40).

Soft energy paths will make primary use of solar energy but they are not to be equated with solar energy. 'The schemes that dominate ERDA's solar research budget – such as making electricity from huge collectors in the desert, or from temperature differences in the oceans, or from Brooklyn Bridge-like satellites in outer space – do not satisfy our criteria' (Lovins, 1977, p. 42). Indeed, through fantasies such as the last of these, human ingenuity can turn this one safe source of energy into an environmental disaster!

Just and sustainable transportation and urban habitat

The decision for soft energy paths will alter national and international life in ways that are more just as well as more sustainable. It will tend to restore opportunities for self-government and initiative to local communities. These paths can be followed much more easily and successfully if the decision to take them is accompanied by others. One of these is in the area of transportation.

For growth-oriented economies modern transportation is of utmost importance. In developing countries networks of roads must be built to bring the exportable products of the hinterland to the ports. Fleets of trucks are required to carry the goods, along with a rail system. The transportation is fuelled chiefly by oil brought by supertankers which are wreaking havoc with the ocean and its coastlines. The cost rises constantly, making the balancing of budgets ever more difficult. At the same time there is a revolution of rising expectations which means that more and more people want and expect automotive transportation. The great cities of the developing countries are clogged with taxis and privately owned motor cars, and all this is the envy of the poor villagers. The motor car creates a further gap between the affluent who own them and can make use of the highways built at public expense and the poor who cannot and are increasingly cut off from the facilities built for those who can. A mere 17 per cent of the world's population own 88 per cent of its cars (Brown, Flavin & Norman, 1980, p. 3). This whole system, based on the increasingly expensive importation of oil, which is being rapidly exhausted, is neither just nor sustainable.

Despite its problems, few have given serious consideration to an alternative. Many assume that progress implies an ever-increasing ownership of cars along with an ever-increasing building of highways to accommodate them. They would wish to shift from oil to other fuels but not reconsider the fundamental problems involved. Others recognise that, at least within cities, private automobiles are not the answer, and they are trying to develop public transportation systems that include buses and rapid transit trains, while discouraging cars at least in some sections of the city. Nevertheless, around the world, the use of the private motor car continues to increase. Motor car manufacturing is the world's largest manufacturing industry. Some 100 000 new cars roll off the assembly lines each working day (Brown *et al.*, 1980, p. 1).

Ivan Illich (1974) is one person who has thought fundamentally about transportation in the context of basic energy policy. He sees three directions in which nations can go, thinking especially of those in the developing world. The first two are widely discussed. They are maximisation of energy consumption and maximisation of efficiency of energy consumption; the third option is 'the least possible use of mechanical energy by the most powerful member of society' (p. 4). But this

> third option is barely noticed. While people have begun to accept ecological limits on maximum *per capita* energy use as a condition for physical survival, they do not yet think about the use of minimum feasible power as the foundation of any of various social orders that would be both modern and desirable. Yet only a ceiling on

> energy use can lead to social relations that are characterized by high levels of equity. The one option that is presently neglected is the only choice within the reach of all nations (p. 5).

Thus far it has been widely assumed that the goal of transportation technology is to move persons and goods as rapidly as possible. On this assumption the Concorde was hailed as a great achievement. It reduced the time taken for travel from Paris to New York by several hours. But this is accomplished at an enormous cost. Even with a 20 per cent surcharge to passengers, the Concorde cannot pay for itself. It can fly only with government subsidies. In short, a few members of the elite save a few hours of their time at the expense of direct financial support by the general public. It is a system whereby the poor pay for the luxuries of the rich, while at the same time the exhaustible resources of the earth are used inefficiently and unnecessary pollution is added to the environment. It is neither sustainable nor just.

The Concorde is, in these respects, only an extreme case of the general tendency of modern transportation. Motorised vehicles, in general, place a heavy burden on society at large, even though in most countries they are driven only by the few. National resources are devoted to the streets and highways they require and to the importation of the necessary fuel as well as the cars themselves. The existence of cars deprives those who do not own them by blocking the streets and polluting the air. But the more people who own cars the less are the advantages anyone gets. The car is a luxury that cannot be 'democratized' (Gorz, 1980, p. 69 *et seq.*).

The bicycle, in contrast, is an eminently appropriate means of transportation especially for developing countries. According to Illich (1974),

> bicycles are not only thermodynamically efficient, they are also cheap. With his much lower salary, the Chinese acquires his durable bicycle in a fraction of the working hours an American devotes to the purchase of his obsolescent car. The cost of public utilities needed to facilitate bicycle traffic versus the price of an infrastructure tailored to high speeds is proportionately even less than the price-differential of the vehicles used in the two systems (p. 61). The model American puts in 1600 hours to get 7500 miles: less than five miles per hour. In countries deprived of a transportation industry, people manage to do the same, walking wherever they want to go, and they allocate only three to eight per cent of their society's time budget to traffic instead of 28 per cent (p. 19).

In the United States the most important single item after housing is transportation. Americans own many cars they could do without. But in other instances even a conventional family with one income needs two cars. There is no other practical way for the husband to get to work. And without a car his wife is unable to carry out her daily chores, much less take part in the life of the community. If the husband's after-tax income is $12 000, it is not unlikely that one-quarter of this may go to car payments, upkeep and other expenses directly or indirectly related to the two cars. If he is also paying for a house bought on today's market at today's interest rates, he will indeed be hard-pressed to survive! Instead of accepting a cut in income so that others may share the work, he is more likely to seek additional part-time work at weekends or in the evening, or ask his wife to do so. It is no wonder that 65 per cent of American households have two wage earners!

The motor car which plays so large a role in family budgets is expensive to the city as well. It has been estimated that 80 per cent of the space in downtown Los Angeles is devoted to the car. The accompanying sprawl adds greatly to the costs of utilities and services. It separates people into age groups, encouraging the institutional solution for the sick and the old. Suburbs take up large sections of what was once prime agricultural land. The car plays a damaging role also in the American economy. It is a major reason for the negative balance of payments that has upset the money markets of the world.

The oft-proposed solution is to improve public transportation. Unfortunately this will help less than expected. A public transportation system works well only when there is a sufficient density of population. The capital expensive Bay Area Rapid Transit (BART) system in San Francisco is capable of serving only five per cent of the area's population (Brown *et al.*, 1980, p. 71). American suburbs have grown up on the basis of the private automobile. Most Americans would still have to drive to the bus or train. Since their destinations are also scattered, many would have to change buses as well and still walk some distance. For many persons 30 minutes commuting by car would become 90 minutes commuting each way by public transport.

For the truly poor who simply cannot support a car, life in a city built around the car is difficult. They are segregated in the least desirable areas, have poor access to many of the city's facilities, and are handicapped in their efforts to obtain employment. They are invisible to the more affluent and perceived only as the reason for high taxes for welfare. In short, in the sprawling city there is no community among persons of differing classes and conditions.

If cities were designed so that home, work, school, shops and recreation facilities were all within easy walking distance of each other, the car would become a luxury which could be readily sacrificed. Some trends in contemporary architecture lead in this direction. There are huge clusters of buildings which combine living and working space with other facilities. The trend needs rapid expansion. It could lead to the rebuilding of community and make possible a much less expensive life that would be more comfortable and would allow more time to do the things one wanted to do.

As long as such planned communities are built as suburbs or as units in the downtown areas of cities, the problem is only half-solved. But they could be built still larger such that they would become cities in themselves. By taking advantage of their three-dimensional shape and the substitution of walkways, lifts and escalators for streets, pavements and roads, they could, without crowding, greatly compress the space now covered by a city. Segregation in such a city would lose its sting. These cities could be built in otherwise unusable spaces adjacent to agricultural and recreational areas, so that their inhabitants would have ready access both to all the facilities of the city and to the outdoors.

Paolo Soleri has created beautiful designs for scores of possible cities which he calls architectural ecologies or arcologies. He envisions building arcologies in such a way that they would make maximum use of the sun's direct energy. They could also be surrounded by greenhouses which would provide food while their sloping roofs would channel heated air into the city. Factories would be located underground, and their waste heat would provide energy for the city. In short a city could be built so that its total energy needs would be a fraction of those of our present cities, and these needs could be met by the sun. Such a city could make possible a society that was both just and sustainable.

It may be objected that even developed countries cannot afford to build such cities, but Soleri believes his cities would cost only about half as much to build as conventional cities. They save vast amounts of space, and greatly reduce the consumption of both renewable and non-renewable resources. A citizen of such a city could enjoy life with far less expenditure than a citizen of a conventional city, and perhaps all would be willing to share what work there is with all others in the city who want to work. The architecture would promote a community.

Arcologies have their most obvious relevance for the developed world as ways of reversing the accelerating system of consumption, excessive pressure on time, unemployment, segregation and alienation produced by our present cities, built around the automobile. But what of the developing world? Are new solutions to the urban problem required there too? Or will a focus on the agricultural base be sufficient?

Based on present trends the United Nations projects that by the year 2000 the urban population of the world will nearly double to over 3000 million. Nearly 1500 million additional people must be housed while much of the present housing is replaced. Most of this growth will be in the developing world. The cost of providing urban facilities for such numbers in the conventional way will be prohibitive. Unless present trends are reversed, we can only expect the continuing expansion of appalling slums of near destitute people around the cities. Most of them will be unemployed.

> Looking at the developing countries as a whole, the International Labour Office (ILO) estimates that 24.7 per cent of the total labour force was either out of work or underemployed in 1970. The comparable figure for 1980 is expected to be 30 per cent. Between 1970 and the end of the century, the labour force in the less developed countries is projected by the ILO to expand by 91 per cent, requiring a phenomenal 922 million jobs (Brown, 1978*a*, p. 186).

The situation to which this points will be neither just nor sustainable.

Much of the solution must be to reverse the trends that these projections embody. If developing nations concentrate their attention on village development, give priority to family life and especially justice for women, and reduce their dependence on patterns of international trade, population growth will slow, the villages will be able to support more people, and urban growth will be much less than projected. Secondly, every effort must be made to build many small cities rather than a few huge ones. Towns built in each region can live in much better relation to the agricultural base. The skills and equipment needed for building smaller towns can more often be provided by those who will live in them. Such towns can develop industries more appropriate both to local resources and to local needs. By emphasising labour-intensive industry, such towns or cities can employ larger numbers of people. But all this will not exclude the need of developing countries also to face many of the same problems of urban life that plague the developed world. At this point, for them too, Soleri's arcologies are relevant.

Conclusions

This chapter has approached the global situation, its problems and possibilities from the perspective of the ecological understanding of life. This view of life leads to concern about individual living things in communities of living things in all their interconnectedness. It leads to interest in agriculture and the communities of people who farm. It encourages a view of the global problem from the bottom up, so to speak, instead of from the top down. Agriculture is not to be seen in global economic categories, in terms of maximisation of production in the short run. Instead it is to be seen more concretely in terms of the communities of plants and animals and human beings and how they can thrive together. Attention is focused on husbands and wives and children and how their life together can be enriched while it enriches the wider community and on how energy and transportation can serve the real needs of masses of people while empowering them to control their own destiny. Cities can become architectural ecologies, that is contexts for human life, and also organic communities related to their environments.

This understanding of life, and of the kind of world in which life would be more free, has implications for personal life. Concerned persons should examine carefully the kinds of impact that their present life-styles, especially middle class life-styles, are having upon other human beings and upon the living environment. But what can people do? They can opt out of the system to some extent. Most can live more frugally than they do. They can be more intentional about avoiding support for some of the worst abuses of the system. They can experiment with new, more communal, life-styles; with more true partnership in living together; with more support for the current movements of liberation of women and minorities; with the sun or wind as sources of energy; with organic gardening; with driving cars less or even doing without them; and with spending more time in supporting grass-roots movements they believe in instead of making money or establishing status in the present system. These life-style gestures will have little direct effect in relation to our great global concerns, but they can help to create that new climate in which the crises of the future can be met constructively instead of only angrily and defensively. They can enable people to husband their financial resources and use them to support causes in their nation and around the world which may have more direct and positive effects. They can help people inwardly to disengage from the unjust and unsustainable system on which they depend for so many of the things they have been taught to prize. In short people can be more alive and more ready for the surprises that Life will bring.

It is not yet too late to shift basic paradigms and act in terms of a new sense of reality. To do that will be very much an act of living and will lead to new possibilities of living. The dominant paradigms of the recent past have not expressed life, they have even attempted to explain life in terms of what is not alive. The liberation of thought about life from these lifeless forms can help to liberate life all over the planet from the present deadening policies. That is why this book is called *The Liberation of Life*. It is not optimistic because it does not underestimate the power of death in established patterns of thought, economics, agriculture, family life, energy production, transportation and urban development. There is enormous potential for death in a global commitment to nuclear energy with its attendant problems of proliferation of nuclear armaments. It is easy to understand why people become despondent at the enormity of the political and economic problems of a divided world bent on pursuing clever means to no clear end.

But Life has strategies still untried and therein lies hope. To trust Life is to be sensitive to the possibilities it offers. It is to be receptive to Life's values ever pressing in on us from all sides and only blocked by us. It is to be open to the compassionate and tender response and to follow one's intuitions. Trust in Life releases human energy that makes it possible to transcend life as it is in order to make life as it could be.

Faith in Life's power to renew life exposes exploitation and injustice as denials of life. It calls for revolutionary commitment against them. But unless that struggle goes hand in hand with actual liberation there is no liberation in the struggle. Otherwise liberation movements take the stamp of their opponents in their struggle. Even when they succeed they can lead to new oppressions and new alienations.

The greatest power in the world is the power that comes from faith in Life and the ideas this faith brings. The need now is for the critical ideas, the new impulses and the new enthusiasm whose time has come. 'There is a tide in the affairs of men, which taken in the flood leads on' (Shakespeare, Julius Caesar, Act 4, Scene 3). There is a yearning in the world, not only within developed Western countries but in the developing world and behind the Iron Curtain. If it is to be met it will not be through the discovery of new energy sources or new battles or a new upsurge of economic growth, but by a silent working of Life in the hearts of men and women. 'Great ideas come into the world on dove's feet,' said Camus (1954). 'If we listen closely we will distinguish amidst the empires and nations the gentle whisper of life and hope.'

References

Abbott, J.C. (1972). The efficient use of world protein supplies. *FAO Monthly Bulletin of Agricultural Economics & Statistics*. vol. 21, no. 6.

Abrecht, P. (ed.) (1979). *Faith, Science and the Future*. Geneva: World Council of Churches.

AIRAC (Australian Ionising Radiation Advisory Council) (1979). *Radioactive Waste Management*. AIRAC no. 6. Canberra: Australian Government Publishing Service.

Alland, A. (1972). *The Human Imperative*. New York: Columbia University Press.

Allee, W.C., Emerson, A.E., Park, O., Park, T. & Schmidt, K.P. (1949). *Principles of Animal Ecology*. Philadelphia: W.B. Saunders & Co.

Altman, D. (1980). *Rehearsals for Change: Politics and culture in Australia*. Melbourne: Fontana/Collins.

Andrewartha, H.G. & Birch, L.C. (1954). *The Distribution and Abundance of Animals*. University of Chicago Press.

Ardrey, R. (1961). *African Genesis: A personal investigation into the animal origins and nature of man*. London: Collins.

Ardrey, R. (1967). *The Territorial Imperative: A personal inquiry into the animal origins of property and nations*. London: Collins.

Ardrey, R. (1970). *The Social Contract: A personal enquiry into the evolutionary sources of order and disorder*. London: Collins.

Ardrey, R. (1976). *The Hunting Hypothesis*. London: Collins.

Armstrong, E.A. (1969). Aspects of the evolution of man's appreciation of bird song. In *Bird Vocalizations: Their relations to current problems in biology and psychology* (ed. R.A. Hinde), pp. 343–65. Cambridge University Press.

Ayala, F.J. & Dobzhansky, Th. (1974). *Studies in the Philosophy of Biology: Reduction and related problems*. London: Macmillan.

Barash, D.P. (1977). *Sociobiology and Behaviour*. New York: Elsevier.

Beckerman, W. (1975). *Two Cheers for the Affluent Society: A spirited defense of economic growth*. New York: St Martin's Press.

Berger, P.L. (1977). *Pyramids of Sacrifice: Political ethics and social change*. Harmondsworth, Middlesex: Penguin Books.

Bergson, H. (1911). *Creative Evolution*. New York: Henry Holt & Co.

Best, J.B. & Rubenstein, I. (1962). Environmental familiarity and feeding in a planarian. *Science* **135**, 916–18.

Birch, C. (1975*a*). *Confronting the Future: Australia and the world the next 100 years*. Australia: Penguin Books.

Birch, C. (1975*b*). Genetics and moral responsibility. In *Genetics and the Quality of Life* (ed. C. Birch & P. Abrecht), pp. 6–20. Australia: Pergamon Press.

Bohm, D. (1969). Some remarks on the notion of order. In *Towards a Theoretical Biology. 2. Sketches* (ed. C.H. Waddington), pp. 18–40. University of Edinburgh Press.

Bohm, D. (1973). Quantum theory as an indication of a new order in physics. *Foundations of Physics* **111**, 139–68.

Bohm, D. (1977). The implicate or enfolded order – a new order for physics. In *Mind in Nature* (ed. J.B. Cobb & D.R. Griffin), pp. 37–42. Washington, DC: University Press of America.

Bohm, D. (1978). The implicate order: A new order for physics. *Process Studies* **8**, 73–102.

Boserup, E. (1970). *Woman's Role in Economic Development.* New York: St Martin's Press.

Boulding, K. (1971). The economics of the coming spaceship earth. In *Global Ecology* (ed. J.H. Holdren & P.R. Ehrlich), pp. 180–7. New York: Harcourt Brace.

Brandt, W. (1980). *North-South: A programme for survival.* London: Pan Books.

Braverman, H. (1974). *Labor and Monopoly Capitalism.* New York: Monthly Review Press.

Brewer, T.H. (1971). A physician on disease and social class. In *The Social Responsibility of the Scientist* (ed. M. Brown), pp. 149–62. New York: Free Press.

British Council for Science & Society. (1976). *Superstar Technologies.* Report of Working Party for Science and Society. London: Barry Rose (Publishers) Ltd.

Brown, L.R. (1974). *By Bread Alone.* New York: Praeger Publications.

Brown, L.R. (1978*a*). *The Twenty-ninth Day: Accommodating human needs and numbers to the earth's resources.* New York: W.W. Norton & Co. Inc.

Brown, L.R. (1978*b*). *The Worldwide Loss of Cropland.* Worldwatch Paper 24. Washington, DC: Worldwatch Institute.

Brown, L.R. (1979). *Resource Trends and Population Policy: A time for reassessment.* Worldwatch Paper 29. Washington, DC: Worldwatch Institute.

Brown, L.R., Flavin, C. & Norman, C. (1980). *Running on Empty: The future of the automobile in an oil short world.* New York: W.W. Norton & Co.

Burnet, F.M. (1978). *Endurance of Life: The implications of genetics for human life.* Melbourne University Press.

Bush, G.L. (1974). The mechanism of sympatric host race formation in the true fruit flies (Tephritidae). In *Genetic Mechanisms of Speciation in Insects* (ed. M.J.D. White), pp. 3–23. Artarmon, Australia. Australia and New Zealand Book Co. Pty. Ltd.

Buvinič, M. (1976). A critical review of some research-concepts and concerns. In *Women and World Development* (ed. I. Tinker, M.B. Bramsen and M. Buvinič), pp. 224–43. New York: Praeger Press.

Calow, P. (1976). *Biological Machines: A cybernetic approach to life.* London: Edward Arnold.

Campbell, K.O. (1979). *Food for the Future.* University of Nebraska Press, Lincoln.

Camus, A. (1954). L'artiste et son temps. A speech given to the Association Culturale Italiana in its 1954/55 season. Quoted by H. Cleveland and T.W. Wilson (1979) in *Human Growth. An essay on growth and the quality of life.* Princeton: Aspen Institute of Humanistic Studies.

Capra, F. (1975). *The Tao of Physics.* London: Wildwood House.

Cardan, P. (1974). *Modern Capitalism and Revolution.* London: Solidarity.

Caughley, G. (1976). Wildlife management and the dynamics of ungulate populations. In *Applied Biology*, vol. 1 (ed. T.H. Coaker), pp. 183–246. London: Academic Press.

Chatwin, B. (1979). Variations on an idée fixe. *New York Review of Books* **26**(19), 8–9.

Clark, S.R.L. (1977). *The Moral Status of Animals.* Oxford: Clarendon Press.

Clarke, C.A. (1969). Problems raised by developments in genetics. In *Biology and Ethics* (ed. T.J. Ebling), p. 93. London: Academic Press.

Cleveland, H. & Wilson, T.W. (1979). *Human Growth: An essay on growth and the quality of life.* Princeton: Aspen Institute of Humanistic Studies.

Cobb, J.B. (1972). *Is It Too Late? A theology of ecology.* Beverly Hills, California: Bruce.

Cocoyoc Declaration. (1974). Development dialogue. *Journal of International Development* **2**, 88–96.

Commoner, B. (1976). *The Poverty of Power*. New York: Alfred A. Knopf Inc.

Corning, W.C., Dyal, J.A. & Willows, A.O.D. (1973). *Invertebrate Learning*, vols. 1–3. New York: Plenum Press.

Council on Environmental Quality (1979). *Environmental Quality*: Tenth annual report of the Council. US Government Printing Office.

Cox, H. (1977). Eastern cults and western culture. *Psychology Today*, July 1977, pp. 36–42.

Crosnier, J. (1974). From artificial kidney to transplant. *South African Outlook*, July 1974, pp. 107–8.

Daly, H.E. (1973). *Towards a Steady State Economy*. San Francisco: W.H. Freeman.

Daly, H.E. (1977). *Steady State Economics: The economics of biophysical equilibria and moral growth*. San Francisco: W.H. Freeman.

Daly, H.E. (1978). On thinking about energy – the future. *Natural Resources Forum* **3**, 9–16.

Daly, H.E. (ed.) (1980). *Economics, Ecology, Ethics: Essays toward a steady-state economy*. San Francisco: W.H. Freeman.

Darwin, C. (1859). *On the Origin of Species by Means of Natural Selection*. John Murray, London, 1st edn. A facsimile of the First Edition with an introduction by Ernst Mayr. Cambridge, Mass: Harvard University Press (1964).

Darwin, C. (1871). The Descent of Man and Selection in Relation to Sex. In: *The Origin of Species and the Descent of Man*. New York: The Modern Library.

Delgado, J.M.R. (1969). *Physical Control of the Mind: Toward a psychocivilized society*. New York: Harper & Row.

Dobzhansky, Th. (1956). *The Biological Basis of Human Freedom*. New York: Columbia University Press.

Dobzhansky, Th. (1967). *The Biology of Ultimate Concern*. New York: New American Library.

Easlea, B. (1973). *Liberation and the Aims of Science: An essay on obstacles to the building of a beautiful world*. London: Chatto & Windus.

Easlea, B. (1974). Who needs the liberation of nature? *Science Studies* **4**, 89.

Eccles, J.C. (1979). *The Human Mystery*. The Gifford Lectures, University of Edinburgh, 1977–78. Berlin: Springer International.

Eckholm, E. (1976). *Losing Ground: Environmental stress and world food prospects*. Washington, DC: Worldwatch Institute.

Eckholm, E. (1977). *The Picture of Health: Environmental sources of disease*. New York: W.W. Norton & Co. Inc.

Eckholm, E. (1978). *Disappearing Species: The social challenge*. Worldwatch Paper 22. Washington, DC: Worldwatch Institute.

Eckholm, E. & Brown, L.R. (1977). *Spreading Deserts – The Hand of Man*. Worldwatch Paper 13. Washington DC: Worldwatch Institute.

Egerton, F.N. (1972). Changing concepts of the balance of nature. *Quarterly Review of Biology* **48**, 322–50.

Ehrlich, P.R., Ehrlich, A.H. & Holdren, J.P. (1977). *Ecoscience: Population, resources and environment*. San Francisco: W.H. Freeman.

Ehrlich, P.R. & Ehrlich, A.H. (1981). *On the Extinction of Species*. New York: Random House.

Elsasser, W.M. (1966). *Atom and Organism: A new approach to theoretical biology*. New Jersey: Princeton University Press.
Elton, C. (1930). *Animal Ecology and Evolution*. Oxford University Press.
Epstein, S.S. (1978). *The Politics of Cancer*. New York: Sierra Club, Ballantine Books.
Flannery, R. (1979). Treating soil like dirt. *Not Man Apart* **9**(7), 19.
Forman, P. (1971). Weimar culture, causality and quantum theory, 1918–1927: Adaptation by German physicists and mathematicians to a hostile intellectual environment. *Historical Studies in the Physical Sciences* **3**, 1–114.
Frankl, V.E. (1964). *Man's Search for Meaning: An introduction to logotherapy*. London: Hodder & Stoughton.
Frith, H.J. (1973). *Wildlife Conservation*. Sydney: Angus & Robertson.
Fukuoka, Masanobu. (1978). *The One Straw Revolution: An introduction to natural farming*. Emmaus, Pennsylvania: Rodale Press.
Gause, G.F. (1934). *The Struggle for Existence*. Baltimore: Williams & Wilkins Co.
Gaylord, W. (1947). *As You Sow: Three studies in the social consequences of agribusiness*. Glencoe, Illinois: The Free Press.
Geertz, C. (1965). The impact of the concept of culture on the concept of man. In *New Views of the Nature of Man* (ed. J.R. Platt), pp. 93–118. University of Chicago Press.
Georgescu-Roegen, N. (1980). Entropy law and the economic problem. In *Economics, Ecology, Ethics: Essays toward a steady-state economy* (ed. H.E. Daly), pp. 49–81. San Francisco: W.H. Freeman.
Ghiselin, M.T. (1974). *The Economy of Nature and the Evolution of Sex*. Berkeley: University of California Press.
Godfrey-Smith, W. (1979). The value of wilderness. *Environmental Ethics* **1**, 309–19.
Godlovitch, R. (1971). Animals and morals. In *Animals, Men and Morals* (ed. R. Godlovitch, S. Godlovitch & J. Harris), pp. 156–72. London: Victor Gollancz.
Goodall, J. van Lawick. (1971). *In the Shadow of Man*. Boston: Houghton Mifflin Co.
Goodfield, J. (1977). *Playing God: Genetic engineering and the manipulation of life*. New York: Random House.
Gorz, A. (1980). *Ecology as Politics*. Boston: South End Press.
Gould, S.J. & Lewontin, R.C. (1979). The spandrels of San Marco and the Panglossian paradigm: a critique of the adaptationist programme. *Proceedings of the Royal Society of London*, B **205**, 581–98.
Grant, J.P. (1977). The world can and must afford it. *Development Forum* **5**(3), 1–2.
Grant, J.P. (1979). *Meeting essential human needs in a sustainable world by the year 2000*. Private paper circulated by the American Association of the Club of Rome.
Gray, E. Dodson. (1979). *Why the Green Nigger? Re-mything Genesis*. Wellesley, Mass.: Roundtable Press.
Griffin, D. (1976). *The Question of Animal Awareness: Evolutionary continuity of mental experience*. New York: The Rockefeller University Press.
Grobstein, C. (1964). *The Strategy of Life*. San Francisco: W.H. Freeman.
Gulland, J.A. (1976). Production and catches of fish in the sea. In *The Ecology of the Seas* (ed. D.H. Cushing and J.J. Walsh), pp. 283–314. Oxford: Blackwell Scientific Publications.
Haeckel, E. (1870). Ueber Entwickelungsgang u. Aufgabe der Zoologie. *Jenaische Zeitschrift* **5**, 353–70.
Hardy, A.C. (1965). *The Living Stream: A restatement of evolution theory and its relation*

to the spirit of man. Gifford Lectures (1963–65), part I. London: William Collins.
Hardy, A.C. (1966). *The Divine Flame: An essay towards a natural history of religion*. Gifford Lectures (1963–65), part II. London: William Collins.
Hardy, A.C. (1975). *The Biology of God: A scientist's study of man, the religious animal*. London: Jonathan Cape.
Harris, M. (ed.) (1972). *Ethical Problems in Human Genetics: Early diagnosis of genetic defects*. Washington, DC: Fogarty International Center Publication Proceedings no. 6.
Harrison, R. (1964). *Animal Machines*. London: Stuart.
Hartshorne, C. (1962). *The Logic of Perfection: and other essays in neoclassical metaphysics*. La Salle, Illinois: Open Court.
Hartshorne, C. (1967). *A Natural Theology for Our Time*. La Salle, Illinois: Open Court.
Hartshorne, C. (1973). *Born to Sing: An interpretation and world survey of bird song*. Bloomington: Indiana University Press.
Hartshorne, C. (1977). Physics and Psychics: The place of mind in nature. In *Mind in Nature* (ed. J.B. Cobb & D.R. Griffin), pp. 89–100. Washington, DC: University Press of America.
Hayes, D. (1978). *Repairs, Reuse, Recycling – First steps toward a sustainable society*. Worldwatch Paper 23. Washington, DC: Worldwatch Institute.
Hayes, D. (1979). *Pollution: The neglected dimension*. Worldwatch Paper 27. Washington DC: Worldwatch Institute.
Henderson, H. (1976). Citizen power in the overdeveloped countries. *World Issues* 1(2), 9–12.
Henderson, H. (1978). *Creating Alternative Futures: The end of economics*. New York: Berkeley Publishing Corporation.
Heschel, A.J. (1962). *The Prophets*. New York: Harper & Row.
Hetzel, B.S. (1974). *Health and the Australian Society*. Australia: Penguin Books.
Higginson, J. (1969). Present trends in cancer epidemiology. *Proceedings of the Canadian Cancer Conference* **8**, 40–75.
Higginson, J. (1979). Cancer and environment: Higginson speaks out. *Science* **205**, 1363–66.
Hill, S. (1979). Who knows best? *UNESCO Review (Australia)* **1**, 23–7.
Huxley, T.H. & Huxley, J. (1947). *Touchstone for Ethics*. New York: Alfred A. Knopf Inc.
Illich, I.D. (1974). *Energy and Equity*. New York: Harper & Row.
IUCN (1980). *World Conservation Strategy: Living resources conserved for sustainable development*. International Union for the Conservation of Nature, Documentation Publication, Nairobi, Kenya.
Jackson, W. (1978). Toward a sustainable agriculture. *Not Man Apart* **8**, 4–6.
Jennings, H.S. (1906). *Behavior of the Lower Organisms*. New York: Columbia University Press.
Johnson, W.R. (1975). Should the poor buy no growth? In *The No Growth Society* (ed. M. Olson & H.H. Landsberg), pp. 165–89. New York: W.W. Norton & Co.
Judson, H.F. (1979). *The Eighth Day of Creation: The makers of the revolution in biology*. London: Jonathan Cape.
Kahn, H., Brown, W. & Martel, L. (1976). *The Next 200 Years: A scenario for America and the world*. New York: William Morrow & Co.

Kass, L.R. (1971). The new biology: what price relieving man's estate? *Science* 174, 779–88.
Kawai, M. (1965). Newly acquired pre-cultural behavior of the natural troop of Japanese monkeys on Koshima Island. *Primates* **6**, 1–30.
Keynes, J.M. (1936). *The General Theory of Employment, Interest and Money*. New York: Harcourt Brace & Co.
King, J. (1980). New genetic technologies: Prospects and hazards. In *Faith and Science in an Unjust World*, vol. 1 (ed. R.L. Shinn), pp. 264–72. Geneva: World Council of Churches.
Knox, R. (1980). Nuclear war: what if? *Science 80* 1(4), 32–4.
Koestler, A. (1967). *The Ghost in the Machine*. London: Hutchinson.
Koestler, A. (1971). *The Call Girls: A tragi-comedy with prologue and epilogue*. London: Hutchinson.
Krebs, C.J. (1978). *Ecology: The experimental analysis of distribution and abundance* (2nd edn). New York: Harper & Row.
Kuenzler, E.J. (1961). Phosphorus budget of a mussel population. *Limnology & Oceanography* **6**, 400–15.
Kuhr, M.D. (1975). Doubtful benefits of Tay-Sachs screening. *New England Journal of Medicine* **292**, 371.
Lappé, F.M. & Collins, J. (1976). More food means more hunger. *Development Forum* 4, 1–2.
Leach, G. (1979). Do we need nuclear energy at all? *Anticipation* **26**, 68–70.
Leghorn, L. & Roodkowsky, M. (1977). *Who Really Starves: Women and world hunger*. New York: Friendship Press.
Leakey, R.E. & Lewin, R. (1977). *Origins: What new discoveries reveal about the emergence of our species and its possible future*. London: Macdonald & Jane's.
Leontief, W.W., Carter, A.P., & Petri, P. (1977). *The Future of the World Economy: A United Nations study*. New York: Oxford University Press.
Leopold, A. (1933). The conservation ethic. *Journal of Forestry* **31**, 634–43.
Levine, C. (1977). Ethics, justice and international health. *The Hastings Center Report* 7(2), 5–7.
Lewis, J. & Towers, B. (1969). *Naked Ape or* Homo sapiens? London: Garnstone Press.
Lewis, K.N. (1979). The prompt and delayed effects of nuclear war. *Scientific American* **241**(1), 27–39.
Lewis, W.A. (1955). *The Theory of Economic Growth*. Homewood, Illinois: Richard D. Irwin.
Lewis, W.A. (1965). A review of economic development. *Manchester School* 55(2), 1–16.
Lewis, W.A. (1978). *The Evolution of the International Economic Order*. New Jersey: Princeton University Press.
Lewontin, R.C. (1979). Adaptation. In *Evolution*. (Scientific American Book), pp. 114–25. San Francisco: W.H. Freeman.
Lillie, R.S. (1937). Directive action and life. *Philosophy of Science* 4, 202–26.
Lillie, R.S. (1945). *General Biology and Philosophy of Organism*. University of Chicago Press.
Linzey, A. (1976). *Animal Rights: A Christian assessment of man's treatment of animals*. London: SCM Press.
Lockeretz, W. (1978). The lessons of the dust bowl. *American Scientist* **66**, 560–70.

Loechler, E., McLennan, T., Park, R., Shore, D., Thacher, S. & Youderian, P. (1978). Social and political issues in genetic engineering. In *Genetic Engineering* (ed. A.M. Chakrabarty), pp. 165–84. West Palm Beach, Florida: CRC Press Inc.

Lorenz, K. (1943). Die angeborenen Formen möglicher erfahrung. *Zeitschrift fur Tierpsychologie* 5(2), 235–409.

Lorenz, K. (1966). *On Aggression*. New York: Harcourt, Brace and World Inc.

Lovins, A.B. (1977). *Soft Energy Paths: Toward a durable peace*. Harmondsworth, Middlesex: Penguin Books.

Lovins, A.B., Lovins, L. & Ross, L. (1980). Nuclear power and nuclear bombs. *Foreign Affairs* 58(5), 1137–77.

Maglen, L.R. (1977). Non-renewable resources and the limits to growth: another look. *Search* 8, 158–66.

Maher, M. (1980). Women at the top. *Development Forum* 8(5), 3.

Marais, E. (1969). *The Soul of the Ape*. New York: Atheneum.

Marcovič, M. (1976). Marxist philosophy in Yugoslavia: The Praxis group. In *Marxism and Religion in Eastern Europe* (ed. R.T. De George & J.P. Scanlan), pp. 63–89. Dordrecht, Holland/Boston: D. Reidel Pub. Co.

Martin, P.S. (1967). Pleistocene overkill. *Natural History* 76(10) 32–8.

Maslow, A.H. (1970). *Religious Values and Peak Experiences*. New York: Viking Press.

Maslow, A.H. (1973). *The Farther Reaches of Human Nature*. New York: Viking Press.

Meadows, D.H. (1975). A look at the future. In *Notes for the Future: An alternative history of the past decade* (ed. R. Clarke), pp. 53–61. London: Thames & Hudson.

Meadows, D.H., Meadows, D.L., Randers, J. & Behrens, W.W. (1972). *The Limits to growth.* A report for the Club of Rome's project on the predicament of mankind. New York: New American Library.

Medawar, P.B. (1957). *The Uniqueness of the Individual*. London: Methuen.

Medawar, P.B. & Medawar, J.S. (1977). *The Life Science: Current ideas of biology*. London: Wildwood House.

Medvedev, Z.A. & Medvedev, R.A. (1974). *A Question of Madness*. Harmondsworth, Middlesex: Penguin Books.

Mendelsohn, E. (1977). The social construction of scientific knowledge. *Society and the Sciences* 1, 3–26.

Midgley, M. (1978). *Beast and Man: The roots of human nature*. Ithaca, New York: Cornell University Press.

Mill, J.S. (1857). *Principles of Political Economy*, vol. 2. London: J.W. Parker.

Mink, P.T. (1980). Help for Third World women. *Ada World* (Americans for democratic action, Washington, DC) 35(3), 3–4.

Moltmann, J. (1979). *The Future of Creation*. London: SCM Press.

Monod, J. (1974). *Chance and Necessity: An essay on the natural philosophy of modern biology*. London: Fontana/Collins.

Morawetz, D. (1977). *Twenty-five Years of Economic Development 1950–1975*. Washington, DC: The World Bank.

Morgan, C.L. (1923). *Emergent Evolution*. The Gifford Lectures 1922. London: Williams & Norgate.

Morris, D. (1967). *The Naked Ape*. London: Jonathan Cape.

Morris, R. & Fox, M.W. (eds) (1978). *On the Fifth Day*. Washington, DC: Acropolis Books.

Narveson, J. (1977). Animal rights. *Canadian Journal of Philosophy* 7, 161–78.

National Academy of Sciences. (1977). *Energy and Climate*. Washington, DC.

Newland, K. (1979). Women's health. *Environment* **21**, 14–20, 35–37.

Newsome, A.E. (1980). The eco-mythology of the red kangaroo in central Australia. *Mankind* **12**(4) (In press).

Niebuhr, R. (1941). *The Nature and Destiny of Man*. New York: Charles Scribner's Sons, New York.

Norman, C. (1978). *Soft Technologies, Hard Choices*. Worldwatch Paper 21. Washington, DC.: Worldwatch Institute.

Odhiambo, T.R. (1980). Perspectives in developing countries – An African perspective. In *Faith and Science in an Unjust World*, vol. 1 (ed. R.L. Shinn), pp. 159–66. Geneva: World Council of Churches.

Oelhaf, R.C. (1979). *Organic Agriculture: Economic and ecological comparisons with conventional methods*. Montclair, New Jersey: Allanheld, Osmun & Co. Inc.

Okun, A.M. (1975). *Equality and Efficiency: The big tradeoff*. Washington, DC: Brookings Institution.

Omo-Fadaka, J. (1977). What can be done about third world poverty? *PHP*, November 1977, 21–32.

Pala, A.A. (1977). Definitions of women and develpment: An African perspective. *Signs: Journal of Women in Culture & Society* **3**, 9–13.

Papanek, H. (1977). Development planning for women. *Signs: Journal of Women in Culture & Society* **3**, 14–21.

Parmar, S.L. (1975). *Some thoughts on human development*. World Council of Churches 5th Assembly document, SV1–2, pp. 1–5.

Passmore, J. (1974). *Man's Responsibility for Nature*. London: Duckworth & Co.

Passmore, J. (1975). The treatment of animals. *Journal of the History of Ideas* **36**, 195–218.

Perelman, M. (1977). *Farming for Profit in a Hungry World*. Montclair, New Jersey: Allanheld, Osmun & Co. Inc.

Pflug, F. (1978). Practical paths to plant power. *Ceres* **3**(5), 19–25.

Podrabinek, A. (1980). *Punitive Medicine*. Ann Arbor: Karoma.

Popper, K.R. (1972). *Objective Knowledge: An evolutionary approach*. Oxford: Clarendon Press.

Popper, K.R. & Eccles, J.C. (1977). *The Self and Its Brain*. Berlin: Springer-Verlag International.

Proceedings (1977). *The World Food Conference of 1976*. Ames, Iowa: Iowa State University Press.

Pulliam, H.R. and Dunford, C. (1980). *Programmed to Learn: An essay on the evolution of culture*. New York: Columbia University Press.

Ramey, J.W. (1976). *Intimate Friendships*. Englewood Cliffs, New Jersey: Prentice Hall Inc.

Randers, J. (1977). *How to Stop Individual Growth with Minimal Pain*. Paper presented to Alternatives to Growth '77 Conference. Houston, Texas.

Ravetz, J. (1971). *Scientific Knowledge and its Social Problems*. Oxford: Clarendon Press.

Rawls, H. (1972). *A Theory of Justice*. Cambridge, Mass.: Harvard University Press.

Regan, T. (1976). Do animals have a right to life? In *Animal Rights and Human Obligations* (ed. T. Regan & P. Singer), pp. 197–204. Englewood Cliffs, New Jersey: Prentice Hall Inc.

Rifkin, J. (1980). *Entropy: A new world view*. New York: The Viking Press.

Rifkin, J. & Howard, T. (1979). *The Emerging Order: God in the age of scarcity*. New York: G.P. Putnam's Sons.

Risser, J. (1978). *Soil Erosion Creates a Problem Down on the Farm*. Conservation Foundation Letter. Washington, DC: Conservation Foundation.

Robertson, J. (1978). *The Sane Alternative: Signposts to a self-fulfilling future*. London: Villiers Publications.

Robinson, M.A. (1980). World fisheries to 2000 – supply, demand and management. *Marine Policy* 4(1), 19–32.

Rockefeller, J.D. (1978). Population growth: The role of the developed world. *Population and Development Review* 4, 509–16.

Rogers, M. (1977). *Biohazard*. New York: Alfred A. Knopf Inc.

Rose, S. (1976). *The Conscious Brain*. Harmondsworth, Middlesex: Penguin Books.

Rossel, J. (1980). The social risks of large scale nuclear energy programmes. In *Faith and Science in an Unjust World*, vol. 1 (ed. R.L. Shinn), pp. 253–60. Geneva: World Council of Churches.

Roszak, T. (1975). Where the wasteland ends. In *Notes for the Future: An alternative history of the past decade* (ed. R. Clarke), pp. 225–30. London: Thames and Hudson.

Ruether, R. (1975). *New Woman – New Earth: Sexist ideologies and human liberation*. New York: Seabury Press.

Ruse, M. (1979). *Sociobiology: Sense or nonsense?* Dordrecht, Holland: D. Reidel Pub. Co.

Russell, E.S. (1945). *The Directiveness of Organic Activities*. Cambridge University Press.

Ryder, R.D. (1975). *Victims of Science: The use of animals in research*. London: Davis-Poynter.

Safa, H.I. (1977). Changing modes of production. *Signs: Journal of Women in Culture & Society* 3, 22–100.

Sagan, C. (1978). *The Dragons of Eden: Speculations on the evolution of human intelligence*. London: Hodder & Stoughton.

Sahlins, M. (1977). *The Use and Abuse of Biology: An anthropological critique of sociobiology*. London: Tavistock Publications.

Sant, M. (1979). *The Least-cost Energy Strategy*. Publication of the Energy Productivity Centre of the Carnegie Mellon Institute, Arlington, Va.

SCEP (1970). *Man's Impact on the Global Environment*. Report of the Study of Critical Environmental Problems (SCEP). Cambridge, Massachusetts: MIT Press.

Schopenhauer, A. (1890). *Religion: A dialogue, and other essays* (2nd edn). London: Swan Sonnenschein & Co.

Schrödinger, E. (1962). *What is Life? The physical aspects of the living cell*. Cambridge University Press.

Schweitzer, A. (1933). *Out of my Life and Thought*. London: Allen & Unwin.

Schweitzer, A. (1949). *Civilization and Ethics*. London: Adam & Charles Black.

Shepard, P.S. (1959). Reverence for life at Lambarene. *Landscape* 8, 26–9.

Shepard, P.S. (1973). *The Tender Carnivore and the Sacred Game*. New York: Charles Scribner's, Sons.

Shepard, P.S. (1978). *Thinking Animals: Animals and the development of human intelligence*. New York: The Viking Press.

Shore, M.F. (1975). Psychological issues in counselling the genetically handicapped. In *Genetics and the Quality of Life* (ed. C. Birch & P. Abrecht), pp. 161–72. Australia: Pergamon Press.

Simpson, G.G. (1952). How many species? *Evolution* **6**, 342–3.
Singer, P. (1973). Review of Animals, Men and Morals (ed. S. Godlovitch, R. Godlovitch & J. Harris). *New York Review of Books* **20**(5), 17–21.
Singer, P. (1976). *Animal Liberation: A new ethics for our treatment of animals*. London: Jonathan Cape.
Sinnott, E.W. (1950). *Cell and Psyche: The biology of purpose*. Chapel Hill, North Carolina: University of North Carolina Press.
Skolimowski, H. (1980). *Eco-philosophy: Designing new tactics for living*. London: Marion Boyers.
Sociobiology Study Group of Science for the people (1976). Sociobiology – another biological determinism. *BioScience* **26**, 182–85.
Söderlund, R. & Svensson, B.H. (1976). The global nitrogen cycle. In *Nitrogen, Phosphorus and Sulphur – Global Cycles* (ed. B.H. Svensson & R. Söderlund), pp. 23–74. Stockholm: Scope Report 7, Ecological Bulletin.
Somers, A.R. (1976). Violence, television and the health of American youth. *New England Journal of Medicine* **294**, 811–17.
Sperry, R.W. (1977). Absolute Values: Problems of the ultimate frame of reference. In *The Search for Absolute Values: Harmony among the sciences*. Proceedings of the Fifth International Conference on the Unity of the Sciences (vol. II), pp. 689–94. New York: The International Cultural Foundation Press.
Steiner, G. (1971) *In Bluebeard's Castle: Some notes towards the redefinition of culture*. New Haven: Yale University Press.
Stobaugh, R. & Yergin, D. (1979). *Energy Future*. New York: Random House.
Storr, A. (1966). *The Integrity of Personality*. Harmondsworth, Middlesex: Penguin Books.
St Vincent Millay, E. & Ellis, N.M. (1967). From Sonnet CXXXVII Collected Poems. New York: Harper & Row.
Szent-Györgyi, A. (1972). *The Living State: With observations on cancer*. New York: Academic Press.
Teilhard de Chardin, P. (1959). *The Phenomenon of Man*. London: Collins.
Thielicke, H. (1970). The doctor as judge of who shall live and who shall die. In *Who Shall Live?* (ed. K. Vaux), pp. 146–94. Philadelphia: Fortress Press.
Thomas, L. (1974). *The Lives of a Cell: Notes of a biology watcher*. New York: Viking Press.
Thomas, L. (1979). *The Medusa and the Snail: More notes of a biology watcher*. New York: Viking Press.
Thorpe, W.H. (1961). *Bird-Song: The biology of vocal communication and expression in birds*. Cambridge University Press.
Thorpe, W.H. (1974). *Animal Nature and Human Nature*. London: Methuen & Co.
Thorpe, W.H. (1977). The frontiers of biology: does process thought help? In *Mind in Nature* (ed J.B. Cobb & D.R. Griffin), pp. 1–12. Washington, DC: University Press of America.
Thorpe, W.H. (1978). *Purpose in a World of Chance: A biologist's view*. Oxford University Press.
Tillich, P. (1949). The depth of existence. In *The Shaking of the Foundations*, P. Tillich, pp. 52–63. London: SCM Press.
Tillich, P. (1955). *The New Being*. New York: Charles Scribner's Sons.
Tinker, I. (1976). The adverse impact of development on women. In *Women and World*

Development (ed. I. Tinker, M.B. Bramsen & M. Buvinič), pp. 22–34. New York: Praeger Publications.

Train, R. (1978). Reverence for life. *Frontiers* **42**, 38–9.

UNEP (United Nations Environment Programme) (1977). The state of the environment: selected topics – 1977. Fifth session, Governing Council, Nairobi. UNEP/GC/88: pp. 1–33.

van den Bosch, R. (1978). *The Pesticide Conspiracy*. New York: Doubleday.

van den Bosch, R. (1979). The pesticide problem. *Environment* **21**(4) 13–42.

Waddington, C.H. (1957). *The Strategy of the Genes: A discussion of some aspects of theoretical biology*. London: Allen & Unwin.

Waddington, C.H. (1960). *The Ethical Animal*. London: Allen & Unwin.

Waddington, C.H. (1975). *The Evolution of an Evolutionist*. Edinburgh University Press.

Wade, N. (1977). *The Ultimate Experiment*. New York: Walker & Co.

Wade, N. (1978). New vaccine may bring man and chimpanzee into tragic conflict. *Science* **200**, 1027–30.

Wallace, A.R. (1869). *The Malay Archipelago: The land of the orang-utan, and the bird of paradise*, Paradisea regia. New York: Harper.

Ward, B. (1979). *Progress for a Small Planet*. Harmondsworth, Middlesex: Penguin Books.

Watts, A. (1976). *Nature, Man and Woman*. London: Abacus Press.

Weinberg, S. (1978). *The First Three Minutes: A modern view of the origin of the universe*. New York: Basic Books Inc.

Weitz, R. (1971). *From Peasant to Farmer*. New York: Columbia University Press.

Westman, W.E. (1977). How much are nature's services worth? *Science* **197**, 960–64.

White, Lynn T. (1975). Christians and Nature. *Pacific Theological Review* 7, 6–11.

White, R.B. (1969). Translator of Victor Hugo, The Relationship Between Man and Animal. *The Ark* (Magazine of the Catholic Study Circle for Animal Welfare) **32**(2), 116.

Whitehead, A.N. (1911). *An Introduction to Mathematics*. Oxford University Press.

Whitehead, A.N. (1926*a*). *Science and the Modern World*. New York: Macmillan.

Whitehead, A.N. (1926*b*). *Religion in the Making*. New York: Macmillan.

Whitehead, A.N. (1929*a*). *The Aims of Education*. New York: Macmillan.

Whitehead, A.N. (1929*b*). *The Function of Reason*. Princeton University Press.

Whitehead, A.N. (1933). *Adventures of Ideas*. New York: Macmillan.

Whitehead, A.N. (1978). *Process and Reality* (corrected edition, ed. D.R. Griffin & D.W. Sherburne). New York: Free Press (original edition, 1929, London & New York: Macmillan.)

Wieman, H.N. (1929). *Methods of Private Religious Living*. New York: Macmillan.

Wieman, H.N. (1946). *The Source of Human Good*. University of Chicago Press.

Wilson, E.O. (1975*a*). *Sociobiology: The new synthesis*. Cambridge, Massachusetts: The Belknap Press of Harvard University Press.

Wilson, E.O. (1975*b*). The origins of human social behaviour. *Harvard Magazine* 77, 21–6.

Wilson, E.O. (1976). Academic vigilantism and the political significance of sociobiology. *BioScience* *26*, 187–90.

Wilson, E.O. (1978). *On Human Nature*. Cambridge, Massachusetts: Harvard University Press.

Woodger, J.H. (1929). *Biological Principles: A critical study*. London: Routledge & Kegan Paul.

Woodwell, G.M. (1971). Toxic substances and ecological cycles. In *Man and the Ecosphere* (ed. P.R. Ehrlich, J.P. Holdren & R.W. Holm), pp. 127–35. San Francisco: W.H. Freeman.

Woodwell, G.M. (1978). The carbon dioxide question. *Scientific American* **238**(1), 34–43.

World Council of Churches (1974). Science and Technology for Human Development. Report of 1974 World Conference in Bucharest. *Anticipation* **19**, 1–43.

Würsig, B. (1979). Dolphins. *Scientific American* **240**(3) 108–19.

Wynder, E.L. & Gori, G.B. (1977). Contributions of the environment to cancer incidence: an epidemiologic exercise. *Journal of the National Cancer Institute* 58, 825–32.

Young, J.Z. (1978). *Programmes of the Brain*. Gifford Lectures 1975–7. Oxford University Press.

Index